RHEOLOGY OF PARTICULATE
DISPERSIONS AND COMPOSITES

SURFACTANT SCIENCE SERIES

RHEOLOGY OF PARTICULATE DISPERSIONS AND COMPOSITES

Rajinder Pal
University of Waterloo
Ontario, Canada

CRC Press
Taylor & Francis Group
Boca Raton   London   New York

CRC Press is an imprint of the
Taylor & Francis Group, an informa business

CRC Press
Taylor & Francis Group
6000 Broken Sound Parkway NW, Suite 300
Boca Raton, FL 33487-2742

© 2007 by Taylor & Francis Group, LLC
CRC Press is an imprint of Taylor & Francis Group, an Informa business

First issued in paperback 2019

No claim to original U.S. Government works

ISBN 13: 978-0-367-45334-3 (pbk)
ISBN 13: 978-1-57444-520-6 (hbk)

Library of Congress Cataloging-in-Publication Data

Pal, Rajinder.
    Rheology of particulate dispersions and composites / Rajinder Pal.
        p. cm. -- (Surfactant science series ; v. 136)
    Includes bibliographical references and index.
    ISBN 1-57444-520-0 (alk. paper)
    1. Granular materials--Fluid dynamics. 2. Inhomogeneous materials--Fluid dynamics. I. Title. II. Series.

TA418.78.P324 2006
660'.294514--dc22   2006045549

Visit the Taylor & Francis Web site at
http://www.taylorandfrancis.com

and the CRC Press Web site at
http://www.crcpress.com

Dedication

To the memory of my mother, Smt Karma-Bhari, and to my father, Shri Khushal Chand

Preface

Particulate dispersion (or simply *dispersion*) consists of fine insoluble particles distributed throughout a continuous *liquid* phase. The particulate phase of the dispersion may be solid, liquid, or gas. When the particulate phase is *solid*, the dispersions are referred to as *suspensions*. When the particulate phase is *liquid (liquid droplets)*, the dispersion is called an *emulsion*. The dispersion with *gaseous* particulate phase (bubbles) is referred to as *bubbly liquid* or *bubbly suspension* when the volume fraction of bubbles in the dispersion is less than ϕ_m, the maximum packing concentration of undeformed bubbles; the dispersion is called *foam* when the volume fraction of bubbles exceeds ϕ_m (ϕ_m is 0.637 for random close packing of uniform spheres).

The particles of the dispersion are not always composed of a single phase (solid, liquid, or gas). For instance, the particles may consist of a liquid core encapsulated with a uniform layer of solid material. Such types of complex particles are referred to as *capsules*. Examples of other types of complex particles are *solid core–liquid shell* particles, which consist of a solid inner core encapsulated with a uniform layer of liquid that is immiscible in the continuous phase, and *liquid core–liquid shell* particles (also referred to as *double-emulsion droplets*), which consist of a liquid inner core completely engulfed by a second liquid that is immiscible in both the inner liquid and the continuous phase.

Particulate dispersions form a large group of materials of industrial importance. Both aqueous and nonaqueous dispersions are encountered in industrial applications. Some of the fields in which dispersions play a vital role are ceramics; computers and particulate recording media; nanotechnology; biomedicine; biotechnology; paints; environment (marine oil spills); food; pipeline transportation of materials; petroleum production; cosmetics and toiletries; and geology (volcanic eruptions).

In the formulation, handling, mixing, processing, storage, and pumping of particulate dispersions, knowledge of the rheological properties is required for the design, selection, and operation of the equipment involved.

Unlike particulate dispersions, which are generally fluidic in nature, *composites* are solid heterogeneous materials composed of two or more phases. Many composites of practical interest are composed of just two phases: the dispersed phase and the matrix (continuous phase). Composite materials can be classified into two broad groups: *particle-reinforced (particulate)* composites and *fiber-reinforced* composites. Particulate composites usually consist of isometric (same dimension in all directions) particles. Fiber-reinforced composites, as the name implies, consist of fibers of high aspect ratio as the dispersed phase. The commercial and industrial applications of composite materials are many. For example, they are used widely as dental restorative materials, as tool materials for high-speed cutting of difficult-to-machine materials, and as solid propellants in aerospace propulsion. Composite materials also play an important role in aircraft, automotive, sport, marine, plastics, and electronics industries.

Knowledge of the rheological properties of composites is required in the analysis and design of structures made from composite materials. Rheology also plays a vital role in the development of new composite materials that meet specific strength or stiffness requirements. In order to develop tailor-made composites with specific mechanical properties, rheological models are needed to predict the mechanical properties of composites from the properties and volume fractions of the individual components.

A number of excellent books are available on the rheology of materials. But none of them provide a systematic and comprehensive coverage of the fundamental rheology of particulate dispersions and two-phase solid composites. This book covers a wide variety of dispersed systems with major emphasis on the exact rheological constitutive equations based on the fundamental laws of mechanics. Empirical results are generally avoided.

The book is divided into six parts. Part I, titled "Introduction," consists of three chapters. Chapter 1 discusses the applications of particulate dispersions and composites. Chapter 2 reviews the fundamental aspects of fluid mechanics, solid mechanics, and interfacial mechanics. Chapter 3 introduces concepts such as bulk stress in dispersions and composites, dipole strength of a particle, and constitutive equations for particulate dispersions and composites. Part II, titled "Rheology of Dispersions of Rigid Particles," consists of Chapter 4 to Chapter 6. Both spherical and nonspherical particles are considered. The rheology of dispersions of rigid dipolar (magnetic) particles, spherical and nonspherical, in the presence of an external magnetic field is also described. Part III, titled "Rheology of Dispersions of Nonrigid Particles," consists of Chapter 7 to Chapter 11. Chapter 7 deals with dispersions of soft (deformable) solid particles. The rheology of emulsions (dispersions of liquid droplets) in the absence and presence of surfactant is discussed in Chapter 8. Chapter 9 describes the rheology of bubbly suspensions. The dispersion of capsules is dealt with in Chapter 10. The rheology of dispersions of core–shell particles is described in Chapter 11. The types of core-shell particles considered are solid core–porous shell, solid core–liquid shell, and liquid core–liquid shell. Part IV, titled "Rheology of Composites," consists of two chapters (Chapter 12 and Chapter 13) dealing with the elastic properties of solid composite materials. Chapter 12 describes particulate composites in this respect and Chapter 13 deals with fiber-reinforced composites. Part V, titled "Linear Viscoelasticity of Particulate Dispersions and Composites," consists of Chapter 14 to Chapter 16. Chapter 14 covers some of the basic aspects of linear viscoelasticity. The dynamic viscoelastic behavior of particulate dispersions and composites are dealt with in Chapter 15 and Chapter 16, respectively. Part VI, titled "Appendices," consists of four useful appendices.

This book provides both an introduction to the subject for newcomers and sufficient in-depth coverage for those involved with the rheology of dispersed systems. Scientists and engineers from a broad range of fields will find the book an attractive and comprehensive source of information on the fundamental rheology of a wide variety of particulate dispersed systems. It could also serve as a textbook for a graduate-level course on rheology.

I wish to thank the late consulting editor of the Surfactant Science Series, Dr. Martin Schick, for inviting me to write a book in my area of specialization, that is, the rheology of dispersed systems. For their love and support, I am grateful to my wife, Archana, and my children, Anuva and Arnav.

Rajinder Pal

About the Author

Rajinder Pal is a professor of chemical engineering at the University of Waterloo, Ontario, Canada. He received his B.Tech degree (1981) in chemical engineering from the Indian Institute of Technology, Kanpur, and a Ph.D. degree (1987) in chemical engineering from the University of Waterloo. The author of more than 100 refereed journal publications in the areas of rheology of dispersed systems (emulsions, suspensions, foams, and particulate composites), pipeline flow behavior of particulate dispersions, and emulsion liquid membranes, Dr. Pal is a fellow of the Chemical Institute of Canada. In recognition of his distinguished contributions in chemical engineering before the age of 40, he received the Syncrude Canada Innovation Award in 1998 from the Canadian Society for Chemical Engineering. In 2001, he received the Teaching Excellence Award of the Faculty of Engineering, University of Waterloo. Dr. Pal served as associate editor of the *Canadian Journal of Chemical Engineering* from 1992 to 2004. He is a registered professional engineer in the province of Ontario.

Table of Contents

PART I *Introduction*

PART II   *Rheology of Dispersions of Rigid Particles*

PART III   *Rheology of Dispersions of Nonrigid Particles*

PART IV   Rheology of Composites

PART V   Linear Viscoelasticity of Particulate Dispersions and Composites

PART VI   Appendices

Part I

Introduction

Chapter 1 deals with the industrial and commercial applications of particulate dispersions and composites. Although the applications of such dispersed systems are numerous, only some important ones are highlighted in this chapter. Industries in which dispersed systems are of importance include ceramics, particulate recording media, nanotechnology, biomedicine, biotechnology, paints, environment, food, pipeline transportation, petroleum, cosmetics and toiletries, geology, cutting tools, aerospace, plastics, dentistry, metals and metal alloys, automobile, electronics, polymers, aircraft, sports, and marine.

Chapter 2 presents a brief review of the mechanics of fluids, solids, and interfaces. The basic concepts of continuum mechanics such as the continuum hypothesis, kinematics, stress vector, stress tensor, and constitutive equation are introduced. Using the principles of conservation of mass and momentum, the continuity equation and equation of motion are derived. In the special case of Newtonian fluids of constant viscosity and density, the equation of motion is shown to reduce to the celebrated Navier–Stokes equation. The section on the mechanics of fluids concludes with creeping flow and creeping flow equations. The mechanics of solids is introduced with the equilibrium equation valid at every point of a solid material subjected to surface and body forces. The constitutive equation for a Hookean solid is discussed. Using the Hookean constitutive equation together with the equilibrium equation, the well-known Navier–Cauchy equation is derived. The last section of the chapter is devoted to the mechanics of interfaces. After introducing some important definitions and the surface divergence theorem, the force balance on interfacial film is carried out. The section concludes with a discussion on interfacial rheology.

Chapter 3 defines the bulk or average fields (stress, velocity, rate of strain, etc.) in dispersions and composites. The expressions for bulk stress in dispersions and composites, in terms of the dipole strength of particles, are derived. The equations for the dipole strength of rigid (spherical and nonspherical) particles and nonrigid droplets with interfacial film are derived. The chapter concludes with a section on the effects of nonhydrodynamic forces (such as Brownian, electrostatic, steric, and van der Waals forces) on the bulk stress of dispersions.

1 Applications of Particulate Dispersions and Composites

1.1 PARTICULATE DISPERSIONS

In this book, particulate dispersions are defined as systems consisting of fine insoluble particles (solid, liquid, or gas) distributed throughout a continuous liquid phase. The particles distributed within the continuous liquid phase are collectively referred to as the *particulate* or *dispersed phase*. The continuous phase of the dispersion is sometimes referred to as the dispersion medium, external phase, or matrix. Dispersions with a solid particulate phase are called *suspensions,* those with a liquid particulate phase are termed *emulsions,* and those with a gaseous particulate phase are referred to as *bubbly liquids* or *bubbly suspensions*. Particulate dispersions have so many commercial and industrial applications that it is not possible to list them all. Only some important ones are highlighted here.

1.1.1 CERAMICS

Ceramics are a class of materials different from plastics (organics) and metals. They are inorganic nonmetal materials with unique properties such as high hardness (harder than steel), high heat and corrosion resistance (higher than polymers or metals), and low density (lower than most metals). Traditional ceramics are composed of materials such as clays and silica. Modern or advanced ceramics are composed of materials such as pure metallic oxides (Al_2O_3) and nonoxide metallic compounds such as carbides, nitrides, borides, and silicides [1–3]. Ceramic materials have a great diversity of applications. They are used in bricks, glassware, dinnerware, home electronics, watches, automobiles, space shuttles, airplanes, cutting tools, filters, electrical resistors and insulators, bearings, fuel cells, dental restoration products, bone implants, orthopedic joint replacement, etc.

The steps involved in the manufacture of ceramics are: (1) preparation of particulate dispersion (suspension) by mixing raw ceramic powder homogeneously in a liquid dispersion medium (water), (2) compaction of particulate dispersion to a high-volume fraction of particles by removing most of the liquid, (3) a shape-forming process, and (4) a heat-treatment process called *firing* or *sintering* to produce a rigid, final product. A good understanding of the rheology of particulate dispersion is vital for the successful completion of steps (1) to (3). Furthermore, online measurement of the rheological properties of particulate dispersion can be used for quality control purposes so as to minimize batch-to-batch variations [4,5].

1.1.2 PARTICULATE RECORDING MEDIA

Floppy disks, audio and video tapes, and magnetic stripes are all referred to as particulate recording media as they are manufactured using a suspension of magnetic particles [6–8]. Figure 1.1 shows a schematic diagram of the process used for the production of particulate recording media (magnetic tapes). A suspension of single-domain rodlike ferromagnetic particles (length of the order of 0.5 μm or less, aspect ratio about 8 to 10) is prepared by dispersing the particles in a liquid dispersion medium consisting of solvent, polymeric binder, surfactant, and other additives in trace amounts. The volume fraction of the magnetic particles (iron oxide, chromium oxide, or barium ferrite) in the suspension is usually 1 to 5%. The suspension is allowed to spread onto the moving polymeric film substrate, also referred to as a web, which is usually made from polyester (polyethylene tetraphthalate) and is about 25 μm thick. In the next stage of the process, the magnetic particles in the film are aligned in the direction of the length of the tape using a magnetic field while the film is still in liquid dispersion form. Following the alignment of the magnetic particles, the web is heated to drive off the volatile solvent and to cross-link the polymeric binder. The cross-linked binder holds the magnetic particles together and attaches them to the supporting substrate. The structure of the magnetic film or layer formed on the substrate is similar to that of grapes-filled Jell-O, with Jell-O representing the binder and grapes representing the magnetic particles. The tape is then rolled to reduce the surface roughness. The thickness of the top magnetic layer of the tape in the finished product is about 3 to 5 μm.

The magnetic particles of the suspension are prone to flocculation as strong attractive forces exist between them. Flocculation of the particles adversely affects the coating and particle-alignment processes. A poor-quality suspension with a high

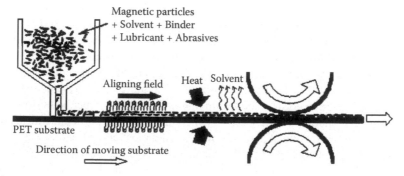

FIGURE 1.1 Production of particulate recording media (magnetic tapes). (From Applied Alloy Chemistry Group, Magnetic Recording, University of Birmingham, http://www.aacg.bham.ac.uk/magnetic_materials/magnetic_recording.htm.)

degree of flocculation of particles results in increased surface roughness and recording noise in the final product. As the rheological properties of magnetic suspension are strongly affected by the degree of aggregation of particles, suspension rheology can be used as a tool to monitor the quality of the suspension [6].

1.1.3 NANOTECHNOLOGICAL APPLICATIONS

Nanotechnology can be defined as a branch of engineering in which dimensions in the range of nanometers (usually 0.1 to 100 nm) play a critical role. The *nanomaterials* subdivision of nanotechnology includes dry powder of nanoparticles as well as dispersion of nanoparticles in liquid. According to a recent market analysis of nanotechnology, the total world market for nanoparticulate materials was $492.5 million in 2000 and $900.1 million in 2005 [9].

Nanomaterials in the form of dispersions of nanoparticles in liquids, also referred to as *nanodispersions*, have many practical applications. One important application of such systems is in the area of *drug delivery* [10,11]. Most drugs, especially those that are poorly soluble, perform better in nanoparticulate form. A significant number of the new drugs discovered today by the pharmaceutical companies are poorly soluble. The bioavailability of the poorly soluble drugs can be enhanced significantly by reducing the size of the drug particles to the nanometer size range and thus increasing the specific surface area of the drug. Nanoparticulate drugs can be formulated either in the dry tablet (capsule) form or in the form of a nanodispersion. However, most production techniques currently utilized to manufacture drug nanoparticles generally yield nanodispersion as the final product. In one production technique, the coarse drug powder is dispersed in a surfactant solution and the dispersion is subjected to high-pressure homogenization. The forces in the homogenizer are so strong that the coarse drug particles disintegrate into drug nanoparticles (nanocrystals) [11]. In another production technique, the drug is first dissolved in some appropriate lipid at a temperature above the melting point of the lipid. The drug–lipid solution is dispersed in a hot aqueous surfactant solution by a stirrer. The coarse emulsion thus formed is homogenized in a high-pressure homogenizer while the temperature is maintained above the melting point of the lipid. The resulting dispersion of nanodroplets is finally cooled down to room temperature. The cooling process leads to crystallization of the lipid, and the final product is a nanodispersion of drug-loaded solid lipid nanoparticles (also referred to as SLN) [10].

The applications of nanodispersions are not restricted to drug delivery alone. Some other areas in which nanodispersions are significant are cosmetics, biomedicine, inks, and paints.

1.1.4 BIOMEDICAL APPLICATIONS

The biomedical applications of particulate dispersions are many. However, the most promising ones involve the use of so-called *ferrofluids* [12–16]. Ferrofluids are stable dispersions of single-domain magnetic nanoparticles, typically

10 nm in diameter. In the absence of an external magnetic field, the ferrofluid has no net magnetization, as the magnetic moments of the particles are randomly distributed owing to thermal agitation (Brownian motion). In the presence of an external magnetic field, however, the magnetic moments of the particles align with the field and the dispersion acts as a magnet. As the single-domain magnetic particles contain several thousand atomic magnetic moments, ferrofluids often behave like a *superparamagnetic* material — magnetizing strongly under the application of an external field but retaining no permanent magnetization upon the removal of the field. The nanoparticles of ferrofluids (usually iron oxide) are stabilized with a layer of surfactant molecules attached to the surface of the particles. The surfactant layer provides steric repulsion between the particles and, consequently, agglomeration or flocculation of particles is prevented.

Ferrofluids have many potential applications in biomedical as well as non-biomedical areas [12–18]. In the biomedical area, ferrofluids are gaining significance in applications involving *targeted drug delivery* and *hyperthermia*. They are also being considered as possible contrast agents for magnetic resonance imaging (MRI).

In targeted drug delivery applications, drug molecules are first adsorbed onto the surface of the magnetic nanoparticles by mixing the particles with the drug solution. The drug-loaded nanoparticles are then released at the intended treatment site. Figure 1.2 shows schematically how the particles can be delivered to the intended site [14]. The dispersion of drug-loaded magnetic nanoparticles is injected into an arterial feed to the intended tissue or organ with the help of a catheter. A powerful magnetic field is then applied near the target site. The magnetic

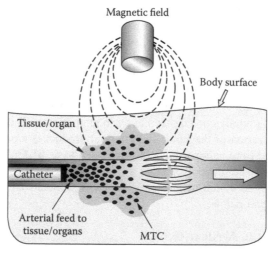

FIGURE 1.2 Delivery of drug-loaded magnetic nanoparticles to the intended tissue or organ. (From Z.M. Saiyed, S.D. Telang, C.N. Ramchand, Application of magnetic techniques in the field of drug discovery and biomedicine, *Biomagn. Res. Technol.* 1: 2, 2003.)

field causes the magnetic particles (MTC–magnetic targeted carriers) to extravasate through the capillary bed into the targeted organ or tissue where the drug is released. Targeted drug delivery is particularly important in cancer treatment. Chemotherapy is known to have many unpleasant side effects; the drugs used to kill cancerous cells also damage healthy cells. These side effects can be prevented by targeting the drug directly at the cancerous cells.

Hyperthermia is used for the treatment of cancer. It involves heating the cancerous tissue to a temperature of about 45°C so as to reduce the viability of cancerous cells and enhance their sensitivity to chemotherapy. Ferrofluids are being studied as potential hyperthermia-causing agents. The magnetic nanoparticles of ferrofluids are first concentrated in the cancerous tissue. Then an external AC magnetic field is applied near the targeted tissue. The magnetic nanoparticles oscillate with the external magnetic field and produce heat.

1.1.5 BIOTECHNOLOGICAL APPLICATIONS

Biotechnology involves the manipulation of microorganisms (bacteria, molds, yeast, etc.), mammalian cells, or plant cells to produce industrial chemicals and pharmaceuticals. Dispersions of microbial organisms or cells in a liquid medium, also referred to as *broths*, are grown inside a bioreactor. As the organisms or cells grow, they also produce the desired chemical. Table 1.1 gives the morphological characteristics (shape, size, etc.) of various microbial organisms and cells [19]. The table indicates that the particulate phase of the broth often consists of aggregates of organisms or cells.

Broth rheology is an active area of research [19–25]. Most broths exhibit non-Newtonian behavior, especially at high biomass concentrations. The design and operation of bioreactor- and downstream-processing equipment require a good understanding of the rheology of broths.

TABLE 1.1
Morphological Characteristics of Microbial Organisms and Cells

Type	Shape	Size (μm)	Cell Wall	Aggregates
Plant cells	Spherical	100–500 (dia)	Yes	Yes
	Cylindrical	20–50 (length)		
Animal cells	Spherical	10–20 (dia)	No	No
Bacteria	Spherical	< 1 (dia)	Yes	Yes/No
	Cylindrical	< 5 (length)		
Yeasts	Spherical	5–10 (dia)	Yes	Yes/No
Moulds	Mycelial	5–10 (dia)	Yes	Yes/No
		< 100 (length)		

Source: From P.M. Kieran, P.F. MacLoughlin, D.M. Malone, *J. Biotechnol.* 59: 39–52, 1997.

1.1.6 PAINTS

Paint is a particulate dispersion composed of pigments and extenders as the particulate phase and binder solution or latex (referred to as the *vehicle*) as the continuous phase. Pigments are fine insoluble solid particles that give the desired color to paint. Titanium dioxide is probably the most widely used pigment in the paint industry. Extenders are also fine insoluble solid particles, but they provide no color to paint; they are basically used as inexpensive fillers to help control the paint rheology and to improve the properties of paint film. Some common extenders are calcium carbonate (whitewash), aluminum silicate (clay), magnesium silicate (talc), and silica [26–29].

The continuous phase or vehicle of paint is either a true liquid solution of film-forming polymeric binder in some appropriate solvent or latex (emulsion) of nanosize binder particles and water. The latex is used in water-based paints and the binder or solvent solution is used in oil-based paints. The solvents used in oil-based paints are low molecular weight organic liquids such as aromatic and aliphatic hydrocarbons, alcohols, esters, and ketones. The most widely used polymeric binders in oil paints are alkyds; they are highly branched, oil-modified polyesters [26]. The function of the binder is to cement (bind) the pigment and extender particles into a uniform continuous film and to make the pigment and extender particles adhere to the surface. In the case of water-based paints, the binder consists of nanosize polymer particles usually made of vinyl acetate (polyvinyl acetate) or 100% acrylic. When the water-based paint is applied on a surface, the water evaporates, forcing the pigment, extender, and binder particles to come closer together; the latex particles eventually coalesce and bind the pigment and extender particles in a continuous film. The film-forming ability of the polymeric latex particles, as well as their ability to bind the pigment and extender particles together, is dependent on factors such as the size of the latex particles and the glass transition temperature (T_g) of the polymer. Small particle size and low T_g (the lower the T_g, the softer the polymer) are better for film forming.

1.1.7 MARINE OIL SPILLS

Marine oil spills are often caused by oil tanker accidents (collisions, hull failure, and groundings) [30,31]. In the well-known *Exxon Valdez* oil spill that occurred in March 1989, the *Exxon Valdez* struck the rocks of Bligh Reef in Prince William Sound [32]. Nearly 11 million gallons of crude oil were spilled within 5 h of the accident. Oil tanker accidents are not the only cause of marine oil spills; they can occur owing to accidents on the offshore oil and gas production facilities (oil rigs) as well. Oil spills can have a major effect on sea animals and humans. The diving and surface-dwelling populations of seabirds are known to be sensitive to oiling; severe oiling almost always results in death. There is some evidence to suggest that oil spills can have long-term effects on sea animals [30].

The oil starts to spread rapidly over the sea surface as soon as it is spilled. At the same time, it starts to lose the lighter components due to evaporation and, consequently, the viscosity of the remaining oil tends to rise. A significant degree of emulsification also occurs simultaneously because of wave action and wind agitation [31,33–35]. Emulsification is a process by which one liquid is dispersed in the form of small droplets in a continuum of another immiscible liquid. In the present situation, water is dispersed in the form of small droplets in the continuous oil phase. Such particulate dispersions are referred to as *water-in-oil* (W/O) emulsions. With certain types of crude oils, highly stable W/O emulsions containing up to 80% by volume sea water can form. Highly concentrated W/O emulsions are generally reddish-brown in color and look like a gel. They are often nicknamed "chocolate mousse" as they resemble this dessert.

The formation of a W/O emulsion makes cleanup operations more difficult. With the addition of a large amount of water droplets to the oil, there occurs a dramatic increase in the viscosity. In fact, the emulsion at high-volume fractions of water droplets is highly non-Newtonian. It often exhibits a large value of yield stress. Furthermore, the formation of an emulsion substantially increases the volume of the spill.

There are a number of different methods proposed to deal with oil floating on the sea [35,36], such as: (1) mechanical cleanup, (2) burning, (3) bioremediation, and (4) treatment with chemical dispersants. Among these, mechanical cleanup is probably the most commonly used method for mitigation of oil spills. Usually, the first step is to limit the spread of oil (the W/O emulsion) on the water surface with the help of so-called booms. Skimmers, which are floating mechanical devices, are then used to remove oil from the water surface. A wide variety of skimmers are available [33], and they seem to function best when the oil or water-in-oil emulsion layer is thick. Pumps are used to transfer oil or the emulsion from skimmers to storage tanks of the recovery vessel. The rheological properties of the oil or W/O emulsion play an important role in the selection and successful operation of the skimmers and pumps.

1.1.8 FOOD APPLICATIONS

Many natural and processed food products are particulate dispersions. A vast majority of them are emulsions of oil and water with additives such as sugar, salts, vitamins, minerals, food-grade surfactants, proteins, gums, colors, flavors, etc. Examples of food emulsions are milk, butter, margarine, mayonnaise, coffee whiteners, and salad dressings [37–39]. A brief description of the composition of these products is given in Table 1.2.

Emulsion rheology plays a critical role not only in the processing of emulsion-based foods, but also in the acceptance of these products by the consumer [39]. Therefore, it is not surprising to note that it is a very active area of research at present.

1.1.9 PIPELINE TRANSPORTATION OF MATERIALS

Pipeline transportation of solids and highly viscous liquids in the form of particulate dispersions [40,41] is not only feasible but has many advantages over other modes of material transportation such as trucks, trains, or sea vessels.

TABLE 1.2
Composition of Emulsion-Based Food Products

Food Product	Particulate Phase	Matrix Phase	Volume Percentage of Particulate Phase
Milk	Oil droplets	Water	3 to 4
Butter	Water droplets	Liquid oil and fat crystals	About 16
Margarine	Water droplets	Liquid oil and fat crystals	16 to 50
Mayonnaise	Oil droplets	Water	≥65
Coffee whiteners	Oil droplets	Water	10 to 15
Salad dressings	Oil droplets	Water	≥30

In addition to long-term economic benefits, pipelines have an aesthetic advantage over other transport options in that they are buried underground and are out of sight for the most part. As they do not cause noise and air pollution, they are also environmentally friendly. A wide variety of industrial materials (solids and liquids) have been transported in particulate dispersion form through pipelines across a range of distances. Examples of solid materials transported in the form of solids-in-water suspensions are: coal, kaolin, limestone, and various metallic ores (copper, iron, zinc, nickel, etc.). Examples of liquids transported in the form of oil-in-water (O/W) emulsions are waxy crude oils, bitumen, and highly viscous heavy oils [41,42].

For economically successful pipeline transport, it is important that a high level of particulate loading is achieved without a substantial increase in the viscosity of the particulate dispersion. One possible method of achieving this is to optimize the particle size distribution (PSD). Many studies have been conducted on the effects of PSD on the rheology of particulate dispersions [43–46]. It is generally concluded that at high levels of particulate loading, a particulate dispersion with bimodal or multimodal PSD has a much lower viscosity than a particulate dispersion with uniform-size particles, when compared at the same volume fraction of the dispersed phase.

1.1.10 PETROLEUM PRODUCTION

A significant portion of the world's crude oil is produced in the form of emulsions, that is, dispersions of oil and water [47]. Emulsions can be classified into two broad groups, namely, W/O emulsions and O/W emulsions. W/O emulsions consist of water droplets dispersed in a continuum of oil phase, whereas O/W emulsions have a reverse arrangement, that is, oil droplets are dispersed in a continuum of water phase. In the petroleum industry, W/O emulsions are more commonly encountered.

In primary oil production where the reservoir's natural energy (pressure) is used to produce the oil, the source of water is the connate water [48] that is naturally present along with the oil in the formation. It is believed that the emulsification of oil and water occurs in the formation near the well bore, where the velocity gradients are very high. The high shear in the production facilities such as pumps, valves, chokes, and turns also favors the formation of the emulsion [48,49]. The source of emulsifying agents in these emulsions is the crude oil itself, which may contain a variety of surface-active chemicals such as long-chain fatty acids and polycyclics [47,48]. Although W/O emulsions are more commonly encountered in primary oil production operations, O/W emulsions are also produced from some oil fields; O/W emulsions frequently occur when a given oil field becomes old and produces an increased amount of water [50].

When the reservoir's natural energy is no longer capable of pushing the oil from the formation into the oil well, a fluid is often injected into the formation through injection wells. A variety of fluids are used for this purpose, such as water, aqueous polymer solution, caustic solution, aqueous surfactant solutions, microemulsion, and steam. Most of these fluids not only push the oil from the formation into the producing well but also help to recover oil that adheres to the walls of the pores (because of strong capillary forces). Very often, however, channeling and breakthrough of fluid into the producing well occurs, causing emulsions to be formed [48].

In some situations where the crude oil or W/O emulsion being produced from a given oil well is highly viscous, an aqueous solution of surface-active chemicals is often injected into the production well bore to convert the high-viscosity oil (or W/O emulsion) to a low-viscosity O/W emulsion [51,52]. This increases the oil production rate substantially by increasing the rod-dropping rate and by lowering the flow line pressure drop. This process, whereby aqueous solutions of surface-active chemicals are pumped down into the producing well to enhance the oil production rate, is referred to as *downhole emulsification*.

Oil-well drilling is another important application in which particulate dispersions (emulsions as well as solids-in-liquid suspensions) are used [53,54]. During the drilling process, particulate dispersions called *drilling muds* are pumped down the drill pipe and out of the nozzles of the drill bit. The mud acts as a coolant and lubricant for the drill bit. When the drilling mud returns to the surface through the annular region between the rotating drill pipe and the borehole wall, it also brings cuttings such as rocks and sand from the hole with it. Drilling muds are either water-based or oil-based. The water-based drilling muds are solids-in-liquid suspensions with clay (usually bentonite) and a weighting agent (usually barium sulfate) as the particulate phase and freshwater or seawater as the liquid phase. The weighting agent is added to increase the density of the mud to overcome formation pressures and to keep oil and other reservoir fluids in place. To control the rheological, fluid-loss, and shale-stabilizing properties of the drilling muds, a wide variety of water-soluble polymers are also used. Although water-based drilling muds are used more often, some situations require the use of oil-based drilling muds. The oil-based drilling muds are basically W/O emulsions with some solids, such as weighting agents and organophilic clay.

1.1.11 COSMETICS AND TOILETRIES

Toothpaste is a particulate dispersion [55–57]. The dispersed phase of a toothpaste consists of solid particles of some mildly abrasive agent (usually dicalcium phosphate) to help remove stains and polish teeth. The weight percentage of solids in the dispersion is about 10 to 50. The continuous phase is an aqueous solution consisting of water, humectant (moisture controller to prevent the toothpaste from becoming dry and firm) such as glycerol, a binder or thickener such as sodium carboxymethylcellulose to prevent the solid particles of the polishing agent from settling out, detergent or surfactant such as lauryl sulphate to provide foaming action and to enhance the cleaning ability of the toothpaste, therapeutic agents such as sodium monofluorophosphate to prevent tooth decay, and minor amounts of flavors, preservatives, and coloring agents.

Most *roll-on antiperspirants* currently sold in the market are suspensions of solid particles of some active ingredient (such as aluminum chlorohydrates) in volatile silicone liquid such as cyclomethicone [58]. To prevent the solid particles of the active ingredient from settling out, additives such as organophilic clay are added to the continuous phase to thicken the product. Rheology plays a key role in the success of roll-on antiperspirants. A properly formulated product should have the right consistency. It should be thick enough so that it is not runny or spillable and the solid particles of the active ingredient do not settle out at the bottom of the container. At the same time, the product should thin down enough during the application to allow a uniform coating to be applied on the skin.

Sunscreen lotions are particulate dispersions available commercially either in the form of an emulsion or in the form of a solids-in-liquid suspension [59–64]. Emulsion-based sunscreen lotions are usually chemical absorbers of ultraviolet radiation (UVR); the chemical ingredient (aromatic compound conjugated with a carbonyl group) that absorbs UVR is incorporated into the oil phase of the emulsion. Suspension-based sunscreen lotions are generally physical blockers of UVR; the particulate phase of the suspension (usually unagglomerated nanosize particles of zinc oxide or titanium dioxide) reflects or scatters UVR.

A large number of *skincare* and *makeup* creams marketed today are in the form of an emulsion [56,57,65–69]. Both O/W and W/O emulsion creams are manufactured commercially, although a majority of the cosmetic creams available in the market are O/W type with oil droplets as the dispersed phase and aqueous solution as the continuous phase. The W/O-type emulsion creams are often used as barrier creams to lock moisture in the skin over a period of time by forming a protective hydrophobic film on the skin. Although the O/W-type emulsion creams are used for various reasons, one important function of these creams is to prevent dryness of the skin by replacing lost moisture and by keeping the skin hydrated over a period of time.

1.1.12 GEOLOGICAL APPLICATIONS

Dispersions of gas bubbles in magma (molten matter of silicate composition), also referred to as *magmatic emulsions* in the literature, play an important role in volcanic eruptions [70–79]. During an eruption, as the magma rises through a volcanic conduit toward the earth's surface, gas bubbles are formed due to ex-solution of volatiles (mainly water and carbon dioxide) that were initially dissolved in magma at high pressures. The bubbles also tend to grow in size as the magma rises toward the surface. The growth of bubbles is governed by decompression and diffusional processes.

The vesiculation of the rising magma drastically changes the physicochemical properties and rheology of magma within the conduit and, thus, the ascent rate of magma. The changes in magma rheology occur partly because of a change in the melt rheology (owing to loss of volatiles) and partly owing to the introduction of bubbles in the melt.

Figure 1.3 summarizes the main processes occurring in a volcanic conduit. A major portion of the conduit consists of a two-phase "bubbly flow" region in which bubbles nucleate and grow. At the end of the bubbly section, the magmatic emulsion fragments and turns into a gas consisting of small particles (ash), which may be liquid or partially solid.

Rheology of magmas and magmatic emulsions is an active area of research. A good understanding of the rheology of bubble-bearing magmas, over a wide range of bubble volume fractions, is important in the analysis of ascent, eruption, and emplacement of magmas [70–77].

1.2 COMPOSITES

Composites are solid heterogeneous materials composed of at least two phases: the continuous phase is termed the *matrix*, and the phase that is dispersed in the matrix as discrete units is called the *dispersed phase*. The mechanical properties of composites depend on factors such as: (1) the volume fraction of the dispersed phase, (2) the geometry of the dispersed phase (shape, size and size distribution,

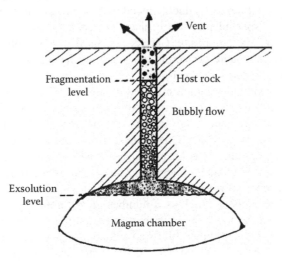

FIGURE 1.3 Processes occurring in a volcanic conduit.

and orientation of the dispersed units), and (3) the properties of the constituent phases. Composite materials can be roughly classified into two broad groups, namely *particle-reinforced (particulate)* composites and *fiber-reinforced* composites. Particulate composites generally consist of nearly isometric (same dimensions in all directions) particles, whereas fiber-reinforced composites consist of fibers of high aspect ratio as the dispersed phase.

1.2.1 PARTICULATE COMPOSITES AND THEIR APPLICATIONS

Particulate composites can be subdivided into two groups: coarse-particle composites and fine-particle composites. Coarse-particle composites consist of particles significantly larger than 1 μm, and the volume fraction of the dispersed phase in these composites is often high. Examples of coarse-particle composites are ceramic-metal composites, solid propellants, wood-plastic composites, and traditional dental composites. Fine-particle composites consist of particles in the submicron or nanometer range and the volume fraction of the dispersed phase is usually small. Examples of fine-particle composites are dispersion-hardened metal alloys, carbon black–filled rubbers and plastics, and clay-polymer nanocomposites.

1.2.1.1 Coarse-Particle Composites

Ceramic-metal particulate composites, also called *cermets*, are widely used as a tool material for high-speed cutting of materials that are difficult to machine, such as hardened steels [80,81]. The ceramic material alone, although hard enough to provide the cutting surface, is extremely brittle whereas the metal alone, although tough, does not possess the requisite hardness. The combination of these two materials (ceramic and metal) in the form of a particulate composite overcomes the limitations of the individual materials. In cermets used for cutting tools, the particles of ceramic (tungsten carbide, titanium carbide, Al_2O_3, etc.) are embedded in a matrix of a ductile metal (cobalt, nickel, etc.). A large volume fraction of the dispersed particulate phase is generally used to maximize the abrasive action of the composite. Cermets are used in many other applications such as: (1) thermocouple protection tubes, (2) mechanical seals, (3) valve and valve seats, and (4) turbine wheels.

The solid propellants commonly used in aerospace propulsion are particulate composites consisting of particles of solid oxidizer (usually ammonium perchlorate NH_4ClO_4) and metal fuel (usually aluminum) dispersed in a polymeric binder (usually polybutadiene). The fuel combines with oxygen provided by the oxidizer to produce gas for propulsion. The volume fraction of particles in solid propellants is typically high [82]. The composite is a rubberlike material with the consistency of a rubber eraser.

The plastics industry employs a number of different types of particulate fillers to improve the mechanical properties of the plastic. In several applications, less expensive particulate fillers are added to expensive plastic materials, mainly to lower the cost. Wood-derived fillers are receiving a lot of attention these days. According to some estimates, wood-plastic composites (WPCs) are the fastest growing construction materials today [83–87]. Wood filler, most often used in a particulate form referred to as wood flour, has several advantages over the traditional inorganic fillers. It is derived from a renewable resource, is lighter, and is less expensive. Also, it is less abrasive to processing equipment as compared to traditional fillers. Commercially produced wood-flour filler generally consists of large-size particles (>100 μm). The weight fraction of wood in WPCs is typically 0.5, although some WPCs contain a much larger amount of wood (as high as 70% by weight), and others contain only a small amount of wood (as low as 10% by weight). Both thermoset plastics and thermoplastics are used as matrix materials for WPCs, although most WPCs are currently manufactured with thermoplastics such as polyethylene, polypropylene, and polyvinylchloride as the matrix.

Dental composites consist of a polymerizable resin matrix, usually urethane dimethacrylate (UDMA) or ethylene glycol dimethacrylate (Bis-GMA), glass particulate fillers, and a silane coupling agent [88,89]. Polymerization of the resin matrix is either light activated or chemically initiated. The silane coupling agent (usually 3-methacryloxypropyltrimethoxy silane) coats the surface of the hydrophilic filler particles, allowing them to couple with the hydrophobic resin matrix. The purpose of fillers in dental composites is to reduce shrinkage (the resin tends to shrink while it sets) and to improve mechanical properties (wear resistance, fracture resistance) of the material. A wide range of particle sizes are used in the manufacture of commercial dental composites [89,90]. Based on the particle size of the filler, dental composites can be classified roughly into four broad groups: (1) traditional composites with filler particle size in the micron range (>> 1 μm), (2) microfilled composites with filler particle size close to 1 μm ($\approx$1 μm), (3) nanocomposites with filler particle size in the nanometer range (< 100 nm), and (4) hybrid composites consisting of a bimodal mixture of very fine and large particles. The volume fraction of the filler in the composite is usually high, somewhere in the range of approximately 0.4 to 0.80.

1.2.1.2 Fine-Particle Composites

Metal alloys and metals are often strengthened and hardened by a uniform dispersion of nanosize particles (usually between 10 to 100 nm) of a very hard and inert material such as thoria (ThO_2). The volume fraction of the dispersed phase in these dispersion-hardened metals and metal alloys is generally small (rarely exceeding 5%) [81].

Automobile tires are made from carbon-black-reinforced rubber [91]. Nearly a quarter of the total weight of a tire is made up of carbon black. The nanosize carbon black particles (usually between 20 and 50 nm) enhance the resistance to wear and tear and increase stiffness and tensile strength of the tires. It is of interest to note that the primary nanosize particles of carbon black aggregate to form chains with various degrees of branching in the rubber matrix. It is the network of these chains of primary particles that provides the reinforcement mechanism in carbon-black-reinforced rubbers.

Plastics filled with fine *conductive particles*, also called conductive plastics, have many practical applications [92]. Plastics are electrically insulating materials. However, the dispersion of conductive filler such as carbon black in the plastic matrix imparts conductive properties to the plastic system. The conductive carbon black particles when dispersed in a plastic matrix form a network of chainlike aggregates. Although the matrix is nonconductive, current can still flow through the network of conductive particles. Carbon-black-filled conductive plastics are used as antistatic materials in the electronic industry. Plastics, widely used as insulators, readily pick up electrostatic charges, especially under low humidity conditions. When earthed, the plastics (carrying electrostatic charge) discharge, and in the process damage electronic circuitry and equipment. To overcome these problems, conductive plastics are used. Conductive plastics are used in many other applications such as EMI (electromagnetic interference) shielding, heating devices, video discs, wire and cable, etc.

Polymer-clay nanocomposites are a rapidly growing class of nano-engineered materials [93,94]. They are composed of nanometer-sized clay particles dispersed uniformly in a polymeric matrix. Note that clays (usually aluminosilicates), in their natural form, consist of aggregates of primary platelike particles, whereas polymer-clay nanocomposites consist of an exfoliated structure, that is, the aggregates are completely separated and the individual primary particles are dispersed uniformly in the polymer matrix (Figure 1.4). The thickness of clay platelets is approximately 1 nm, and their diameter can vary anywhere from tens of nanometers to hundreds of nanometers. As the clay particles in their natural form are generally hydrophilic, they need to be treated to make them "organophilic" and compatible with hydrophobic polymers. Usually, only small quantities (less than 6% by weight) of clay are incorporated in polymer-clay nanocomposites. A small quantity of clay in exfoliated form is often enough to provide a large improvement in mechanical and other desired properties.

1.2.2 FIBER-REINFORCED COMPOSITES AND THEIR APPLICATIONS

Fiber-reinforced composites, as the name implies, consist of fibers of high aspect ratio as the dispersed phase. They are the most widely used composite materials at present [95–97]. The fibers bear the major portion of an applied load, and the function of the matrix is to distribute and transmit load to the fibers. The main types of fiber materials used in these composites are glass, aramid (Kevlar), carbon, boron, and silicon [97]. The diameter of the fibers varies from a few microns to as much as 200 μm. The volume fraction of the dispersed phase (fibers) in the composite can be anywhere in the range of about 0.1 to 0.7. Fiber-reinforced composites can be classified into two broad groups: continuous (long) fiber composites and discontinuous (short) fiber composites. The orientation of the fibers relative to one another is an important factor in determining the strength of the composite. The fibers may all be aligned in a single direction or their orientation may be random.

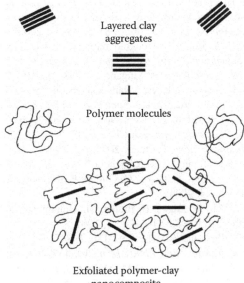

Layered clay
aggregates

+

Polymer molecules

Exfoliated polymer-clay
nanocomposite

FIGURE 1.4 Polymer-clay nanocomposite.

Fiber-reinforced composite materials have many commercial and industrial applications [95–97]. They are widely used in aircraft, space, automotive, and sports industries. Weight reduction is often the primary reason for selecting fiber-reinforced composites in many applications. The specific modulus (modulus per unit mass) and specific strength (strength per unit mass) of fiber-reinforced composites are significantly higher than those of traditional materials such as metals [96]. Therefore, it is possible to manufacture components having lower weight without sacrificing the stiffness and strength of the material.

REFERENCES

1. D.W. Richerson, *Modern Ceramic Engineering*, Marcel Dekker, New York, 1982.
2. G.C. Phillips, *A Concise Introduction to Ceramics*, Van Nostrand Reinhold, New York, 1991.
3. W.D. Kingery, H.K. Bowen, D.R. Uhlmann, *Introduction to Ceramics*, John Wiley & Sons, New York, 1976.
4. R.A. Terpstra, P.P.A.C. Pex, A.H. de Vries, *Ceramic Processing*, Chapman & Hall, London, 1995.
5. L. Bergstrom, in *Surface and Colloid Chemistry in Advanced Ceramics Processing*, R.J. Pugh, L. Bergstrom, Eds., Marcel Dekker, New York, 1994.
6. T.E. Karis, *J. Appl. Polym. Sci.*, 59: 1405–1416, 1996.
7. Applied Alloy Chemistry Group, magnetic recording, University of Birmingham, http://www.aacg.bham.ac.uk/magnetic_materials/magnetic_recording.htm.

8. T.M. Kwon, P.L. Frattini, L.N. Sadani, M.S. Jhon, *Colloids Surf. A*, 80: 47–61, 1993; also see *Colloids Surf. A*. 80: vii–viii, 1993.
9. M.N. Rittner, *Am. Ceramic Soc. Bull.*, 81(3), March 2002.
10. R.H. Muller, S.A. Runge, Solid lipid nanoparticles (SLN) for controlled drug delivery, in *Submicron Emulsions in Drug Targeting and Delivery*, S. Benita, Ed., Harwood Academic Publishers, Singapore, 1998, pp. 219–234, chap. 9.
11. R.H. Muller, C. Jacobs, O. Kayser, Nanosuspensions for the formulation of poorly soluble drugs, in *Pharmaceutical Emulsions and Suspensions*, F. Nielloud, G. Marti-Mestres, Eds., Marcel Dekker, New York, 2002, pp. 383–407, chap. 12.
12. I. Safarik, M. Safarikova, *Monatsh Chem.*, 133: 737–759, 2002.
13. P. Gould, *Materials Today*, February 2004, pp. 36–43.
14. Z.M. Saiyed, S.D. Telang, C.N. Ramchand, *Biomagn. Res. Technol.*, 1: 2, 2003.
15. K. Buscher, C.A. Helm, C. Gross, G. Glockl, E. Romanus, W. Weitschies, *Langmuir*, 20: 2435–2444, 2004.
16. D.K. Kim, W. Voit, W. Zapka, B. Bjelke, M. Muhammed, K.V. Rao, *Mater. Res. Soc. Symp. Proc.* 676: Y8.32.1–Y8.32.6, 2001.
17. J. Popplewell, *Phys. Technol.*, 15: 150–156, 1984.
18. R.E. Rosensweig, *Annu. Rev. Fluid Mech.*, 19: 437–463, 1987.
19. P.M. Kieran, P.F. MacLoughlin, D.M. Malone, *J. Biotechnol.*, 59: 39–52, 1997.
20. D.G. Allen, C.W. Robinson, *Chem. Eng. Sci.*, 45: 37–48, 1990.
21. M. Rodriguez-Monroy, E. Galindo, *Enzyme Microb. Technol.*, 24: 687–693, 1999.
22. G. Trejo-Tapia, A. Jimenez-Aparicio, L. Villarreal, M. Rodriguez-Monroy, *Biotechnol. Lett.*, 23: 1943–1946, 2001.
23. E.S. Olsvik, B. Kristiansen, *Biotechnol. Bioeng.*, 40: 375–387, 1992.
24. E.H. Dunlop, P.K. Namdev, M.Z. Rosenberg, *Chem. Eng. Sci.*, 49: 2263–2276, 1994.
25. E.T. Papoutsakis, *Trends Biotechnol.*, 9: 427–437, 1991.
26. A. Doroszkowski, Paints, in *Technical Applications of Dispersions*, R.B. McKay, Ed., Marcel Dekker, New York, 1994, chap. 1.
27. C.R. Martens, *Technology of Paints, Varnishes, and Lacquers*, Krieger, Malabar, FL, 1974.
28. G.P.A. Turner, *Introduction to Paint Chemistry and Principles of Paint Technology*, Chapman & Hall, London, 1980.
29. H.P. Preuss, *Paint Additives*, Noyes Data Corp., Park Ridge, NJ, 1970.
30. G. Peet, International cooperation to prevent oil spills at sea: not quite the success it should be, in *Green Globe Yearbook of International Cooperation on Environment and Development*, H.O. Bergesen, G. Parmann, Eds., Oxford University Press, Oxford, 1994, pp. 41–54.
31. M.L. O'Brien, At-sea recovery of heavy oils — a reasonable response strategy, in *3rd R&D Forum on High Density Oil Spill Response*, The International Tanker Owners Pollution Federation Limited, London, 2002, pp. 1–20.
32. Alaska Oil Spill Association, Spill: The Wreck of the Exxon Valdez, final report, the State of Alaska, 1990.
33. M. Fingas, *The Basics of Oil Spill Cleanup*, Lewis Publishers, Boca Raton, FL, 2001.
34. ITOPF, Fate of Marine Oil Spills — Technical information paper No. 2, The International Tanker Owners Pollution Federation Limited (ITOPF), London, 1–8, 2002.
35. P. Hepple, Ed., *Water Pollution by Oil*, The Institute of Petroleum, London, 1971.
36. J.R. Clayton, J.R. Payne, J.S. Farlow, *Oil Spill Dispersants — Mechanisms of Action and Laboratory Tests*, C.K. Smoley Inc., Boca Raton, FL, 1993.

37. M.J. Lynch, W.C. Griffin, Food emulsions, in *Emulsions and Emulsion Technology — Part 1*, K.J. Lissant, Ed., Marcel Dekker, New York, 1974, chap. 5.

38. N. Krog, T.H. Riisom, K. Larsson, Applications in the food industry — I, *Encyclopedia of Emulsion Technology: Applications*, Vol. 2, P. Becher, Ed., Marcel Dekker, New York, 1985, chap. 5.

39. E.N. Jaynes, Applications in the food industry — II, *Encyclopedia of Emulsion Technology*, Vol. 2, P. Becher, Ed., Marcel Dekker, New York, 1985, chap. 6.

40. M.C. Roco, C.A. Shook, *Slurry Flow*, Butterworth-Heinemann, Oxford, 1991.

41. S.S. Marsden, K. Ishimoto, L. Chen, *Colloids Surf.*, 29: 133–146, 1988.

42. J.L. Zakin, R. Pinaire, M.E. Borgmeyer, *J. Fluids Eng.*, 101: 100, 1979.

43. J.S. Chong, E.B. Christiansen, A.D. Baer, *J. Appl. Polym. Sci.*, 15: 2007, 1971.

44. L.Y. Sadler, K.G. Sim, *Chem. Eng. Prog.*, 87(3): 68–71, 1991.

45. R.L. Hoffman, *J. Rheol.*, 36: 947–965, 1992.

46. R. Pal, *AIChE J.*, 42: 3181–3190, 1996.

47. F. Steinhauff, *Petroleum*, 25: 294–296, 1962.

48. K.J. Lissant, in *AIChE Symposium on Improved Oil Recovery by Surfactant and Polymer Flooding*, D.O. Shah, P.S. Schechter, Eds., AIChE, New York, 1978, pp. 93–100.

49. R. Raghavan, S.S. Marsden, *SPEJ.*, 11: 153–161, 1971.

50. F.W. Jenkins, Oil in water emulsions: causes and treatment, in *Water Problems in Oil Production — An Operator's Manual*, 2nd ed., L.C. Case, Ed., Petroleum Publishing Co. Tulsa, OK, 1977, chap. 13.

51. R. Simon, W.G. Poynter, Down-hole emulsification for improving viscous crude production. *J. Pet. Technol.*, 20(12): 1349–1353, 1968.

52. A.H. Beyer, D.E. Osborn, Down-Hole Emulsification for Improving Paraffinic Crude Production, presented at the 44th Annual Fall Meeting of the SPE and AIME, Denver, CO, 1968; paper SPE 2676.

53. T.G.J. Jones, T.L. Hughes, Drilling fluid suspensions, in *Suspensions: Fundamentals and Applications in the Petroleum Industry*, L.L. Schramm, Ed., American Chemical Society, Washington, D.C., 1996, chap. 10.

54. J.L. Lummus, Water-In-Oil Emulsion Drilling Fluid, U.S. Patent. 2,661,334, 1953.

55. M. Pader, Dentrifice rheology, in *Rheological Properties of Cosmetics and Toiletries*, D. Laba, Ed., Marcel Dekker, New York, 1993, chap. 7.

56. T. Hargreaves, *Chemical Formulation — An Overview of Surfactant-Based Preparations Used in Everyday Life*, The Royal Society of Chemistry, Cambridge University Press, Cambridge, 2003.

57. J.S. Douglas, *Making Your Own Cosmetics*, Pelham Books, London, 1979.

58. D. Laba, Antiperspirant/deodorant rheology, in *Rheological Properties of Cosmetics and Toiletries*, D. Laba, Ed., Marcel Dekker, New York, 1993, chap. 6.

59. M. Mitchnick, *Drug Cosmet. Ind.*, 153: 38–43, August 1993.

60. M. Mitchnick, *Cosmet. Toiletries* 107: 111–116, 1992.

61. J.P. Hewitt, *Drug Cosmet. Ind.*, 151: 26–29, September 1992.

62. U. Heinrich, H. Tronnier, D. Kockott, R. Kuckuk, H.M. Heise, *Int. J. Cosmet. Sci.*, 26: 79–89, 2004.

63. S.A. Wissing and R.H. Muller, *Int. J. Cosmet. Sci.*, 23: 233–243, 2001.

64. M. Turkoglu, S. Yener, *Int. J. Cosmet. Sci.*, 19: 193–201, 1997.

65. P. Becher, *Emulsions: Theory and Practice*, Krieger Publishing Co., Malabar, FL, 1977.

66. H. Bennett, J.L. Bishop, M.F. Wulfinghoff, *Practical Emulsions: Applications*, Vol. II, Chemical Publishing Co., New York, 1968, chap. 4.
67. R.G. Harry, J.B. Wilkinson, *Harry's Cosmeticology*, George Godwin Ltd., London, 1976.
68. J.S. Jellinek, *Formulation and Function of Cosmetics*, Wiley-Interscience, New York, 1970.
69. M.M. Breuer, Cosmetic emulsions, in *Encyclopedia of Emulsion Technology: Applications*, Vol. 2, P. Becher, Ed., Marcel Dekker, New York, 1985, chap. 7.
70. A.M. Lejeune, Y. Bottinga, T.W. Trull, P. Richet, *Earth Planet. Sci. Lett.*, 166: 71–84, 1999.
71. F.J. Spera, D.J. Stein, *Earth Planet. Sci. Lett.*, 175: 327–331, 2000.
72. D.J. Stein, F.J. Spera, *J. Volcanol. Geotherm. Res.*, 49: 157–174, 1992.
73. M. Manga, J. Castro, K.V. Cashman, M. Loewenberg, *J. Volcanol. Geotherm. Res.*, 87: 15–28, 1998.
74. M. Manga, M. Loewenberg, *J. Volcanol. Geotherm. Res.*, 105: 19–24, 2001.
75. A.C. Rust, M. Manga, *J. Non-Newtonian Fluid Mech.*, 104: 53–63, 2002.
76. D.J. Stein, F.J. Spera, *J. Volcanol. Geotherm. Res.*, 113: 243–258, 2002.
77. R. Pal, *Earth Planet. Sci. Lett.*, 207: 165–179, 2003.
78. R.S.J. Sparks, *J. Volcanol. Geotherm. Res.*, 105: 19–24, 2001.
79. G. Macedonio, A. Neri, J. Marti, A. Folch, *J. Volcanol. Geotherm. Res.*, 143: 153–172, 2005.
80. J.M. Berthelot, *Composite Materials*, Springer-Verlag, New York, 1999.
81. M.M. Schwartz, *Composite Materials Handbook*, McGraw-Hill, New York, 1992.
82. A. Gocmez, C. Erisken, U. Yilmazer, F. Pekel, S. Ozkar, *J. Appl. Polym. Sci.*, 67: 1457–1464, 1998.
83. B. Dawson-Andoh, L.M. Matuana, J. Harrison, *J. Vinyl Additive Technol.*, 10: 179–186, 2004.
84. C. Clemons, *Forest Products J.*, 52: 10–18, June 2002.
85. B. Shah, L.M. Matuana, *J. Vinyl Additive Technol.*, 10: 121–128, 2004.
86. H. Jiang, D.P. Kamdem, *J. Vinyl Additive Technol.*, 10: 59–69, 2004.
87. G. Pritchard, *Reinforced Plastics*, 48(6): 26–29, June 2004.
88. D.W. Jones, *J. Can. Dent. Assoc.* 64: 732–734, 1998.
89. M. Zhou, J.L. Drummond, L. Hanley, *Dent. Mater.*, 21: 145–155, 2005.
90. Y. Li, M.L. Swartz, R.W. Phillips, B.K. Moore, T.A. Roberts, *J. Dent. Res.* 64: 1396–1401, 1985.
91. D. Parkinson, *Br. J. Appl. Phys.* 2: 273–280, 1951.
92. J. Markarian, *Plastics Additives Compounding*, 7(1): 26–30, 2005.
93. T.J. Pinnavaia, G.W. Beall, Eds., *Polymer-Clay Nanocomposites*, John Wiley & Sons, New York, 2000.
94. F. Gao, Clay/polymer composites: the story, *Materials Today*, 7(11): 50–55, 2004.
95. P.K. Mallick, *Fiber-Reinforced Composites-Materials, Manufacturing, and Design*, Marcel Dekker, New York, 1988.
96. D. Hull, *An Introduction to Composite Materials*, Cambridge University Press, Cambridge, 1981.
97. D. Gay, S.V. Hoa, S.W. Tsai, *Composite Materials — Design and Applications*, CRC Press, Boca Raton, FL, 2003.

2 Brief Review of Mechanics of Fluids, Solids, and Interfaces

2.1 BASIC CONCEPTS OF CONTINUUM MECHANICS

2.1.1 CONTINUUM MODEL OF MATTER

Matter is known to be discrete or discontinuous at the microscopic level. All materials are composed of molecules or atoms with empty space between them. Solids, liquids, and gases differ in the average spacing of molecules. The molecules of solids are more closely packed compared to those of liquids. In gases, the average spacing between the molecules is much larger compared to solids and liquids.

In the continuum model of matter, the discrete molecular structure of matter is ignored, and it is treated as a continuous medium. This hypothesis that matter is a continuum is adopted in many branches of science. At each point of this continuum, one can assign unique values of kinematical and dynamical variables. Thus, field variables are considered to be continuous functions of space coordinates and time. At any particular point in space, the values of field variables can be regarded as the average values of the various molecules that occupy a small volume in the neighborhood of that point.

The continuum hypothesis is not always valid. For this hypothesis to be valid, it is necessary that the average spacing between the molecules be very small in comparison with the characteristic size of the flow system. In most situations encountered in engineering and science, this condition is readily met.

2.1.2 KINEMATICS

Kinematics is the branch of mechanics that deals with the analysis and description of deformation and flow of matter without referring to the forces that produce the deformation or motion. It deals with physical quantities such as displacement, velocity, acceleration, deformation, and rotation of material particles.

2.1.2.1 Lagrangian and Eulerian Descriptions of Deformation and Flow

There are two basic methods of describing the deformation or flow of materials: Lagrangian and Eulerian methods.

In the Lagrangian description (also referred to as material description), one essentially follows the history of individual particles as the body (collection of matter) undergoes deformation or flow. Each particle of the continuum is labeled, usually by specifying its spatial coordinates at some initial time. Then, the path, velocity, etc. of each individual particle are traced as time passes.

Suppose that at some initial time $t = t_o$, the particle P occupies a position described by its position vector $\vec{R}$ measured from the origin of some fixed frame of reference. Let the position vector of the particle some time later, at time t, be $\vec{r}$. Then the equation of the form:

$$\vec{r} = \vec{r}(\vec{R},t) \text{ with } \vec{r}(\vec{R},t_o) = \vec{R} \tag{2.1}$$

describes the path of every particle, which, at time $t = t_o$ is located at $\vec{r} = \vec{R}$. Obviously, $\vec{R}$ is different for different particles.

Equation 2.1 can also be written in component form if we take $\vec{R}$ as

$$\vec{R} = X\hat{i} + Y\hat{j} + Z\hat{k} \tag{2.2}$$

and $\vec{r}$ as

$$\vec{r} = x\hat{i} + y\hat{j} + z\hat{k} \tag{2.3}$$

where $\hat{i}, \hat{j}$, and $\hat{k}$ are unit vectors in the three coordinate directions. Thus,

$$x = x(X,Y,Z,t) \tag{2.4a}$$

$$y = y(X,Y,Z,t) \tag{2.4b}$$

$$z = z(X,Y,Z,t) \tag{2.4c}$$

In Equation 2.4, the coordinates (X,Y,Z) serve to identify the different particles of the continuum body and are known as material coordinates. Equation 2.1 or Equation 2.4 is said to define the motion of a continuum. An important point to note is that in the Lagrangian, or material description, the independent variables are: X,Y,Z, and t, where X,Y,Z are the coordinates of the particle at some reference time $t = t_o$.

As the body undergoes deformation or flow, physical quantities (other than the position vector already discussed) associated with the individual particles change with time. All these physical quantities can be expressed as functions of time and material coordinates X,Y,Z of the individual particles. Thus,

$$Q = Q(X,Y,Z,t) \qquad (2.5)$$

where Q is any physical quantity (scalar, vector, or tensor) associated with the particle.

EXAMPLE

Consider the motion $\vec{r} = \vec{R} + ktY\hat{i}$, that is, $x = X + ktY$, $y = Y$, $z = Z$, where the material coordinates (X,Y,Z) give the position of a particle at $t = 0$. A material body, at $t = 0$, has the shape of a cube of unit sides as shown in the figure. Sketch the configuration of the body at time t if it undergoes the given motion.

Solution

At $t = 0$, the positions of the particles located at A, B, C, and D are given as: $(X,Y,Z)_{\text{particle at A}} = (0, 0, 0)$, $(X,Y,Z)_{\text{particle at B}} = (1, 0, 0)$, $(X,Y,Z)_{\text{particle at C}} = (1, 1, 0)$, $(X,Y,Z)_{\text{particle at D}} = (0, 1, 0)$. Owing to the given motion, the position of the particles at time t are: $(x,y,z)_{\text{particle at A}} = (0, 0, 0)$ — the particle at "A" does not move, $(x,y,z)_{\text{particle at B}} = (1, 0, 0)$ — the particle at "B" does not move, $(x,y,z)_{\text{particle at C}} = (1 + kt, 1, 0)$ — the particle at "C" moves a distance "kt" in the x-direction to a new location C', $(x,y,z)_{\text{particle at D}} = (kt, 1, 0)$ — the particle at "D" moves a distance "kt" in the x-direction to a new location D'. Thus, the configuration of the body at time t is ABC'D', as shown in the figure. This type of motion is known as *simple shearing motion*.

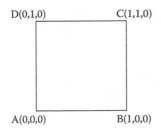

Configuration of the body at t = 0

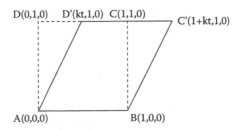

Configuration of the body at time t

In contrast to the Lagrangian approach, the *Eulerian* method does not trace individual particles when the body undergoes deformation or flow. Rather, it describes the state of affairs at every fixed point in the space occupied by the material continuum as a function of time. For example, the velocity field in Eulerian description can be written as: $\bar{u} = \bar{u}(x, y, z, t)$. The independent variables here are space coordinates x, y, and z , and time t. As any given spatial position is occupied by different particles at different times, the Eulerian description (also referred to as *spatial description*) does not provide direct information regarding the changes in particle properties as it moves about. It describes only the change of quantities at a fixed location as a function of time.

EXAMPLE

Given the motion of a body in Lagrangian description to be

$$x = X/(1+tX); \quad y = Y; \quad z = Z$$

Find the spatial and material descriptions of velocity.

Solution

The Lagrangian description of velocity can be expressed as

$$u_x = \left[\frac{\partial x}{\partial t}\right]_{(X,Y,Z-fixed)} = -\frac{X^2}{\left(1+tX\right)^2}$$

$$u_y = \left[\frac{\partial y}{\partial t}\right]_{(X,Y,Z-fixed)} = 0$$

$$u_z = \left[\frac{\partial z}{\partial t}\right]_{(X,Y,Z-fixed)} = 0$$

Solving for X and substituting in the material description of velocity, one can obtain the spatial description of velocity. Thus,

$$X = x/(1-tx); \quad u_x = -x^2; \quad u_y = 0; \quad u_z = 0$$

Note that the functional form of u_x in Eulerian description ($u_x = -x^2$) is different from the functional form in Lagrangian description, that is, $u_x = -X^2/(1+tX)^2$.

2.1.2.2   Displacement Gradient Tensors

Suppose a body having a particular configuration at some reference time t_o changes to another configuration at time t. The neighboring particles that occupy points P_o and Q_o before deformation move to points P and Q, respectively, in the deformed configuration (see Figure 2.1). The position vectors of points P_o and Q_o with respect to the origin are $\vec{X}$ and $\vec{X} + d\vec{X}$, respectively. The position vectors of points P and Q with respect to the origin are $\vec{x}$ and $\vec{x} + d\vec{x}$, respectively. The particle at point P_o undergoes a displacement $\vec{\xi}$, whereas the particle at point Q_o undergoes a displacement $\vec{\xi} + d\vec{\xi}$. The displacement gradient tensor $(\bar{\bar{J}})$ is defined as:

$$\bar{\bar{J}} = \frac{\partial \vec{\xi}}{\partial \vec{X}} = \left(\nabla \vec{\xi}\right)^T \tag{2.6}$$

where $\nabla \vec{\xi}$ is the dyadic product of ∇ and $\vec{\xi}$, and $(\nabla \vec{\xi})^T$ is the transpose of $\nabla \vec{\xi}$. The displacement gradient tensor so defined with respect to the initial configuration (with $\vec{X}$ as independent coordinates) is also referred to as the *material/Lagrangian* displacement gradient tensor. The *spatial/Eulerian* displacement gradient tensor $(\bar{\bar{K}})$ is defined as:

$$\bar{\bar{K}} = \frac{\partial \vec{\xi}}{\partial \vec{x}} \tag{2.7}$$

where the independent coordinates are $\vec{x}$, which refer to the final deformed configuration.

Information about the deformation and rotation of the material is embodied in the displacement gradient tensor. It can readily be shown that:

$$d\vec{x} = d\vec{X} + \bar{\bar{J}} \bullet d\vec{X} \tag{2.8}$$

If $\bar{\bar{J}}$ is zero, $d\vec{x} = d\vec{X}$ and the motion of the body is just a rigid body translation.

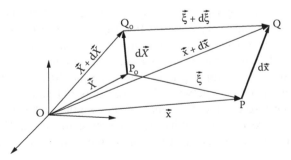

FIGURE 2.1 Displacement of two neighboring particles upon deformation of a material.

2.1.2.3 Deformation Gradient Tensors

The material line element P_0Q_0 shown in Figure 2.1 undergoes stretching and rotation as the body deforms. The change in the material line element PQ, that is $d\vec{x}$, with respect to the undeformed line element P_0Q_0, that is $d\vec{X}$, can be expressed as:

$$d\vec{x} = \bar{\bar{F}} \bullet d\vec{X} \qquad (2.9)$$

where $\bar{\bar{F}} = \partial \vec{x}/\partial \vec{X} = (\nabla \vec{x})^T$ is called the *deformation gradient tensor*. As the independent coordinates in $\bar{F}$ refer to the initial configuration, the tensor $\bar{F}$ is also called the *material (Lagrangian)* deformation gradient tensor. The *spatial (Eulerian)* deformation gradient tensor $\bar{\bar{H}}$ is defined as:

$$\bar{\bar{H}} = \frac{\partial \vec{X}}{\partial \vec{x}} \qquad (2.10)$$

where the independent coordinates are $\vec{x}$, which refer to the final deformed configuration.

The deformation gradient tensor, similar to the displacement gradient tensor, contains information about stretching and rotation of material elements. The deformation gradient tensors are related to displacement gradient tensors as follows:

$$\bar{\bar{J}} = \bar{\bar{F}} - \bar{\bar{\delta}} \qquad (2.11)$$

$$\bar{\bar{K}} = \bar{\bar{\delta}} - \bar{\bar{H}} \qquad (2.12)$$

where $\bar{\bar{\delta}}$ is a unit tensor.

2.1.2.4 Finite Strain Tensors

When a material undergoes a rigid body motion (translation and/or rotation), the length of each and every material line segment remains constant. If and only if the distance between any two material points of the body changes, the body is said to have undergone strain. One way to distinguish between rigid body motion and straining is to determine the difference of the square of the differential lengths between line segments PQ and P_0Q_0 (see Figure 2.1). One could also use the difference of the lengths, rather than the difference of the square of the lengths, as a measure of strain. However, it is mathematically more convenient to compute the difference of the square of the lengths.

The square of the line element PQ of length $d\vec{x}$ is:

$$(\text{length PQ})^2 = (dx)^2 = d\vec{x} \bullet d\vec{x} = d\vec{X} \bullet \bar{\bar{G}} \bullet d\vec{X} \qquad (2.13)$$

where $\bar{\bar{G}} = \bar{\bar{F}}^T \bullet \bar{\bar{F}}$ is called *Green's deformation tensor.*
The square of the line element P_oQ_o of length $d\vec{X}$ is:

$$(\text{length } P_oQ_o)^2 = (dX)^2 = d\vec{X} \bullet d\vec{X} = d\vec{x} \bullet \bar{\bar{C}} \bullet d\vec{x} \qquad (2.14)$$

where $\bar{\bar{C}} = \bar{\bar{H}}^T \bullet \bar{\bar{H}}$ is known as *Cauchy's deformation tensor.*
Note that if $\bar{\bar{G}} = \bar{\bar{C}} = \bar{\bar{\delta}}$, then there is no change in the length of the material line elements and, hence, no deformation takes place.
The difference $(dx)^2 - (dX)^2$ can now be expressed as:

$$(dx)^2 - (dX)^2 = d\vec{X} \bullet (\bar{\bar{G}} - \bar{\bar{\delta}}) \bullet d\vec{X} = d\vec{X} \bullet 2\bar{\bar{\ell}} \bullet d\vec{X} \qquad (2.15)$$

where $\bar{\bar{\ell}} = (1/2)(\bar{\bar{G}} - \bar{\bar{\delta}}) = (1/2)(\bar{\bar{F}}^T \bullet \bar{\bar{F}} - \bar{\bar{\delta}})$ is called the *Lagrangian finite strain tensor.* It is a measure of the strain of the body. If $\bar{\bar{\ell}} = 0$, then no change occurs in the lengths of the material line segments and the body undergoes no strain.
The difference $(dx)^2 - (dX)^2$ can also be expressed as:

$$\left(dx\right)^2 - \left(dX\right)^2 = d\vec{x} \bullet \left(\bar{\bar{\delta}} - \bar{\bar{C}}\right) \bullet d\vec{x} = d\vec{x} \bullet 2\bar{\bar{e}} \bullet d\vec{x} \qquad (2.16)$$

where $\bar{\bar{e}} = (1/2)(\bar{\bar{\delta}} - \bar{\bar{C}}) = (1/2)(\bar{\bar{\delta}} - \bar{\bar{H}}^T \bullet \bar{\bar{H}})$ is called the *Eulerian finite strain tensor.*
The finite strain tensors can also be expressed in terms of the displacement gradient tensors as follows:

$$\bar{\bar{\ell}} = \frac{1}{2}\left[\bar{\bar{J}} + \bar{\bar{J}}^T + \bar{\bar{J}}^T \bullet \bar{\bar{J}}\right] = \frac{1}{2}\left[(\nabla\vec{\xi}) + (\nabla\vec{\xi})^T + (\nabla\vec{\xi}) \bullet (\nabla\vec{\xi})^T\right] \qquad (2.17)$$

$$\bar{\bar{e}} = \frac{1}{2}\left[\bar{\bar{K}} + \bar{\bar{K}}^T - \bar{\bar{K}}^T \bullet \bar{\bar{K}}\right] = \frac{1}{2}\left[(\nabla\vec{\xi}) + (\nabla\vec{\xi})^T - (\nabla\vec{\xi}) \bullet (\nabla\vec{\xi})^T\right] \qquad (2.18)$$

The independent coordinates in $\bar{\bar{\ell}}$ are the initial coordinates $\vec{X}$, whereas the independent coordinates in $\bar{\bar{e}}$ are the final coordinates $\vec{x}$.

Note that for small deformation, the displacement gradient is small compared to unity and the product terms $(\nabla\vec{\xi}) \bullet (\nabla\vec{\xi})^T$ are negligible. As a consequence:

$$\bar{\bar{\ell}} = \frac{1}{2}\left[\bar{\bar{J}} + \bar{\bar{J}}^T\right] = \frac{1}{2}\left[(\nabla\vec{\xi}) + (\nabla\vec{\xi})^T\right] \tag{2.19}$$

$$\bar{\bar{e}} = \frac{1}{2}\left[\bar{\bar{K}} + \bar{\bar{K}}^T\right] = \frac{1}{2}\left[(\nabla\vec{\xi}) + (\nabla\vec{\xi})^T\right] \tag{2.20}$$

These forms of strain tensors are referred to as *infinitesimal strain tensors*. When both the displacements and the displacement gradients are sufficiently small, the Lagrangian and Eulerian infinitesimal strain tensors are equal.

2.1.2.5   Material Time Derivative of a Spatial Function

Let $Q = Q(x,y,z,t)$ be a spatial (Eulerian) function representing some scalar, vector, or tensor property of a continuum. The material time derivative of the spatial function, denoted by DQ/Dt, is given as:

$$\frac{DQ}{Dt} = \frac{\partial Q}{\partial t} + \vec{u} \bullet \nabla Q \tag{2.21}$$

The material time derivative DQ/Dt describes the time rate of change of Q for a particle instantaneously located at (x,y,z). The term $\partial Q/\partial t$ gives the local rate of change of Q at a given location and is zero under steady-state conditions. The term $\vec{u} \bullet \nabla Q$ gives the convective rate of change of Q resulting from the movement of a particle from one position in space to another. The convective term is zero only if Q is space invariant.

The material time derivative is also referred to as the co-moving, substantial, hydrodynamic or particle time derivative. Although the concept of this derivative is Lagrangian in that it gives the time rate of change of Q for a material particle, the derivative itself is Eulerian. Many important flow field variables require the determination of a material time derivative. For example, the acceleration of a material particle instantaneously located at (x,y,z) is simply the material time derivative of the velocity field $\vec{u}(x,y,z,t)$.

Note that the material time derivative operates on quantities referred to a fixed coordinate system, unlike the convected time derivative which operates on quantities referred to a convected coordinate system.

EXAMPLE

The spatial distribution of velocity is described as: $u_x = x/(1+t)$, $u_y = 2y/(1+t)$, $u_z = 3z/(1+t)$. Determine the acceleration of the particles.

Solution

The acceleration of a particle is the time rate of change of velocity of the particle. Thus, it is the material time derivative of velocity given as:

$$\vec{a} = \frac{D\vec{u}}{Dt} = \frac{\partial \vec{u}}{\partial t} + \vec{u} \bullet \nabla \vec{u}$$

Using the preceding expression for $\vec{a}$ and the given spatial distribution of velocity, it can be shown that:

$$a_x = 0, \quad a_y = 2y/(1+t)^2, \quad a_z = 6z/(1+t)^2$$

Thus, the acceleration of the particles is:

$$\vec{a} = \left[2y/(1+t)^2 \right] \hat{j} + \left[6z/(1+t)^2 \right] \hat{k}$$

2.1.2.6   Material Time Derivative of a Volume Integral (Reynolds Transport Theorem)

The fundamental laws of mechanics are stated for a material system, defined as an arbitrary quantity of mass of fixed identity. We need to be able to transform these laws into an Eulerian formulation. This can be achieved by using the material derivative of a volume integral, also referred to as *Reynolds Transport Theorem*.

Let $b = b(x, y, z, t)$ be some scalar, vector, or tensor property per unit mass of a continuum given in spatial (Eulerian) coordinates. Consider an arbitrary volume V fixed in space and bounded by surface S. The total amount of the property in V, denoted by B, is then the volume integral

$$B = \int_V \rho b \, dV \qquad (2.22)$$

where ρ is the density of the material. The material time derivative of the volume integral, denoted by DB/Dt, is given as:

$$\frac{DB}{Dt} = \frac{D}{Dt} \int_V \rho b \, dV = \frac{\partial}{\partial t} \int_V \rho b \, dV + \int_S \rho b (\hat{n} \bullet \vec{u}) dS \qquad (2.23)$$

where $\hat{n}$ is an outer unit normal vector of S. This equation, known as the *Reynolds Transport Theorem*, states that the time rate of change of the property B in that portion of the continuum instantaneously occupying the volume V is equal to the time rate of change of the property within the volume V fixed in space, plus the net flux of B passing through the bounding surface S.

2.1.2.7   Velocity Gradient Tensor

Consider a material continuum undergoing flow. Let P and Q be two neighboring points. The velocity of the material at P is $\vec{u}(\vec{x},t)$, and the simultaneous velocity at a neighboring position Q is $\vec{u}+d\vec{u}$. The relative velocity at Q with respect to P, that is $d\vec{u}$, is given by:

$$d\vec{u} = \left(\frac{\partial \vec{u}}{\partial \vec{x}}\right)\bullet d\vec{x} = \left(\nabla\vec{u}\right)^{T}\bullet d\vec{x} \qquad (2.24)$$

where $(\partial\vec{u}/\partial\vec{x})$ is the velocity gradient tensor, given as the transpose of the dyadic product of the differential operator ∇ and velocity vector $\vec{u}$. The velocity gradient tensor, denoted by $\bar{\bar{\Gamma}}$, is a measure of the steepness of velocity variation in the flow field as one moves from one location to another at a given instant in time. The velocity gradient tensor $\bar{\bar{\Gamma}}$ contains all the information on rotation and deformation of the material elements.

2.1.2.8   Rate of Deformation Tensor and Vorticity Tensor

The velocity gradient tensor $\bar{\bar{\Gamma}}$ can be split into two tensors, one symmetric and one anti-symmetric, as follows:

$$\bar{\bar{\Gamma}} = \bar{\bar{E}} + \bar{\bar{\Omega}} \qquad (2.25)$$

where $\bar{\bar{E}}$ and $\bar{\bar{\Omega}}$ are given as:

$$\bar{\bar{E}} = \frac{1}{2}\left(\bar{\bar{\Gamma}}+\bar{\bar{\Gamma}}^{T}\right) = \frac{1}{2}\left[\nabla\vec{u}+\left(\nabla\vec{u}\right)^{T}\right] \qquad (2.26a)$$

$$\bar{\bar{\Omega}} = \frac{1}{2}\left(\bar{\bar{\Gamma}}-\bar{\bar{\Gamma}}^{T}\right) = \frac{1}{2}\left[\left(\nabla\vec{u}\right)^{T}-\nabla\vec{u}\right] \qquad (2.26b)$$

The tensor $\bar{\bar{E}}$, which is symmetric, is called the *rate of deformation tensor.* The rate of deformation tensor, also referred to as *rate of strain tensor* or *stretching tensor,* describes the rate at which neighboring material particles move with respect to each other independently of superimposed rigid rotations. Thus, all the information about deformation of material elements is contained in $\bar{\bar{E}}$. In general, the material elements undergo *linear* and *shear* strains. The linear strain (also called *extensional* or *normal strain*) is defined as the ratio of the change in length to the original length of a material line segment. The shear strain is defined as the change in angle between two line segments from the undeformed state to the deformed state. The linear strain rate of a material line segment of length dl is related to $\bar{\bar{E}}$ as follows:

$$\frac{1}{dl}\left[\frac{D\,(dl)}{Dt}\right] = \hat{b} \bullet \bar{\bar{E}} \bullet \hat{b} \tag{2.27}$$

where $\hat{b}$ is a unit vector parallel to the material line segment. The linear strain may produce changes in volume of the material elements; the *volumetric* strain rate (Δ), defined as the rate of change of volume per unit volume, is given by the trace of the rate of deformation tensor, that is:

$$\Delta = \underset{V(t)\to 0}{Lim}\frac{1}{V(t)}\frac{DV(t)}{Dt} = tr\left(\bar{\bar{E}}\right) = \nabla \bullet \vec{u} \tag{2.28}$$

The shear (angular) strain rate between two material line segments with unit vectors $\hat{a}$ and $\hat{b}$ (originally at right angles, i.e., $\hat{a} \bullet \hat{b} = 0$) is related to $\bar{\bar{E}}$ as follows:

$$\frac{D\theta}{Dt} = -\hat{a} \bullet 2\bar{\bar{E}} \bullet \hat{b} \tag{2.29}$$

The tensor $\bar{\bar{\Omega}}$ in Equation 2.25 is called the *vorticity (spin) tensor*. It is anti-symmetric, as can be seen from Equation 2.26b. The vorticity tensor contains all the information about the local rates of rotation of material elements. Interestingly, the relative velocity between two neighboring material particles, separated by a small distance $d\vec{x}$, can be divided into two parts as:

$$d\vec{u} = \left(\bar{\bar{E}} + \bar{\bar{\Omega}}\right) \bullet d\vec{x} = \left(\bar{\bar{E}} \bullet d\vec{x}\right) + \left(\bar{\bar{\Omega}} \bullet d\vec{x}\right) \tag{2.30}$$

The contribution $(\bar{\bar{E}} \bullet d\vec{x})$ to the relative velocity represents pure straining motion (no rotation) between the neighboring particles, and $(\bar{\bar{\Omega}} \bullet d\vec{x})$ represents a rigid body rotation.

2.1.2.9 Shear Rate Tensor

The shear rate tensor $\overset{=}{\dot{\gamma}}$ is defined as:

$$\overset{=}{\dot{\gamma}} = \nabla \vec{u} + \left(\nabla \vec{u}\right)^T = 2\overset{=}{E} \tag{2.31}$$

2.1.3 STRESS VECTOR AND STRESS TENSOR

Two types of forces act on a material continuum. One of these is the *body force* that acts throughout a volume and is expressed as either force per unit mass or force per unit volume. For example, gravitational force per unit mass of a material continuum is $\vec{g}$ and force per unit volume is $\rho\vec{g}$, where $\vec{g}$ is the acceleration due to gravity and ρ is the material density. The body force acts at a distance and not as a result of direct contact. The other type of force acting on a material continuum is the *surface force* (also referred to as *contact force*). The surface force acts on a surface, whether it is an arbitrary internal surface or the bounding surface of the continuum, and is expressed as force per unit area.

2.1.3.1 Stress Vector

Consider a material continuum occupying a volume V, and imagine a closed surface S within the volume V, as shown in Figure 2.2. The material outside S exerts a surface force on the material inside S. Let $\Delta\vec{F}$ be the total surface force exerted on a small surface element of area ΔS, whose spatial orientation is given by the unit normal vector $\hat{n}$. The force $\Delta\vec{F}$ depends on the location, size of the area element, and the orientation of the area element specified by unit normal vector $\hat{n}$. The average surface force per unit area on ΔS is $\Delta\vec{F}/\Delta S$. As ΔS tends to zero,

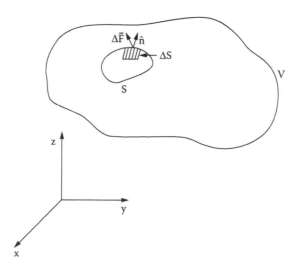

FIGURE 2.2 A material continuum occupying a volume V and subjected to surface forces.

keeping $\hat{n}$ constant, the ratio $\Delta\vec{F}/\Delta S$ approaches a definite limit $d\vec{F}/dS$. The limiting vector $d\vec{F}/dS$ is called the *stress vector* or sometimes the *traction vector*. Thus,

$$\vec{\sigma}_{(\hat{n})} = \underset{\Delta S \to 0}{Lim} \frac{\Delta\vec{F}}{\Delta S} = \frac{d\vec{F}}{dS} \tag{2.32}$$

where $\vec{\sigma}_{(\hat{n})}$ is the stress vector, which is a function of position, surface orientation, and time in general; that is:

$$\vec{\sigma}_{(\hat{n})} = \vec{\sigma}_{(\hat{n})}(\hat{n}, \vec{r}, t) \tag{2.33}$$

where $\vec{r}$ is the position vector and t is time.

2.1.3.2   Stress Tensor

Consider an arbitrary point Q in a material continuum. For each unit normal vector $\hat{n}$ at Q, there exists a corresponding stress vector $\vec{\sigma}_{(\hat{n})}$. The *state of stress* at point Q is defined by the totality of all possible pairs of companion vectors $\vec{\sigma}_{(\hat{n})}$ and $\hat{n}$ at that point. As an infinite number of unit normal vectors can be imagined at Q, each having a different orientation, there exist an infinite number of pairs of $\hat{n}$ and $\vec{\sigma}_{(\hat{n})}$ at point Q. Fortunately every pair of $\hat{n}$ and associated $\vec{\sigma}_{(\hat{n})}$ need not be specified. According to *Cauchy's theorem*, if the stress vectors acting on three mutually perpendicular planes at a given point are known, then all stress vectors at that point can be determined. Cauchy's formula, which is as follows, allows us to express the stress vector at any given location as $\hat{n}$ varies:

$$\vec{\sigma}_{(\hat{n})} = \bar{\bar{\sigma}}(\hat{r}, t) \bullet \hat{n} \tag{2.34}$$

The linear operator $\bar{\bar{\sigma}}$ that transforms the unit normal vector $\hat{n}$ into the stress vector $\vec{\sigma}_{(\hat{n})}$ is called the *stress tensor*. The stress tensor is independent of $\hat{n}$ (orientation of the surface) and has nine components. It is a point function that completely describes the state of stress at a given point. In the absence of body-moment and couple-stress, the stress tensor is symmetric, that is:

$$\bar{\bar{\sigma}} = \bar{\bar{\sigma}}^T \tag{2.35}$$

Consequently, $\bar{\bar{\sigma}}$ has only six independent components.

The total stress tensor $\bar{\bar{\sigma}}$ can be divided into two tensors as follows:

$$\bar{\bar{\sigma}} = -P\bar{\bar{\delta}} + \bar{\bar{\tau}} \tag{2.36}$$

where $\bar{\bar{\tau}}$ is known as the extra or deviatoric stress tensor (also referred to as viscous stress tensor), P is the local thermodynamic pressure, and $\bar{\bar{\delta}}$ is the unit tensor. For incompressible fluids, the thermodynamic pressure P is the same as the mean pressure $\bar{P}$, defined as:

$$\bar{P} = -\frac{1}{3}(tr\bar{\bar{\sigma}}) = -\frac{1}{3}\bar{\bar{\delta}} : \bar{\bar{\sigma}} \tag{2.37}$$

where tr refers to trace of a tensor.

2.1.4 CONSTITUTIVE EQUATION

Constitutive equations are equations that relate dynamical (stress) variables to appropriate kinematical (deformation) variables. Not all materials follow the same form of a constitutive equation. These equations are particular to individual classes of materials, and they serve to distinguish one class of material from another.

The so-called *Newtonian* fluids obey the following form of a constitutive equation:

$$\bar{\bar{\sigma}} = -P\bar{\bar{\delta}} + \lambda(tr\bar{\bar{E}})\bar{\bar{\delta}} + 2\eta\bar{\bar{E}} \tag{2.38}$$

where P is the thermodynamic pressure, η and λ are the first and second viscosity coefficients, respectively, of the fluid, and $\bar{\bar{E}}$ is the rate of strain tensor. The second viscosity coefficient λ is related to the first viscosity coefficient η (shear viscosity) and dilational (bulk) viscosity η^k as follows:

$$\lambda = \eta^k - \frac{2}{3}\eta \tag{2.39}$$

Consequently, the constitutive equation for Newtonian fluids can be written as:

$$\bar{\bar{\sigma}} = -P\bar{\bar{\delta}} + \left(\eta^k - \frac{2}{3}\eta\right)(tr\bar{\bar{E}})\bar{\bar{\delta}} + 2\eta\bar{\bar{E}} \tag{2.40}$$

The bulk viscosity η^k is related to viscous stresses produced only because of volumetric (dilational) strain of fluid elements. It is zero for monoatomic gases. Even for polyatomic gases, it is rather small and can frequently be neglected.

It should be noted that, in general, the mean pressure $\bar{P}$ and the thermodynamic pressure P are not the same. From Equation 2.37 and Equation 2.40, it follows that:

$$\bar{P} = P - \eta^k(\nabla \bullet \vec{u}) \tag{2.41}$$

However, for *incompressible* fluids (liquids are often assumed to be incompressible), $tr\bar{\bar{E}} = \nabla \bullet \vec{u} = 0$, and therefore, the mean and thermodynamic pressures are the same. As $tr\bar{\bar{E}} = 0$ for incompressible fluids, the constitutive equation (Equation 2.40) reduces to:

$$\bar{\bar{\sigma}} = -P\bar{\bar{\delta}} + 2\eta\bar{\bar{E}} \tag{2.42}$$

It is important to note that not all fluids follow the Newtonian constitutive equation. For example, particulate dispersions and polymeric liquids exhibit much more complex rheological behavior.

Another class of materials, the so-called *Hookean* solids (homogeneous and isotropic, linearly elastic solids), obeys the following constitutive equation:

$$\bar{\bar{\sigma}} = \lambda\left(tr\bar{\bar{e}}\right)\bar{\bar{\delta}} + 2G\bar{\bar{e}} \tag{2.43}$$

where λ and G (shear modulus) are the Lamé constants, and $\bar{\bar{e}}$ is the infinitesimal strain tensor. The Lamé constants λ and G are material properties. They vary from one Hookean material to another. Three other elastic constants of interest are Poisson's ratio (ν), bulk modulus (K), and Young's modulus (E); they can all be expressed in terms of λ and G as:

$$\nu = \frac{\lambda}{2(\lambda+G)} \tag{2.44}$$

$$K = \frac{3\lambda+2G}{3} \tag{2.45}$$

$$E = \left(\frac{3\lambda+2G}{\lambda+G}\right)G \tag{2.46}$$

For *incompressible* solids: $\lambda \to \infty$, $K \to \infty$, $\nu = +0.5$, $E = 3G$.

2.2 MECHANICS OF FLUIDS

2.2.1 THE CONTINUITY EQUATION

Consider an arbitrary volume V fixed in space and bounded by surface S. The amount of mass in that portion of the fluid instantaneously occupying the volume V is given by the integral

$$m = \int_V \rho(x,y,z,t)dV \tag{2.47}$$

where $\rho(x,y,z,t)$ is the density of the fluid given in spatial coordinates.

According to the law of conservation of mass, the mass of the fluid instantaneously occupying the volume V must remain constant. Thus, the material time derivative of m is zero:

$$\frac{Dm}{Dt} = \frac{D}{Dt} \int_V \rho(x, y, z, t) \, dV = 0 \tag{2.48}$$

Using the Reynolds transport theorem, Equation 2.48 gives:

$$\frac{\partial}{\partial t} \int_V \rho \, dV + \int_S \rho(\hat{n} \bullet \vec{u}) \, dS = 0 \tag{2.49}$$

Converting the second integral on the left-hand side of Equation 2.49 to a volume integral with the help of the Gauss divergence theorem (see Appendix C), one can write:

$$\frac{\partial}{\partial t} \int_V \rho \, dV + \int_V (\nabla \bullet \rho \vec{u}) \, dV = 0 \tag{2.50}$$

For a fixed volume in space, the operations of differentiation and integration may be interchanged in the first term of the left-hand side of the preceding equation. Thus,

$$\int_V \frac{\partial \rho}{\partial t} dV + \int_V (\nabla \bullet \rho \vec{u}) \, dV = 0 \tag{2.51}$$

or,

$$\int_V \left(\frac{\partial \rho}{\partial t} + \nabla \bullet \rho \vec{u} \right) dV = 0 \tag{2.52}$$

Because this equation holds for an arbitrary volume V, the integrand must also be zero, that is:

$$\frac{\partial \rho}{\partial t} + \nabla \bullet \rho \vec{u} = 0 \tag{2.53}$$

This is the *continuity equation*. If the density of the fluid is constant, the continuity equation reduces to:

$$\nabla \bullet \vec{u} = 0 \tag{2.54}$$

2.2.2 THE EQUATION OF MOTION

Newton's second law of motion states that the time rate of change of linear momentum of a body is equal to the resultant force (sum of forces) acting upon the body. Newton's second law of motion can be generalized for a fluid as follows: the time rate of change of linear momentum of a fluid instantaneously occupying an arbitrary volume V fixed in space is equal to the sum of forces acting upon the fluid occupying V.

The linear momentum of the fluid occupying the spatial volume V is given by:

$$\vec{M} = \int_V \rho \vec{u} dV \tag{2.55}$$

According to Newton's second law of motion, the material time derivative of $\vec{M}$ is

$$\frac{D\vec{M}}{Dt} = \frac{D}{Dt} \int_V \rho \vec{u} dV = \vec{F}_{body} + \vec{F}_{surface} \tag{2.56}$$

where the body force $\vec{F}_{body}$ is

$$\vec{F}_{body} = \int_V \rho \vec{g} dV \tag{2.57}$$

and the surface force $\vec{F}_{surface}$ is

$$\vec{F}_{surface} = \int_S \vec{\sigma}_{(\hat{n})} dS = \int_S \left(\bar{\bar{\sigma}} \bullet \hat{n} \right) dS \tag{2.58}$$

Using Gauss's divergence theorem, $\vec{F}_{surface}$ can be rewritten as

$$\vec{F}_{surface} = \int_V \left(\nabla \bullet \bar{\bar{\sigma}} \right) dV \tag{2.59}$$

Note that $\bar{\bar{\sigma}}$ is assumed to be symmetric. Combining Equation 2.56, Equation 2.57, and Equation 2.59 gives:

$$\frac{D}{Dt} \int_V \rho \vec{u} dV = \int_V \rho \vec{g} dV + \int_V \left(\nabla \bullet \bar{\bar{\sigma}} \right) dV \tag{2.60}$$

Using the Reynolds transport theorem and the Gauss divergence theorem, the material time derivative of $\vec{M}$ can be written as:

$$\frac{D}{Dt}\int_V \rho\vec{u}\,dV = \frac{\partial}{\partial t}\int_V (\rho\vec{u})dV + \int_S (\hat{n}\bullet\rho\vec{u}\vec{u})\,dS$$

$$= \int_V \frac{\partial}{\partial t}(\rho\vec{u})dV + \int_V (\nabla\bullet\rho\vec{u}\vec{u})\,dV \qquad (2.61)$$

Combining Equation 2.60 and Equation 2.61 gives:

$$\int_V \left[\frac{\partial}{\partial t}(\rho\vec{u})+\nabla\bullet\rho\vec{u}\vec{u}-\rho\vec{g}-\nabla\bullet\bar{\bar{\sigma}}\right]dV = 0 \qquad (2.62)$$

Because this equation holds for an arbitrary volume V, the integrand must also be zero, that is:

$$\frac{\partial}{\partial t}(\rho\vec{u})+\nabla\bullet\rho\vec{u}\vec{u} = \rho\vec{g}+\nabla\bullet\bar{\bar{\sigma}} = \rho\vec{g}-\nabla P+\nabla\bullet\bar{\bar{\tau}} \qquad (2.63)$$

This is the *equation of motion*. It can also be written as:

$$\rho\frac{D\vec{u}}{Dt} = \rho\vec{g}+\nabla\bullet\bar{\bar{\sigma}} = \rho\vec{g}-\nabla P+\nabla\bullet\bar{\bar{\tau}} \qquad (2.64)$$

2.2.3 THE NAVIER–STOKES EQUATION

For incompressible Newtonian fluids, the stress tensor $\bar{\bar{\sigma}}$ is given as:

$$\bar{\bar{\sigma}} = -P\bar{\bar{\delta}}+2\eta\bar{\bar{E}} = -P\bar{\bar{\delta}}+\eta\left[\nabla\vec{u}+\left(\nabla\vec{u}\right)^T\right] \qquad (2.65)$$

Substituting this equation for $\bar{\bar{\sigma}}$ into the equation of motion yields:

$$\rho\left[\frac{\partial\vec{u}}{\partial t}+\vec{u}\bullet\nabla\vec{u}\right] = -\nabla P+\rho\vec{g}+\eta\nabla^2\vec{u} \qquad (2.66)$$

This is the celebrated *Navier–Stokes equation*, which, together with the continuity equation, describes the flow of Newtonian fluids of constant density and viscosity.

2.2.4 CREEPING–FLOW EQUATIONS

When the fluid motion is very slow, terms such as $\rho \partial \vec{u}/\partial t$ and $\rho \vec{u} \cdot \nabla \vec{u}$ in the Navier–Stokes equation can be generally neglected. This greatly simplifies the equation. Thus, for very slow fluid motion (with gravity term neglected):

$$\eta \nabla^2 \vec{u} = \nabla P \qquad (2.67)$$

$$\nabla \cdot \vec{u} = 0 \qquad (2.68)$$

These equations are referred to as *creeping flow* or *Stokes equations*. Note that the characteristic Reynolds number (N_{Re}) of flow has to be very small $\left(N_{Re} \rightarrow 0 \right)$ for Stokes equations to be applicable. This follows from the dimensional analysis of the Navier–Stokes equation. Let u_c and l_c be characteristic velocity and length, respectively, of flow. The relevant dimensionless variables can be defined as:

$$\vec{u}^* = \vec{u}/u_c \qquad (2.69)$$

$$x^* = x/l_c, \quad y^* = y/l_c, \quad z^* = z/l_c \qquad (2.70)$$

$$t^* = t/(l_c/u_c) \qquad (2.71)$$

$$P^* = P/(\eta u_c/l_c) \qquad (2.72)$$

Substituting the dimensionless variables into the Navier–Stokes equation (with the gravity term neglected) gives:

$$N_{Re}\left[\frac{\partial \vec{u}^*}{\partial t^*} + \vec{u}^* \cdot \nabla \vec{u}^* \right] = -\nabla P^* + \nabla^{*2} \vec{u}^* \qquad (2.73)$$

where N_{Re}, the characteristic Reynolds number, is defined as:

$$N_{Re} = \frac{\rho u_c l_c}{\eta} \qquad (2.74)$$

When $N_{Re} \to 0$, the left-hand side of the dimensionless Navier–Stokes equation can be neglected. Consequently,

$$0 = -\nabla^* P^* + \nabla^{*2} \vec{u}^* \tag{2.75}$$

This is the dimensionless form of the creeping flow equation, Equation 2.67.

2.3 MECHANICS OF SOLIDS

Consider a solid material subjected to surface and body forces. At equilibrium, the resultant force on the material is zero and the equation of motion (Equation 2.63) gives:

$$\nabla \bullet \overline{\overline{\sigma}} + \rho \vec{g} = 0 \tag{2.76}$$

This is the *equilibrium equation* valid at every point of the material body. As discussed earlier, the constitutive equation for a *Hookean solid* is given as:

$$\overline{\overline{\sigma}} = \lambda \left(tr \overline{\overline{e}} \right) \overline{\overline{\delta}} + 2G \overline{\overline{e}} \tag{2.77}$$

where λ and G are Lamé constants and $\overline{\overline{e}}$ is the infinitesimal strain tensor defined earlier as:

$$\overline{\overline{e}} = \frac{1}{2} \left[\nabla \vec{\xi} + (\nabla \vec{\xi})^T \right] \tag{2.78}$$

$\vec{\xi}$ in this equation is the displacement vector. For constant λ and G, Equation 2.76 to Equation 2.78 lead to the following equation:

$$G\nabla^2 \vec{\xi} + \left(\lambda + G \right) \nabla \left(\nabla \bullet \vec{\xi} \right) + \rho \vec{g} = 0 \tag{2.79}$$

This is the well-known *Navier–Cauchy* equation. The Navier–Cauchy equation governs the equilibrium displacement field (with respect to the natural unstrained state) in a Hookean solid.

2.4 MECHANICS OF INTERFACES

The mechanics of interfaces, or interfacial films, comes into play when one is dealing with the rheology of dispersions of liquid droplets and capsules. A very thin interfacial film (membrane) is always present at the interface between the liquid droplets and the surrounding matrix fluid. The interfacial film, in the absence of any surfactant or other additives, is characterized by a constant interfacial tension. When surfactants and other additives are present at the surface of the droplets, the interfacial film exhibits a more complex rheological behavior. In the case of capsules, the interfacial film is a thin membrane of deformable elastic-solid material.

2.4.1 DEFINITIONS OF IMPORTANT TERMS

2.4.1.1 Surface Unit Tensor and Surface Gradient Operator

The two-dimensional surface unit tensor $\bar{\bar{\delta}}_s$ is defined as:

$$\bar{\bar{\delta}}_s = \bar{\bar{\delta}} - \hat{n}\hat{n} \tag{2.80}$$

where $\bar{\bar{\delta}}$ is the three-dimensional unit tensor, and $\hat{n}$ is the outward unit vector normal to the surface.

The surface gradient operator ∇_s is defined as

$$\nabla_s = \bar{\bar{\delta}}_s \bullet \nabla = \left(\bar{\bar{\delta}} - \hat{n}\hat{n}\right) \bullet \nabla \tag{2.81}$$

where ∇ is the three-dimensional spatial gradient operator.

2.4.1.2 Surface Deformation Gradient Tensor

The surface deformation gradient tensor $\left(\bar{\bar{F}}_s\right)$ is defined as:

$$\bar{\bar{F}}_s = \left(\bar{\bar{\delta}} - \hat{n}\hat{n}\right) \bullet \frac{\partial \vec{x}}{\partial \vec{X}} \bullet \left(\bar{\bar{\delta}} - \hat{N}\hat{N}\right) \tag{2.82}$$

where $\hat{n}$ is the unit vector normal to the interface in the deformed configuration, and $\hat{N}$ is the unit vector normal to the interface in the reference configuration. $\bar{\bar{F}}_s$ is a two-dimensional version of the three-dimensional Lagrangian deformation gradient tensor $\bar{\bar{F}}$ defined as $\partial \vec{x}/\partial \vec{X}$ (Equation 2.9).

2.4.1.3 Surface Strain Tensor

The surface strain tensor $\left(\bar{\bar{\ell}}_s\right)$ is defined as:

$$\bar{\bar{\ell}}_s = \frac{1}{2}\left[\bar{\bar{F}}_s^T \bullet \bar{\bar{F}}_s - \left(\bar{\bar{\delta}} - \hat{N}\hat{N}\right)\right] \tag{2.83}$$

This is a two-dimensional version of the three-dimensional Lagrangian finite strain tensor $\bar{\bar{\ell}}$ (Equation 2.15). In the reference configuration, $\bar{\bar{\ell}}_s = 0$. Note that the Eulerian surface-strain tensor ($\bar{\bar{e}}_s$) can be defined in a similar manner, that is,

$$\bar{\bar{e}}_s = \frac{1}{2}\left[\left(\bar{\bar{\delta}} - \hat{n}\hat{n}\right) - \bar{\bar{H}}_s^T \bullet \bar{\bar{H}}_s\right] \tag{2.84}$$

where $\bar{\bar{H}}_s$, the Eulerian surface-deformation gradient tensor, is a two-dimensional version of the three-dimensional Eulerian deformation gradient tensor $\bar{\bar{H}}$ defined as $\partial \vec{X}/\partial \vec{x}$ (Equation 2.10).

2.4.1.4 Surface Rate of Strain Tensor

The surface rate of strain tensor ($\bar{\bar{E}}_s$) is defined as:

$$\bar{\bar{E}}_s = \frac{1}{2}\left[(\nabla_s \vec{u})\bullet\bar{\bar{\delta}}_s + \bar{\bar{\delta}}_s \bullet (\nabla_s \vec{u})^T\right] \tag{2.85}$$

where $\vec{u}$ is the velocity field imposed on the interface.

2.4.1.5 Surface Stress Vector

Consider a material surface S bounded by curve C as shown in Figure 2.3. Let $\hat{m}$ be a unit vector normal to the differential line segment $d\ell$ and tangent to the surface. The material on the positive side of the line segment $d\ell$ exerts a force $d\vec{F}_s$ on the material on the negative side of the line segment.

The surface stress vector, denoted as $\vec{P}_{s(\hat{m})}$, is defined as:

$$\vec{P}_{s(m)} = \frac{d\vec{F}_s}{d\ell} \tag{2.86}$$

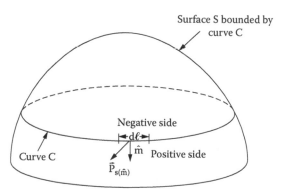

FIGURE 2.3 A surface S bounded by curve C. $\hat{m}$ is a unit vector normal to the differential line segment $d\ell$ and tangent to the surface. $\vec{P}_{s(\hat{m})}$, is the surface stress vector.

The surface stress vector, having the units of force per unit length, is a function of position, orientation of line segment, and time in general.

2.4.1.6 Surface Stress Tensor

The surface stress tensor, denoted as $\overline{\overline{P}}_s$, completely specifies the state of stress at a given point within a material surface. It is defined such that

$$\vec{P}_{s(\hat{m})} = \hat{m} \bullet \overline{\overline{P}}_s \tag{2.87}$$

In general, $\overline{\overline{P}}_s$ consists of six independent components.

2.4.2 Surface Divergence Theorem

The surface divergence theorem allows the line integral to be written as a surface integral. Let S be the surface bounded by curve C, as shown in Figure 2.4. Let z, $\vec{w}$, and $\overline{\overline{T}}$ be some scalar, vector, and tensor fields, respectively, defined at every point of the surface. Then,

$$\int_C mz \, d\ell = \iint_S [-n(\nabla_s \bullet n) + \nabla_s] z \, dS \tag{2.88}$$

$$\int_C (m \bullet \vec{w}) \, d\ell = \iint_S [-n(\nabla_s \bullet n) + \nabla_s] \bullet \vec{w} \, dS \tag{2.89}$$

$$\int_C (m \bullet \overline{\overline{T}}) \, d\ell = \iint_S [-n(\nabla_s \bullet n) + \nabla_s] \bullet \overline{\overline{T}} \, dS \tag{2.90}$$

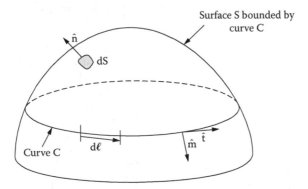

FIGURE 2.4 A surface S bounded by curve C. $\hat{n}$ is the outward unit vector normal to the surface; $\hat{t}$ is the unit vector tangent to the curve, and $\hat{m}$ is the outward unit vector normal to the curve C and tangent to the surface ($\hat{m}$ lies on the surface). $d\ell$ is the differential arc length along the bounding curve C, and dS is the differential area element on the surface S.

2.4.3 FORCE BALANCE ON INTERFACIAL FILM

Consider an interfacial film of area S bounded by curve C (see Figure 2.5). The film separates two immiscible fluids I and II. Let $\bar{\bar{\sigma}}_I$ and $\bar{\bar{\sigma}}_{II}$ be the stress tensors in the bulk fluids below and above the interface and let $\bar{\bar{P}}_s$ be the surface stress tensor present within the interfacial film. The force exerted on the film by fluid I is $\iint_S [(-\hat{n}) \bullet \bar{\bar{\sigma}}_I] dS$. The force exerted on the film by fluid II is $\iint_S (\hat{n} \bullet \bar{\bar{\sigma}}_{II}) \, dS$. The force exerted across the curve C by material toward which $\hat{m}$ is pointing on the interfacial film bounded by curve C is $\int_C \vec{P}_{s(\hat{m})} \, dl$. Assuming that the inertia of the interfacial film is negligible, the sum of all forces acting on the interfacial film must be zero. Thus,

$$\iint_S \left(-\hat{n} \bullet \bar{\bar{\sigma}}_I\right) dS + \iint_S \left(\hat{n} \bullet \bar{\bar{\sigma}}_{II}\right) dS + \int_C \vec{P}_{s(\hat{m})} \, dl = 0 \qquad (2.91)$$

Substituting $\vec{P}_{s(\hat{m})} = \hat{m} \bullet \bar{\bar{P}}_s$ and using the surface divergence theorem, the line integral can be written as $\iint_S \left[-\hat{n}(\nabla_s \bullet \hat{n}) + \nabla_s\right] \bullet \bar{\bar{P}}_s \, dS$. Therefore,

$$\iint_S \left[\hat{n} \bullet \left(\bar{\bar{\sigma}}_{II} - \bar{\bar{\sigma}}_I\right) + \left\{-\hat{n}\left(\nabla_s \bullet \hat{n}\right) + \nabla_s\right\} \bullet \bar{\bar{P}}_s\right] dS = 0 \qquad (2.92)$$

Because this equation holds for an arbitrary surface S, the integrand must also be zero, that is:

$$\hat{n} \bullet \left(\bar{\bar{\sigma}}_{II} - \bar{\bar{\sigma}}_I\right) + \left[-\hat{n}\left(\nabla_s \bullet \hat{n}\right) + \nabla_s\right] \bullet \bar{\bar{P}}_s = 0 \qquad (2.93)$$

or

$$\hat{n} \bullet \left(\bar{\bar{\sigma}}_{II} - \bar{\bar{\sigma}}_I\right) = \left(\nabla_s \bullet \hat{n}\right)\left(\hat{n} \bullet \bar{\bar{P}}_s\right) - \nabla_s \bullet \bar{\bar{P}}_s \qquad (2.94)$$

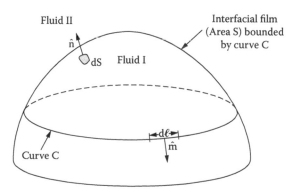

FIGURE 2.5 An interfacial film of area S bounded by curve C. The film separates two immiscible fluids, I and II.

Assuming that the surface stress tensor has no component in the direction of $\hat{n}$, $\hat{n} \bullet \bar{\bar{P}}_s = 0$. Thus, the force balance on interfacial film reduces to:

$$\hat{n} \bullet \left(\bar{\bar{\sigma}}_{II} - \bar{\bar{\sigma}}_I \right) = -\nabla_s \bullet \bar{\bar{P}}_s \qquad (2.95)$$

If interfacial tension γ is the only nonzero mechanical property of the film, then

$$\bar{\bar{P}}_s = \gamma \bar{\bar{\delta}}_s \qquad (2.96)$$

Consequently, the force balance becomes:

$$\hat{n} \bullet (\bar{\bar{\sigma}}_{II} - \bar{\bar{\sigma}}_I) = -\nabla_s \bullet (\gamma \bar{\bar{\delta}}_s)$$
$$= -\gamma (\nabla_s \bullet \bar{\bar{\delta}}_s) \qquad (2.97)$$

As $\nabla_s \bullet \bar{\bar{\delta}}_s = -(\nabla_s \bullet \hat{n}) \hat{n}$, the preceding force balance reduces to:

$$\hat{n} \bullet (\bar{\bar{\sigma}}_{II} - \bar{\bar{\sigma}}_I) = \gamma \hat{n} (\nabla_s \bullet \hat{n}) , \qquad (2.98)$$

where $\nabla_s \bullet \hat{n}$ is the local mean curvature of the interface.

2.4.4 INTERFACIAL RHEOLOGY

The rheological behavior of interfacial films can be purely viscous (inelastic) or viscoelastic, depending on the composition of the film. The presence of polymers, proteins, or large surfactant molecules at the interface often imparts viscoelasticity to the interface.

For a purely viscous Newtonian interface, the surface stress tensor $\bar{\bar{P}}_s$ is related to the surface rate-of-strain tensor $(\bar{\bar{E}}_s)$ as:

$$\bar{\bar{P}}_s = \gamma \bar{\bar{\delta}}_s + 2\eta_s \bar{\bar{E}}_s + \left(\eta_s^k - \eta_s \right) \left(\bar{\bar{\delta}}_s : \bar{\bar{E}}_s \right) \bar{\bar{\delta}}_s \qquad (2.99)$$

where γ is the interfacial tension, η_s is the surface or interfacial shear viscosity, η_s^k is surface or interfacial dilational viscosity, and $\bar{\bar{E}}_s$ is the surface rate-of-strain tensor. In the case of a purely viscous non-Newtonian interface, the interfacial viscosities η_s and η_s^k are not constant independent of the rate of deformation.

More often than not, the surfactant-absorbed interfaces exhibit shear-thinning and dilation-thinning behaviors; that is, η_s and η_s^k decrease with the increase in the rate of deformation.

For a purely elastic interface, the surface stress tensor is related to surface-strain tensor $\bar{\bar{e}}_s$ as:

$$\bar{\bar{P}}_s = \gamma \bar{\bar{\delta}}_s + 2G_s \bar{\bar{e}}_s + (K_s - G_s)(\bar{\bar{\delta}}_s : \bar{\bar{e}}_s) \bar{\bar{\delta}}_s \qquad (2.100)$$

where G_s is the surface-shear modulus and K_s is the surface-dilational modulus.

For a viscoelastic interface subjected to oscillatory strain, the constitutive equation can be expressed as:

$$\bar{\bar{P}}_s^* = \gamma \bar{\bar{\delta}}_s + 2G_s^* \bar{\bar{e}}_s^* + (K_s^* - G_s^*)(\bar{\bar{\delta}}_s : \bar{\bar{e}}_s^*) \bar{\bar{\delta}}_s \qquad (2.101)$$

where $\bar{\bar{P}}_s^*$ is the complex surface-stress tensor, $\bar{\bar{e}}_s^*$ is complex surface-strain tensor, G_s^* is complex surface-shear modulus, and K_s^* is complex surface-dilational modulus. The complex moduli G_s^* and K_s^* are defined as:

$$G_s^* = G_s + j\eta_s \omega \qquad (2.102)$$

$$K_s^* = K_s + j\eta_s^k \omega \qquad (2.103)$$

where G_s, η_s, K_s, and η_s^k are functions of frequency.

NOTATION

$\bar{\bar{C}}$	Cauchy deformation tensor
D/Dt	material time derivative
E	Young's modulus
$\bar{\bar{E}}$	rate of strain (deformation) tensor
$\bar{\bar{E}}_s$	surface rate of strain tensor
$\bar{\bar{e}}$	Eulerian strain tensor
$\bar{\bar{e}}_s$	Eulerian surface strain tensor
$\bar{\bar{e}}_s^*$	complex surface strain tensor
$\bar{\bar{F}}$	material/Lagrangian deformation gradient tensor
$\bar{\bar{F}}_s$	surface deformation gradient tensor

$\vec{F}_{body}$	body force
$\vec{F}_{Surface}$	surface force
G	shear modulus
$\bar{\bar{G}}$	Green deformation tensor
G_s	surface shear modulus
G_s^*	complex surface shear modulus
$\vec{g}$	acceleration due to gravity
$\bar{\bar{H}}$	spatial/Eulerian deformation gradient tensor
$\bar{\bar{H}}_s$	Eulerian surface deformation gradient tensor
$\bar{\bar{J}}$	material/Lagrangian displacement gradient tensor
j	imaginary number
K	bulk (dilational) modulus
$\bar{\bar{K}}$	spatial/Eulerian displacement gradient tensor
K_s	surface dilational modulus
K_s^*	complex surface dilational modulus
$\bar{\bar{\ell}}$	Lagrangian finite strain tensor
$\bar{\bar{\ell}}_s$	surface strain tensor
m	mass
$\hat{m}$	unit vector normal to a line segment and tangent to the surface
$\vec{M}$	linear momentum
$\hat{n}$	outer unit normal vector (normal to the surface)
N_{Re}	Reynolds number
P	thermodynamic pressure
$\bar{P}$	mean pressure
$\bar{\bar{P}}_s$	surface stress tensor
$\bar{\bar{P}}_s^*$	complex surface stress tensor
$\vec{P}_{s(\hat{m})}$	surface stress vector
$\vec{R}$	position vector at some reference time
$\vec{r}$	position vector at time t
S	surface area
t	time
tr	trace of a tensor
$\vec{u}$	velocity vector

u_x, u_y, u_z	velocity components
V	volume
x,y,z	coordinates at time t
X,Y,Z	coordinates at some reference time

GREEK

$\bar{\bar{\Gamma}}$	velocity gradient tensor
γ	interfacial tension
$\bar{\bar{\dot{\gamma}}}$	shear rate tensor
$\bar{\bar{\delta}}$	unit tensor
$\bar{\bar{\delta}}_s$	surface unit tensor
∇	del operator (three-dimensional spatial gradient operator)
∇_s	surface gradient operator
η	shear viscosity
η_s	surface shear viscosity
η^k	dilational (bulk) viscosity
η_s^k	surface dilational viscosity
λ	either first viscosity coefficient or Lamé constant
ν	Poisson's ratio
$\vec{\xi}$	displacement vector
ρ	density
$\bar{\bar{\sigma}}$	stress tensor
$\vec{\sigma}_{(\hat{n})}$	stress vector
$\bar{\bar{\tau}}$	extra or deviatoric stress tensor (also called *viscous stress tensor*)
$\bar{\bar{\Omega}}$	vorticity (spin) tensor
ω	frequency of oscillation

SUPERSCRIPT

T	transpose

SUPPLEMENTAL READING

1. G.E. Mase, *Theory and Problems of Continuum Mechanics*, McGraw-Hill, New York, 1970.
2. Y.C. Fung, *A First Course in Continuum Mechanics*, Prentice-Hall, Englewood Cliffs, NJ.

3. D. Frederick, T.S. Chang, *Continuum Mechanics*, Scientific Publishers, Boston, MA, 1972.
4. W.M. Lai, D. Rubin, E. Krempl, *Introduction to Continuum Mechanics*, Pergamon Press, New York, 1978.
5. J. Salencon, *Handbook of Continuum Mechanics*, Springer-Verlag, New York, 2001.
6. G. Emanuel, *Analytical Fluid Dynamics*, CRC Press, Boca Raton, FL, 2001.
7. R.B. Bird, W.E. Stewart, E.N. Lightfoot, *Transport Phenomena*, John Wiley & Sons, New York, 2002.
8. R.B. Bird, R.C. Armstrong, O. Hassager, *Dynamics of Polymeric Liquids*, Vol. 1, John Wiley & Sons, New York, 1987.
9. P.J. Kundu, I.M. Cohen, *Fluid Mechanics*, Academic Press, San Diego, CA, 2002.
10. R.A. Granger, *Fluid Mechanics*, CBS College Publishing, New York, 1985.
11. S.W. Yuan, *Foundations of Fluid Mechanics*, Prentice-Hall, Englewood Cliffs, NJ, 1967.
12. S. Whitaker, *Introduction to Fluid Mechanics*, Prentice-Hall, Englewood Cliffs, NJ, 1968.
13. T.G.M. Van de ven, *Colloidal Hydrodynamics*, Academic Press, San Diego, CA, 1989.
14. J. Happel, H. Brenner, *Low Reynolds Number Hydrodynamics*, Noordhoff International Publishing, Leyden, 1973.
15. C.W. Macosko, *Rheology Principles, Measurements, and Applications*, John Wiley & Sons, New York, 1994.
16. A. Nadim, *Chem. Eng. Commun.*, 148–150: 391–407, 1996.
17. D. Edwards, H. Brenner, D. Wasan, *Interfacial Transport Processes and Rheology*, Butterworth-Heinemann, Boston, MA, 1991.
18. J.G. Oldroyd, *Proceed. R. Soc. Lond.*, A 232: 567–577, 1955.

3 Bulk Stress in Particulate Dispersions and Composites

3.1 DISPERSED SYSTEM AS A CONTINUUM

Strictly speaking, dispersed systems (dispersions and composites) are heterogeneous systems consisting of discrete inclusions (particles, droplets, bubbles, capsules, etc.) of one phase suspended in a continuum of another phase. Nevertheless, they are generally regarded as continuous (homogeneous) systems when it comes to defining their macroscopic rheological behavior [1–3]. In most situations of practical interest, this hypothesis that the dispersed systems can be regarded as a continuum is sound. But continuum assumption becomes meaningless in certain situations. It is important to consider two length scales before treating dispersions and composites as a continuum. The two length scales are: (1) the average spacing between the centers of adjacent inclusions, ℓ; (2) the characteristic linear dimension L of the apparatus boundaries constraining the material. Provided that $\ell \ll$ L, the dispersed system can be regarded as a continuum with suitably defined bulk or average fields.

3.2 ENSEMBLE VS. VOLUME AVERAGING

For a given set of macroscopic boundary conditions, there are many possible geometric configurations (realizations) of the dispersed systems. The relative positions of the suspended particles are generally time-dependent even when the system undergoes macroscopically steady flow. Velocity, pressure, velocity gradient, and stress all tend to fluctuate with time at any given location $\vec{x}$ in the dispersed system; at some time the position $\vec{x}$ is occupied by a particle and at another time, the same position may be occupied by the matrix. Thus, dispersed systems are determinate only in a statistical sense [3].

A large number of configurations corresponding to the same macroscopic boundary conditions make up an ensemble. An ensemble average of any quantity is defined as the average taken over the various values occurring in these configurations. If q is some field variable (function of position $\vec{x}$ and time t) for some particular realization, ξ, of the process, then the ensemble average of q is

$$\bar{q}(\vec{x},t) = \int_M q(\vec{x},t;\xi)\,dm \qquad (3.1)$$

where *dm* is the probability of realization ξ and *M* is the set (ensemble) of all realizations [4].

Alternatively, the bulk or average fields can be defined as volume averages of the corresponding local fields over a representative volume element of volume *V* whose typical linear dimension, $V^{1/3}$, is large compared to the average spacing between the particles ℓ, but small compared to the apparatus length scale L, that is, $\ell << V^{1/3} << L$. As volume averages of local fields are more convenient for analytical purposes as compared to ensemble averages, they are widely used in the development of constitutive equations for dispersed systems. Note that an ensemble average at any point $\vec{x}$ is identical with the volume average over representative volume *V* for one realization if *V* is taken to be large enough ($V^{1/3} >> \ell$) to contain a statistically significant number of particles [2,3].

Examples of volume averages of certain quantities of interest (velocity, velocity gradient, rate of strain tensor), denoted by angular brackets, are given as follows:

$$\langle \vec{u} \rangle = \frac{1}{V} \int_V \vec{u} \, dV \tag{3.2}$$

$$\langle \nabla \vec{u} \rangle = \frac{1}{V} \int_V (\nabla \vec{u}) \, dV \tag{3.3}$$

$$\langle (\nabla \vec{u})^T \rangle = \frac{1}{V} \int_V (\nabla \vec{u})^T \, dV \tag{3.4}$$

$$\langle \bar{\bar{E}} \rangle = \frac{1}{V} \int_V \bar{\bar{E}} \, dV = \frac{1}{2} \left[\frac{1}{V} \int_V (\nabla \vec{u}) \, dV + \frac{1}{V} \int_V (\nabla \vec{u})^T \, dV \right] \tag{3.5}$$

where $\vec{u}$ is local velocity vector and $\bar{\bar{E}}$ is the local rate of strain tensor.

3.3   BULK STRESS IN DISPERSIONS AND COMPOSITES

The bulk or volume-average stress tensor $\langle \bar{\bar{\sigma}} \rangle$ in dispersed systems (dispersions and composites) can be defined as:

$$\langle \bar{\bar{\sigma}} \rangle = \frac{1}{V} \int_V \bar{\bar{\sigma}} \, dV \tag{3.6}$$

where V is the volume of the element large enough to contain a statistically significant number of particles but small relative to the characteristic macroscale of the system.

The integral in Equation 3.6 can be broken up into integrals over the volume occupied by the continuous-phase (matrix), V_c, and by the dispersed particles, V_d, to yield:

$$\langle \overline{\overline{\sigma}} \rangle = \frac{1}{V} \int_{V_c} \overline{\overline{\sigma}} dV + \frac{1}{V} \int_{V_d} \overline{\overline{\sigma}} dV \qquad (3.7)$$

However, it should be noted that this definition of volume-average stress tensor (Equation 3.6) contains an implicit assumption that the effects of inertia forces on the motion of the matrix and the particles are negligible [3,5]. This is a standard assumption in the study of the rheology of dispersed systems.

3.3.1 Dispersions

Dispersions are defined as systems consisting of fine insoluble particles distributed in a continuous *liquid* phase. The stress tensor in the ambient continuous-phase liquid (assumed Newtonian) is:

$$\overline{\overline{\sigma}} = - P\overline{\overline{\delta}} + 2\eta_c \overline{\overline{E}} \qquad (3.8)$$

where η_c is the continuous-phase viscosity. Using Equation 3.8, Equation 3.7 can be written as:

$$\langle \overline{\overline{\sigma}} \rangle = \frac{1}{V} \int_{V_c} \left(-P\overline{\overline{\delta}} + 2\eta_c \overline{\overline{E}} \right) dV + \frac{1}{V} \int_{V_d} \overline{\overline{\sigma}} dV \qquad (3.9)$$

or

$$\langle \overline{\overline{\sigma}} \rangle = \frac{1}{V} \int_{V_c} \left(-P\overline{\overline{\delta}} + 2\eta_c \overline{\overline{E}} \right) dV + \frac{1}{V} \sum_{i=1}^{N} \int_{V_i} \overline{\overline{\sigma}} dV \qquad (3.10)$$

where V_i is the volume of ith particle and N is the total number of particles in V.

Assuming that the particles are force-free and that there is no interfacial film present at the surface of the particles, the stress tensor within the particles is divergence-free, that is,

$$\nabla \bullet \overline{\overline{\sigma}} = 0 \qquad (3.11)$$

Because the stress tensor within the particles is also symmetric, we have an identity

$$\bar{\bar{\sigma}} = \nabla \bullet \left(\bar{\bar{\sigma}} \, \vec{x} \right) \tag{3.12}$$

Making use of this identity and the Gauss divergence theorem, the second volume integral on the RHS (right-hand side) of Equation 3.10 can be written as:

$$\int_{V_i} \bar{\bar{\sigma}} \, dV = \int_{V_i} \left(\nabla \bullet \bar{\bar{\sigma}} \, \vec{x} \right) dV = \int_{A_i} \left(\bar{\bar{\sigma}} \bullet \hat{n} \, \vec{x} \right) dA \tag{3.13}$$

where A_i is the surface area of ith particle, and $\hat{n}$ is a unit outward normal to the particle surface.

Note that the stress in the surface integral of Equation 3.13 should be taken as the boundary value of the Newtonian stress in the ambient continuous-phase fluid.

Using Equation 3.13, Equation 3.10 can be written as:

$$\left\langle \bar{\bar{\sigma}} \right\rangle = \frac{1}{V} \int_{V_c} \left(-P\bar{\bar{\delta}} + 2\eta_c \bar{\bar{E}} \right) dV + \frac{1}{V} \sum_{i=1}^{N} \int_{A_i} \left(\bar{\bar{\sigma}} \bullet \hat{n} \, \vec{x} \right) dA \tag{3.14}$$

The origin of the position vector $\vec{x}$ present in the surface integral of Equation 3.14 is arbitrary as the particles are force-free [3].

The volume-average velocity gradients $\left\langle \nabla \vec{u} \right\rangle$ and $\left\langle (\nabla \vec{u})^T \right\rangle$ in a dispersion can be expressed as

$$\left\langle \nabla \vec{u} \right\rangle = \frac{1}{V} \int_{V_c} (\nabla \vec{u}) dV + \frac{1}{V} \sum_{i=1}^{N} \int_{V_i} (\nabla \vec{u}) dV \tag{3.15}$$

$$\left\langle (\nabla \vec{u})^T \right\rangle = \frac{1}{V} \int_{V_c} (\nabla \vec{u})^T dV + \frac{1}{V} \sum_{i=1}^{N} \int_{V_i} (\nabla \vec{u})^T dV \tag{3.16}$$

Using the Gauss divergence theorem, the second volume integrals on the RHS of Equation 3.15 and Equation 3.16 can be converted into surface integrals. As a result,

$$\frac{1}{V} \int_{V_c} \left[\nabla \vec{u} + (\nabla \vec{u})^T \right] dV = \left[\left\langle \nabla \vec{u} \right\rangle + \left\langle (\nabla \vec{u})^T \right\rangle \right] - \frac{1}{V} \sum_{i=1}^{N} \int_{A_i} (\hat{n}\vec{u} + \vec{u}\hat{n}) dA \tag{3.17}$$

Using the definitions of the local and average rate-of-strain tensors,

$$\bar{\bar{E}} = \frac{1}{2}\left[\nabla\vec{u} + \left(\nabla\vec{u}\right)^T\right] \qquad (3.18)$$

$$\left\langle\bar{\bar{E}}\right\rangle = \frac{1}{2}\left[\left\langle\nabla\vec{u}\right\rangle + \left\langle\left(\nabla\vec{u}\right)^T\right\rangle\right] \qquad (3.19)$$

Equation 3.17 can be expressed as:

$$\frac{2}{V}\int_{V_c}\bar{\bar{E}}\,dV = 2\left\langle\bar{\bar{E}}\right\rangle - \frac{1}{V}\sum_{i=1}^{N}\int_{A_i}(\hat{n}\vec{u} + \vec{u}\hat{n})\,dA \qquad (3.20)$$

From Equation 3.14 and Equation 3.20, the bulk stress in a dispersion can be written as:

$$\left\langle\bar{\bar{\sigma}}\right\rangle = \frac{1}{V}\int_{V_c}\left[-P\bar{\bar{\delta}}dV + 2\,\eta_c\left\langle\bar{\bar{E}}\right\rangle + \frac{1}{V}\sum_{i=1}^{N}\int_{A_i}\left[\bar{\bar{\sigma}}\bullet\hat{n}\,\vec{x} - \eta_c\left(\hat{n}\vec{u} + \vec{u}\hat{n}\right)\right]dA \qquad (3.21)$$

or

$$\left\langle\bar{\bar{\sigma}}\right\rangle = -\left\langle P\right\rangle\bar{\bar{\delta}} + \frac{1}{V}\int_{V_d}\left(P\bar{\bar{\delta}}\right)dV + 2\,\eta_c\left\langle\bar{\bar{E}}\right\rangle + \frac{1}{V}\sum_{i=1}^{N}\int_{A_i}\left[\bar{\bar{\sigma}}\bullet\hat{n}\,\vec{x} - \eta_c(\hat{n}\vec{u} + \vec{u}\hat{n})\right]dA \qquad (3.22)$$

The pressure P within the particles can be expressed as:

$$P = -\frac{1}{3}\,tr\,(\bar{\bar{\sigma}}) = -\frac{1}{3}(\bar{\bar{\sigma}}:\bar{\bar{\delta}}) \qquad (3.23)$$

where *tr* refers to trace of a tensor. Using Equation 3.23 and the Gauss divergence theorem, one can show that:

$$\frac{1}{V}\int_{V_d}P\bar{\bar{\delta}}\,dV = \frac{1}{V}\sum_{i=1}^{N}\int_{V_i}-\frac{1}{3}(\bar{\bar{\sigma}}:\bar{\bar{\delta}})\bar{\bar{\delta}}\,dV = \frac{1}{V}\sum_{i=1}^{N}\int_{A_i}-\frac{1}{3}(\vec{x}\bullet\bar{\bar{\sigma}}\bullet\hat{n})\bar{\bar{\delta}}dA \qquad (3.24)$$

Combining Equation 3.22 and Equation 3.24 gives the desired expression for the bulk stress in dispersions of force-free particles:

$$\langle \overline{\overline{\sigma}} \rangle = -\langle P \rangle \overline{\overline{\delta}} + 2\eta_c \langle \overline{\overline{E}} \rangle + \frac{1}{V} \sum_{i=1}^{N} \int_{A_i} \left[\overline{\overline{\sigma}} \bullet \hat{n}\, \vec{x} - \frac{1}{3}(\vec{x} \bullet \overline{\overline{\sigma}} \bullet \hat{n}) \overline{\overline{\delta}} - \eta_c \left(\hat{n}\,\vec{u} + \vec{u}\,\hat{n} \right) \right] dA \quad (3.25)$$

Equation 3.25 is often expressed in the form

$$\langle \overline{\overline{\sigma}} \rangle = \langle \overline{\overline{\sigma}}_0 \rangle + \frac{1}{V} \sum_{i=1}^{N} \overline{\overline{S}}_i \quad (3.26)$$

where $\langle \overline{\overline{\sigma}}_0 \rangle$ is the stress tensor in pure matrix material under conditions corresponding to the imposed macroscopic rate-of-deformation tensor $\langle \overline{\overline{E}} \rangle$ on the dispersion, and $\overline{\overline{S}}_i$ is the force "dipole strength" of the ith particle. The stress tensor $\langle \overline{\overline{\sigma}}_0 \rangle$ is given as:

$$\langle \overline{\overline{\sigma}}_0 \rangle = -\langle P \rangle \overline{\overline{\delta}} + 2\eta_c \langle \overline{\overline{E}} \rangle \quad (3.27)$$

The dipole strength $\overline{\overline{S}}$ of a single particle is given by (see Equation 3.25):

$$\overline{\overline{S}} = \int_{A_0} \left[\overline{\overline{\sigma}} \bullet \hat{n}\, \vec{x} - \frac{1}{3}(\vec{x} \bullet \overline{\overline{\sigma}} \bullet \hat{n}) \overline{\overline{\delta}} - \eta_c \left(\hat{n}\,\vec{u} + \vec{u}\,\hat{n} \right) \right] dA \quad (3.28)$$

where A_0 is the surface area of a particle. $\overline{\overline{S}}$ as defined in Equation 3.28 is a traceless tensor. It can be interpreted as a measure of the increase in stress in region V_0 (bounded by a closed surface A_0) resulting from the replacement of matrix material there by particle material [6], keeping the local rate of strain constant. It is also expressed as:

$$\overline{\overline{S}} = \int_{V_0} \left(\overline{\overline{\sigma}} - \overline{\overline{\sigma}}_0 \right) dV \quad (3.29)$$

where $\overline{\overline{\sigma}}$ is the local stress tensor in region V_0, and $\overline{\overline{\sigma}}_0$ is the value the stress tensor would have at that same location if that particle is replaced by matrix material, with the rate of strain held constant. The stress difference $(\overline{\overline{\sigma}} - \overline{\overline{\sigma}}_0)$ is often termed *polarization stress* [7]. Using the Gauss divergence theorem, it can be shown that the expressions for $\overline{\overline{S}}$ given in Equation 3.28 and Equation 3.29 are equivalent.

It should be noted that the dipole strength $\overline{\overline{S}}$, and hence the bulk stress in a dispersion, is symmetric only in the absence of externally applied torque (couple) on the particles. In the presence of externally applied torque on the particles,

$\overline{\overline{S}}$ and the bulk stress tensor of dispersion contain an antisymmetric component directly related to the torque on each particle. This point can be further clarified by expressing the dipole strength $\overline{\overline{S}}$ in terms of symmetric and antisymmetric parts as:

$$\overline{\overline{S}} = \int_{A_0} \left[\overline{\overline{\sigma}} \bullet \hat{n}\, \vec{x} - \frac{1}{3} \left(\vec{x} \bullet \overline{\overline{\sigma}} \bullet \hat{n} \right) \overline{\overline{\delta}} - \eta_c \left(\hat{n}\,\vec{u} + \vec{u}\,\hat{n} \right) \right] dA = \overline{\overline{T}} + \overline{\overline{L}} \quad (3.30)$$

where $\overline{\overline{T}}$, the symmetric part of dipole strength referred to as *stresslet*, is defined as:

$$\overline{\overline{T}} = \frac{1}{2} \int_{A_0} \left[\left(\overline{\overline{\sigma}} \bullet \hat{n}\, \vec{x} + \vec{x}\, \overline{\overline{\sigma}} \bullet \hat{n} \right) - \frac{2}{3} \left(\vec{x} \bullet \overline{\overline{\sigma}} \bullet \hat{n} \right) \overline{\overline{\delta}} - 2\eta_c \left(\vec{u}\,\hat{n} + \hat{n}\,\vec{u} \right) \right] dA \quad (3.31)$$

and $\overline{\overline{L}}$, the antisymmetric part of $\overline{\overline{S}}$ referred to as *couplet*, is defined as:

$$\overline{\overline{L}} = \frac{1}{2} \int_{A_0} \left(\overline{\overline{\sigma}} \bullet \hat{n}\, \vec{x} - \vec{x}\, \overline{\overline{\sigma}} \bullet \hat{n} \right) dA \quad (3.32)$$

The torque $\overline{\mathscr{L}}$ on the particle is related to $\overline{\overline{L}}$ as [8]:

$$\overline{\mathscr{L}} = \overline{\overline{\overline{\epsilon}}} : \overline{\overline{L}} \quad (3.33)$$

where $\overline{\overline{\overline{\epsilon}}}$ is the alternating isotropic triadic (third-order tensor) defined in terms of the permutation symbol ϵ_{ijk} as [9]:

$$\overline{\overline{\overline{\epsilon}}} = \sum_i \sum_j \sum_k \hat{\delta}_i\, \hat{\delta}_j\, \hat{\delta}_k\, \epsilon_{ijk} \quad (3.34)$$

In the absence of externally applied torque, $\overline{\overline{L}} = 0$ and the dipole strength $\overline{\overline{S}}$ is equal to the stresslet $\overline{\overline{T}}$.

3.3.1.1 Dipole Strength of Rigid Particles

The integral of $\vec{u}\,\hat{n} + \hat{n}\,\vec{u}$ over the surface of A_0 of a rigid particle is zero [3]:

$$\int_{A_0} \left(\vec{u}\hat{n} + \hat{n}\,\vec{u} \right) dA = 0 \quad (3.35)$$

Consequently, the dipole strength $\overline{\overline{S}}$ of a rigid particle, as given by Equation 3.28, reduces to:

$$\overline{\overline{S}} = \int_{A_0} \left[\overline{\overline{\sigma}} \bullet \hat{n} \, \vec{x} - \frac{1}{3} \left(\vec{x} \bullet \overline{\overline{\sigma}} \bullet \hat{n} \right) \overline{\overline{\delta}} \right] dA \qquad (3.36)$$

3.3.1.2   Dipole Strength of Droplets with Interfacial Film

A very thin interfacial film (membrane) is always present at the interface between the ambient fluid and the liquid droplet. In the absence of any surfactant or other additives, the interfacial film is characterized by a constant interfacial tension. In the presence of surfactants/additives, the interfacial film could exhibit a more complex rheological behavior.

An important point to note is that in the presence of an interfacial film at the interface between the ambient fluid and the particle, the expression for the dipole strength (Equation 3.28) remains the same except that the surface over which the surface integral is taken lies on the outside of the interfacial film; that is,

$$\overline{\overline{S}} = \int_{A_0^+} \left[\overline{\overline{\sigma}} \bullet \hat{n} \, \vec{x} - \frac{1}{3} \left(\vec{x} \bullet \overline{\overline{\sigma}} \bullet \hat{n} \right) \overline{\overline{\delta}} - \eta_c \left(\hat{n} \vec{u} + \vec{u} \, \hat{n} \right) \right] dA \qquad (3.37)$$

where A_0^+ denotes the outer surface of the interfacial film and $\overline{\overline{\sigma}}$ in this expression is the continuous-phase stress tensor at the outer surface of the interfacial film.

The force balance on the interfacial film gives [10] (see Chapter 2):

$$\hat{n} \bullet \left(\overline{\overline{\sigma}}_c - \overline{\overline{\sigma}}_d \right) = \left(\nabla_s \bullet \hat{n} \right) \left(\hat{n} \bullet \overline{\overline{P}}_s \right) - \nabla_s \bullet \overline{\overline{P}}_s \qquad (3.38)$$

where $\hat{n}$ is the unit normal to the interface directed into the continuous phase, $\overline{\overline{\sigma}}_c$ is continuous-phase stress at the outer surface of the film A_0^+, $\overline{\overline{\sigma}}_d$ is the dispersed-phase (droplet) stress at the inner surface of the film A_0^-, $\overline{\overline{P}}_s$ is the surface-stress tensor (stress tensor existing in the interface itself), and ∇_s is the surface gradient operator defined as $\left(\overline{\overline{\delta}} - \hat{n}\hat{n} \right) \bullet \nabla$. The component of the surface stress tensor in the direction of $\hat{n}$ is usually assumed to be zero. Thus,

$$\hat{n} \bullet \left(\overline{\overline{\sigma}}_c - \overline{\overline{\sigma}}_d \right) = -\nabla_s \bullet \overline{\overline{P}}_s \qquad (3.39)$$

The volume integral of stress for the particle with interfacial film can be written as:

$$\int_{V_0^+} \bar{\bar{\sigma}} \, dV = \int_{V_0^-} \bar{\bar{\sigma}}_d \, dV + \int_{A_0} \bar{\bar{P}}_s \, dA \qquad (3.40)$$

where V_0^+ denotes the volume of the particle bounded by the closed surface A_0^+ on the outer side of the interfacial film, V_0^- is the volume of a particle on the inner side of the interfacial film, bounded by closed surface A_0^-, that is, V_0^- is the volume of a particle without the interfacial film, and A_0 is the interfacial area. (Note that $A_0^+ = A_0^- = A_0$ and $V_0^+ = V_0^- = V_0$ as the film is infinitely thin.)

Because the stress tensor inside the particle (inner side of the interfacial film) is divergence-free and symmetric,

$$\bar{\bar{\sigma}}_d = \nabla \bullet \bar{\bar{\sigma}}_d \, \vec{x} \qquad (3.41)$$

Consequently,

$$\int_{V_0^-} \bar{\bar{\sigma}}_d \, dV = \int_{A_0^-} \left(\hat{n} \bullet \bar{\bar{\sigma}}_d \, \vec{x} \right) dA \qquad (3.42)$$

As $\hat{n} \bullet \bar{\bar{\sigma}}_d = \hat{n} \bullet \bar{\bar{\sigma}}_c + \nabla_s \bullet \bar{\bar{P}}_s$ (see Equation 3.39), Equation 3.42 can be rewritten as:

$$\int_{V_0^-} \bar{\bar{\sigma}}_d \, dV = \int_{A_0^+} \left(n \bullet \bar{\bar{\sigma}}_c \, \vec{x} \right) dA + \int_{A_0} \left(\nabla_s \bullet \bar{\bar{P}}_s \right) \vec{x} \, dA \qquad (3.43)$$

Combining Equation 3.40 and Equation 3.43 gives:

$$\int_{V_0^+} \bar{\bar{\sigma}} \, dV = \int_{A_0^+} \left(\hat{n} \bullet \bar{\bar{\sigma}}_c \, \vec{x} \right) dA + \int_{A_0} \left[\bar{\bar{P}}_s + \left(\nabla_s \bullet \bar{\bar{P}}_s \right) \vec{x} \right] dA \qquad (3.44)$$

The integrand of the second surface integral on the RHS of Equation 3.44 is equal to $\nabla_s \bullet (\bar{\bar{P}}_s \vec{x})$. The surface divergence theorem when applied to a closed surface with no bounding contours gives [10] (see Chapter 2):

$$\int_{A_0} \nabla_s \bullet \left(\bar{\bar{P}}_s \, \vec{x} \right) dA = 0 \qquad (3.45)$$

Thus, Equation 3.44 reduces to:

$$\int_{V_0^+} \overline{\overline{\sigma}} \, dV = \int_{A_0^+} \left(\hat{n} \bullet \overline{\overline{\sigma}}_c \, \vec{x} \right) dA \tag{3.46}$$

The expression for the dipole strength of droplets in the presence of the interfacial film can now be derived. From Equation 3.40 and Equation 3.46:

$$\int_{V_0^+} \overline{\overline{\sigma}} \, dV = \int_{A_0^+} \left(\overline{\overline{\sigma}}_c \bullet \hat{n} \, \vec{x} \right) dA = \int_{V_0^-} \overline{\overline{\sigma}}_d \, dV + \int_{A_0} \overline{P}_s \, dA \tag{3.47}$$

Note that $\hat{n} \bullet \overline{\overline{\sigma}}_c = \overline{\overline{\sigma}}_c \bullet \hat{n}$ as $\overline{\overline{\sigma}}_c$ is a symmetric tensor. The stress tensor in the droplets $\overline{\overline{\sigma}}_d$ is:

$$\overline{\overline{\sigma}}_d = -P\overline{\overline{\delta}} + \eta_d \left[\nabla \hat{u} + \left(\nabla \hat{u} \right)^T \right] \tag{3.48}$$

Therefore,

$$\int_{V_0^-} \overline{\overline{\sigma}}_d \, dV = \int_{V_0^-} -P\overline{\overline{\delta}} \, dV + \eta_d \int_{V_0^-} \left[\nabla \vec{u} + \left(\nabla \vec{u} \right)^T \right] dV \tag{3.49}$$

Using the Gauss divergence theorem and the fact that the velocity $\vec{u}$ is continuous at the interface, Equation 3.49 can be expressed as:

$$\int_{V_0^-} \overline{\overline{\sigma}}_d \, dV = \eta_d \int_{A_0} \left(\hat{n}\vec{u} + \vec{u}\hat{n} \right) dA + \int_{A_0^+} \frac{1}{3} (\vec{x} \bullet \overline{\overline{\sigma}}_c \bullet \hat{n}) \overline{\overline{\delta}} dA \tag{3.50}$$

Substituting Equation 3.50 into Equation 3.47 gives:

$$\int_{A_0^+} \left(\overline{\overline{\sigma}}_c \bullet \hat{n} \, \vec{x} \right) dA = \eta_d \int_{A_0} \left(\hat{n}\vec{u} + \vec{u}\hat{n} \right) dA + \int_{A_0} \overline{P}_s \, dA + \int_{A_0^+} \frac{1}{3} (\vec{x} \bullet \overline{\overline{\sigma}}_c \bullet \hat{n}) \overline{\overline{\delta}} dA \tag{3.51}$$

From Equation 3.37 and Equation 3.51, the dipole strength of droplets with interfacial film is:

$$\overline{\overline{S}} = (\eta_d - \eta_c) \int_{A_0} \left(\hat{n}\vec{u} + \vec{u}\hat{n} \right) dA + \int_{A_0} \overline{\overline{P}}_s \, dA \tag{3.52}$$

For interfacial film characterized solely by a constant interfacial tension γ, the surface stress tensor is given by:

$$\bar{\bar{P}}_s = \gamma \bar{\bar{\delta}}_s = \gamma \left(\bar{\bar{\delta}} - \hat{n}\,\hat{n} \right) \tag{3.53}$$

Consequently, the dipole strength of droplets is:

$$\bar{\bar{S}} = \int_{A_0} \gamma \left(\bar{\bar{\delta}} - \hat{n}\,\hat{n} \right) dA + \left(\eta_d - \eta_c \right) \int_{A_0} \left(\hat{n}\,\vec{u} + \vec{u}\hat{n} \right) dA \tag{3.54}$$

3.3.2 COMPOSITES

Composites are *solid* heterogeneous materials composed of two or more phases. Many composites of practical interest are composed of just two phases — the dispersed phase and the matrix (continuous phase). The bulk stress in two-phase composite materials can be written as:

$$\langle \bar{\bar{\sigma}} \rangle = \langle \bar{\bar{\sigma}}_o \rangle + \frac{1}{V} \sum_{i=1}^{N} \bar{\bar{S}}_i \tag{3.55}$$

where $\langle \bar{\bar{\sigma}} \rangle$ is the bulk (volume-average) stress in the composite material, $\langle \bar{\bar{\sigma}}_o \rangle$ is the stress tensor in pure matrix material under conditions corresponding to the imposed bulk strain $\langle \bar{\bar{e}} \rangle$ on the composite, V is a sample volume whose dimensions are large relative to the average spacing between the particles but small relative to the characteristic macroscale of the system, N is the number of particles within V, and $\bar{\bar{S}}$ is the dipole strength of a particle defined as [11–13]:

$$\bar{\bar{S}} = \int_{A_0} \left[\bar{\bar{\sigma}} \bullet \hat{n}\,\vec{x} - \lambda_c \left(\vec{\xi} \bullet \hat{n} \right) \bar{\bar{\delta}} - G_c \left(\vec{\xi}\hat{n} + \hat{n}\,\vec{\xi} \right) \right] dA \tag{3.56}$$

where A_0 is the surface area of a particle, $\hat{n}$ is a unit outward normal to the particle surface, $\vec{\xi}$ is the surface displacement vector, $\bar{\bar{\sigma}}$ is the stress tensor at the particle surface, $\vec{x}$ is the position vector with respect to a fixed origin, and $\bar{\bar{\delta}}$ is the unit tensor. The matrix material is assumed to be isotropic and linearly elastic with Lamé constants λ_c and G_c. Note that for a linearly elastic matrix, the stress tensor $\langle \bar{\bar{\sigma}}_o \rangle$ is given as [11–14]:

$$\langle \bar{\bar{\sigma}}_o \rangle = \lambda_c \left(tr \langle \bar{\bar{e}} \rangle \right) \bar{\bar{\delta}} + 2G_c \langle \bar{\bar{e}} \rangle \tag{3.57}$$

3.4 CONSTITUTIVE EQUATIONS FOR DISPERSED SYSTEMS

Constitutive equations are equations that relate bulk stress tensor $\langle\bar{\bar{\sigma}}\rangle$ to appropriate kinematical variables. As noted in sections 3.3.1 and 3.3.2, the bulk stress tensor for a dispersed system (dispersion or composite), valid for all concentrations of dispersed phase, can be written as:

$$\langle\bar{\bar{\sigma}}\rangle = \langle\bar{\bar{\sigma}}_o\rangle + \frac{1}{V}\sum_{i=1}^{N}\bar{\bar{S}}_i \tag{3.58}$$

where $\langle\bar{\bar{\sigma}}_o\rangle$ is the stress tensor in pure matrix material and $\bar{\bar{S}}_i$ is the dipole strength of the ith particle. For dispersions, the stress tensor $\langle\bar{\bar{\sigma}}_0\rangle$ is given by Equation 3.27. For composites, the stress tensor $\langle\bar{\bar{\sigma}}_0\rangle$ is given by Equation 3.57.

The dipole strength of a particle $\bar{\bar{S}}_i$ depends on the size, shape, orientation, and internal transport properties of the particle and on the location, size, shapes, and orientations of the other adjoining particles [3,6,15].

For a disperse system of identical particles (same shape, size, and internal constitution), the constitutive equation can be expressed as:

$$\langle\bar{\bar{\sigma}}\rangle = \langle\bar{\bar{\sigma}}_o\rangle + \frac{N}{V}\langle\bar{\bar{S}}\rangle \tag{3.59}$$

where N/V is the number density of the particles in the system and $\langle\bar{\bar{S}}\rangle$ is the average value of $\bar{\bar{S}}$ for a reference particle.

For given shape, size, and internal constitution of the particles, the dipole strength $\bar{\bar{S}}$ of a reference particle is affected by two microstructural factors, namely: (1) the orientation of the reference particle if the particle is nonspherical and (2) the relative positions and the orientations of the other neighboring particles. Consequently, the mean value of $\bar{\bar{S}}$ for a reference particle can be expressed as:

$$\langle\bar{\bar{S}}\rangle = \int \bar{\bar{S}}(\vec{p},\xi)\psi(\vec{p},\xi/O)\,d\vec{p}\,d\xi \tag{3.60}$$

where $\vec{p}$ denotes the orientation vector of the reference particle, ξ denotes the configuration (relative positions and orientation) of the surrounding particles, $\bar{\bar{S}}(\vec{p},\xi)$ is the dipole strength of the reference particle for a given orientation $\vec{p}$ and given configuration of surrounding particles ξ, and $\psi(\vec{p},\xi/O)$ is the probability density function of the configuration ξ with the reference particle at the origin having orientation $\vec{p}$. For *spherical* particles, orientation effects are irrelevant. Consequently,

$$\left\langle \overline{\overline{S}} \right\rangle = \int \overline{\overline{S}}\,(\xi)\,\psi(\xi/O)\,d\xi \qquad (3.61)$$

where the configuration ξ is specified by a set of position vectors $\vec{x}_1, \vec{x}_2, \dots$ giving the positions of the surrounding particles relative to the reference particle at the origin.

3.4.1 DILUTE SYSTEM OF NONSPHERICAL PARTICLES

If the dispersed system is *dilute* so that the particles are far apart, the interactions between the particles is negligible. In such situations, the mean value of the dipole strength $\left\langle \overline{\overline{S}} \right\rangle$ for the reference particle can be calculated assuming that the reference particle alone is present in an infinite matrix. As the value of $\left\langle \overline{\overline{S}} \right\rangle$ no longer depends on the configuration of other particles, the mean dipole strength can be expressed as:

$$\left\langle \overline{\overline{S}}^{o} \right\rangle = \int \overline{\overline{S}}^{o}(\vec{p})\,\psi(\vec{p})\,d\vec{p} \qquad (3.62)$$

where $\overline{\overline{S}}^{o}(\vec{p})$ is the value of the dipole strength of the reference particle for particle orientation $\vec{p}$, evaluated on the basis that the reference particle alone is present in an infinite matrix. Thus, the bulk stress tensor for a dilute dispersed system can be written as:

$$\left\langle \overline{\overline{\sigma}} \right\rangle = \left\langle \overline{\overline{\sigma}}_{o} \right\rangle + \frac{N}{V}\left\langle \overline{\overline{S}}^{o} \right\rangle \qquad (3.63)$$

3.4.2 DILUTE SYSTEM OF SPHERICAL PARTICLES

For a dilute system of spherical particles, the bulk stress tensor can be expressed as:

$$\left\langle \overline{\overline{\sigma}} \right\rangle = \left\langle \overline{\overline{\sigma}}_{o} \right\rangle + \left(\frac{3\phi}{4\pi R^3} \right)\overline{\overline{S}}^{o} \qquad (3.64)$$

where R is the particle radius, ϕ is the volume fraction of particles, and $\overline{\overline{S}}^{o}$ is the dipole strength of a single spherical particle located in an infinite matrix.

3.5 NONHYDRODYNAMIC EFFECTS IN DISPERSIONS

In the discussion thus far, we have made no mention of nonhydrodynamic forces such as electrostatic, steric, van der Waals, and Brownian forces. These forces, also referred to as *thermodynamic* forces, play a significant role in governing the macroscopic rheological behavior of dispersions of fine (submicron) colloidal particles [2,5,6,16–33].

The thermodynamic forces affect the rheology of colloidal dispersions at all levels of particle concentrations (volume fraction of dispersed phase). Consider, for example, the effect of Brownian motion on the rheology of dispersion of nonspherical rigid particles (say rods). Even when the dispersion is dilute so that there is negligible interaction between the particles, Brownian motion affects the bulk rheological behavior of the dispersion in two distinct ways:

1. First, Brownian motion affects the probability distribution of the particle orientation $\psi(\vec{p})$ and, hence, the average force dipole strength $\langle \bar{\bar{s}}^o \rangle$ (see Equation 3.62). The viscous stresses of the imposed flow field tend to align the particles in such a manner so as to alleviate these stresses. This ordering effect of flow causes gradients in the orientation distribution function and thereby generates an opposing-rotational-diffusion process. The rotary Brownian motion tends to randomize the particle orientations, thereby destroying the ordering or alignment of particles. The relative influence of these counteracting effects — alignment by flow and randomizing by rotary Brownian motion — is described by the Fokker–Planck equation for the probability distribution of the particle orientation. This effect of Brownian motion on the bulk stress of a dispersion whereby the orientation distribution function is altered is referred to as the *indirect* effect of Brownian motion in the literature.
2. Brownian motion also makes a "direct" contribution to the bulk stress of a dispersion. The Brownian couple acting upon a suspended particle imparts an angular velocity to the particle. Consequently, the effective rotation rate of the particle is $\vec{\omega} + \vec{\omega}^{Br}$, where $\vec{\omega}$ is the angular velocity in the absence of Brownian motion, and $\vec{\omega}^{Br}$ is the angular velocity due to Brownian couple alone in the absence of bulk motion. The force dipole strength of a particle due to rotation of a particle at $\vec{\omega}^{Br}$ in a fluid at rest at infinity is the direct contribution made to the bulk stress by Brownian motion.

The bulk stress in a dispersion of particles can be written as:

$$\langle \bar{\bar{\sigma}} \rangle = -\langle P \rangle \bar{\bar{\delta}} + 2\eta_c \langle \bar{\bar{E}} \rangle + \langle \bar{\bar{\sigma}}^P \rangle \qquad (3.65)$$

where $\langle \bar{\bar{\sigma}}^P \rangle$ is the particle contribution to the bulk stress. For dilute dispersion of Brownian particles, $\langle \bar{\bar{\sigma}}^P \rangle$ can be expressed as:

$$\langle \bar{\bar{\sigma}}^P \rangle = -(N/V)kT \bar{\bar{\delta}} + (N/V)\left[\langle \bar{\bar{S}}^H \rangle + \langle \bar{\bar{S}}^{Br} \rangle \right] \qquad (3.66)$$

where $-(N/V)kT\bar{\bar{\delta}}$ is just the isotropic stress associated with the thermal kinetic energy of the Brownian particles, $\bar{\bar{\delta}}$ is a unit tensor, N/V is the number density of particles, $\langle \bar{\bar{S}}^H \rangle$ is the dipole strength of a particle which is undisturbed by Brownian motion except to the extent that it governs the orientation distribution $\psi(\bar{p})$, and $\langle \bar{\bar{S}}^{Br} \rangle$ is the direct Brownian contribution to dipole strength arising from particle rotation associated with Brownian couple.

Note that Brownian motion does not affect the bulk (deviatoric) stress of a dilute dispersion of *spherical* particles. In case of a dispersion of *spherical* particles, Brownian motion affects the bulk stress only at nondilute concentrations.

In *nondilute* dispersions, interactions between the particles become significant. The particles can no longer be treated as isolated from one another. Both hydrodynamic and nonhydrodynamic interactions between the particles could come into play. In such situations, the particle contribution to the bulk stress can be separated into hydrodynamic and thermodynamic contributions; hydrodynamic stress $\langle \bar{\bar{\sigma}}^H \rangle$ is the particle stress due to the imposed flow in the absence of nonhydrodynamic interactions, and thermodynamic stress $\langle \bar{\bar{\sigma}}^T \rangle$ is due to nonhydrodynamic interactions between the particles when the dispersion as a whole is at rest.

As an example, let us consider a nondilute dispersion of interacting (hydrodynamically and nonhydrodynamically) rigid particles that are spherical in shape. For spherical particles, orientations are irrelevant and the configuration or the microstructure of the dispersion is specified by spatial distribution of particles. The hydrodynamic contribution to the particle stress tensor can be expressed as:

$$\langle \bar{\bar{\sigma}}^H \rangle = (N/V)\langle \bar{\bar{S}}^H \rangle = (N/V) \int \bar{\bar{S}}^H(\xi) \psi(\xi/O) d\xi \qquad (3.67)$$

where $\langle \bar{\bar{S}}^H \rangle$ is the average value of the dipole strength for a reference particle in the absence of nonhydrodynamic interactions, and $\psi(\xi/O)$ is the probability density function of the configuration ξ with the reference particle at the origin. However, it should be noted that the hydrodynamic stress $\langle \bar{\bar{\sigma}}^H \rangle$ includes the "indirect" effect of nonhydrodynamic forces through their influence on the probability density function $\psi(\xi/O)$, which appears as a weighting function in the integral expression given in Equation 3.67.

The nonhydrodynamic or thermodynamic forces generate bulk stresses by both opposing the change in the configuration and inducing hydrodynamic flow fields. Thus, the thermodynamic forces make a direct contribution to the particle stress in two ways [33].

Let $\psi_N(\vec{x}_1, \vec{x}_2, ..., \vec{x}_N)$ be the probability density of a configuration of N particles defined by the N position vectors $(\vec{x}_1, \vec{x}_2, ..., \vec{x}_N)$. Let $\Lambda_{ij}(r_{ij})$ be the interaction potential between particles i and j, separated by a distance r_{ij}.

The interaction force on the ith particle, $\vec{F}_i$, is given by:

$$\vec{F}_i = -kT\vec{\nabla}_i(\ell n\psi_N) - \vec{\nabla}_i\left(\sum_{j\neq i}\Lambda_{ij}\right) \qquad (3.68)$$

The first term on the RHS of Equation 3.68 is the interaction force due to Brownian motion and the second term on the RHS is the interactive force due to van der Waals, electrostatic, or steric interactions.

The stress caused by the change in the Helmholtz free energy, resulting from the distortion of the microstructure of the dispersion, is given as:

$$\left\langle\overline{\overline{\sigma}}^{T_1}\right\rangle = -\frac{1}{V}\sum_{i=1}^{N}\vec{x}_i\vec{F}_i$$
$$= -(N/V)\left\langle\vec{x}\,\vec{F}\right\rangle \qquad (3.69)$$

The other contribution of thermodynamic forces to the particle stress comes from the hydrodynamic coupling effect. Any force (Brownian or interparticle) acting on a particle causes motion of a particle. The motion of a particle generates a hydrodynamic flow field resulting in an additional stress. This contribution of thermodynamic forces to the particle stress can be expressed as:

$$\left\langle\overline{\overline{\sigma}}^{T_2}\right\rangle = -\frac{1}{V}\sum_{i=1}^{N}\overline{\overline{\overline{C}}}_i\bullet\vec{F}_i$$
$$= -(N/V)\left\langle\overline{\overline{\overline{C}}}\bullet\vec{F}\right\rangle \qquad (3.70)$$

where $\overline{\overline{\overline{C}}}$ is a third-order hydrodynamic tensor that relates stresses to forces [33]. It depends on the entire configuration of all N particles.

The total thermodynamic stress $\langle\overline{\overline{\sigma}}^T\rangle$ is the summation of $\langle\overline{\overline{\sigma}}^{T_1}\rangle$ and $\langle\overline{\overline{\sigma}}^{T_2}\rangle$. Thus,

$$\left\langle\overline{\overline{\sigma}}^T\right\rangle = -\frac{1}{V}\sum_{i=1}^{N}\vec{x}_i\,\vec{F}_i - \frac{1}{V}\sum_{i=1}^{N}\overline{\overline{\overline{C}}}_i\bullet\vec{F}_i$$
$$= -(N/V)\left[\left\langle\vec{x}\,\vec{F}\right\rangle + \left\langle\overline{\overline{\overline{C}}}\bullet\vec{F}\right\rangle\right] \qquad (3.71)$$

The particle contribution to the bulk stress of a dispersion $\langle \bar{\bar{\sigma}}^P \rangle$ (see Equation 3.65) can now be expressed as:

$$\left\langle \bar{\bar{\sigma}}^P \right\rangle = -\left(N/V \right) kT\, \bar{\bar{\delta}} + \left(N/V \right) \left[\left\langle \bar{\bar{S}}^H \right\rangle - \left(\left\langle \vec{x}\,\vec{F} \right\rangle + \left\langle \bar{\bar{C}} \bullet \vec{F} \right\rangle \right) \right] \quad (3.72)$$

where the thermal kinetic energy of the Brownian particles is included as an isotropic stress.

NOTATION

A	surface area
A_i	surface area of ith particle
A_o	surface area of a particle
A_o^-	surface area of a particle on the inner side of the interfacial film
A_o^+	surface area of a particle on the outer side of the interfacial film
$\bar{\bar{\bar{C}}}$	third-order hydrodynamic tensor that relates stresses to forces
$\bar{\bar{E}}$	rate of strain tensor
$\bar{\bar{e}}$	strain tensor
$\vec{F}$	force vector
G_c	shear modulus of matrix material
k	Boltzmann constant
L	characteristic linear dimension of the apparatus
ℓ	average spacing between particles
$\bar{\bar{L}}$	antisymmetric part of $\bar{\bar{S}}$, referred to as *couplet*
$\vec{\mathscr{L}}$	torque acting on a particle
N	number of particles
$\hat{n}$	outer unit normal vector
P	thermodynamic pressure
$\vec{p}$	orientation vector of nonspherical particle
$\bar{\bar{P}}_s$	surface stress tensor
$\bar{\bar{S}}$	dipole strength of a particle
$\bar{\bar{S}}_i$	dipole strength of ith particle
$\bar{\bar{S}}^o$	dipole strength of a single particle located in an infinite matrix
$\bar{\bar{S}}^{Br}$	dipole strength of a particle due to rotation associated with Brownian couple

$\overline{\overline{S}}^H$	dipole strength of a particle in the absence of nonhydrodynamic forces and Brownian motion
$\underset{=}{T}$	temperature (as superscript, T generally refers to transpose)
$\overline{\overline{T}}$	symmetric part of $\overline{\overline{S}}$, referred to as *stresslet*
t	time
$\vec{u}$	velocity vector
V	volume
V_c	volume occupied by continuous phase (matrix)
V_d	volume occupied by dispersed particles
V_i	volume of ith particle
V_o	volume of a particle
V_o^-	volume of a particle on the inner side of the interfacial film
V_o^+	volume of the particle including the interfacial film
$\vec{x}$	position vector

GREEK

γ	interfacial tension
$\overline{\overline{\delta}}$	unit tensor
$\hat{\delta}_1, \hat{\delta}_2, \hat{\delta}_3$	unit vectors
∇	del operator, three-dimensional spatial gradient operator
∇_s	surface gradient operator
$\underset{=}{\overline{\in}}$	alternating isotropic triadic
$\in_{ijk}$	permutation symbol
η_c	continuous-phase (matrix) viscosity
η_d	viscosity of dispersed phase
Λ_{ij}	interaction potential between particles i and j
λ	orientation of a particle
λ_c	Lamé constant of matrix material
ξ	configuration of a system of particles (relative positions and orientation of particles)
$\vec{\xi}$	surface displacement vector
$\overline{\overline{\sigma}}$	stress tensor
$\overline{\overline{\sigma}}_c$	stress tensor in continuous phase
$\overline{\overline{\sigma}}_d$	stress tensor in dispersed phase
$\overline{\overline{\sigma}}_o$	stress tensor in pure matrix material
$\overline{\overline{\sigma}}^H$	hydrodynamic stress (stress tensor due to imposed flow in the absence of nonhydrodynamic forces)
$\overline{\overline{\sigma}}^P$	particle contribution to stress tensor

$\bar{\bar{\sigma}}^T$	thermodynamic stress (stress tensor due to nonhydrodynamic interactions alone)
$\bar{\bar{\sigma}}^{T_1}$	thermodynamic stress tensor due to distortion of the microstructure of dispersion
$\bar{\bar{\sigma}}^{T_2}$	thermodynamic stress tensor due to hydrodynamic coupling effect
ψ_N	probability density function of a configuration of N particles
$\psi(\vec{p}, \xi/O)$	probability density function of configuration ξ with the reference particle at the origin having orientation $\vec{p}$
$\psi(\vec{p})$	probability density of a particle having orientation $\vec{p}$
$\vec{\omega}$	angular velocity of a particle
$\vec{\omega}^{Br}$	angular velocity of a particle due to Brownian couple alone
$\langle X \rangle$	volume average of quantity X

REFERENCES

1. H. Brenner, *Annu. Rev. Fluid Mech.*, 2: 137–176, 1970.
2. H. Brenner, *Prog. Heat Mass Transfer*, 5: 89–129, 1972.
3. G.K. Batchelor, *J. Fluid Mech.*, 41: 545–570, 1970.
4. D.A. Drew, G.S. Arnold, R.T. Lahey, in *Two Phase Flows and Waves*, D.D. Joseph, D.G. Schaeffer, Eds., Springer-Verlag, New York, 1990, pp. 45–56.
5. D.J. Jeffrey, A. Acrivos, *AIChE J.*, 22: 417–432.
6. G.K. Batchelor, *Annu. Rev. Fluid Mech.*, 6: 227–255, 1974.
7. R.W. O'Brien, *J. Fluid Mech.*, 91: 17–39, 1979.
8. K.R. Hase, R.L. Powell, *Phys. Fluids*, 13: 32–44, 2001.
9. J. Happel, H. Brenner, *Low Reynolds Number Hydrodynamics*, Noordhoff International Publishing, Leyden, the Netherlands, 1973.
10. A. Nadim, *Chem. Eng. Commun.*, 148–150: 391–407, 1996.
11. W.B. Russel, A. Acrivos, *J. Appl. Math. Phys. (ZAMP)*, 23: 435–464, 1972.
12. H.S. Chen, A. Acrivos, *Int. J. Solids Struct.*, 14: 331–348, 1978.
13. H.S. Chen, A. Acrivos, *Int. J. Solids Struct.*, 14: 349–364, 1978.
14. G.E. Mase, *Continuum Mechanics*, McGraw-Hill, New York, 1970, chap. 6.
15. G.K. Batchelor, J.T. Green, *J. Fluid Mech.*, 56: 401–427, 1972.
16. E.J. Hinch, L.G. Leal, *J. Fluid Mech.*, 52: 683–712, 1972.
17. L.G. Leal, E.J. Hinch, *J. Fluid Mech.*, 55: 745–765, 1972.
18. E.J. Hinch, L.G. Leal, *J. Fluid Mech.*, 57: 753–767, 1973.
19. G.K. Batchelor, in *Theoretical and Applied Mechanics*, W.T. Koiter, Ed., North-Holland Publishing Co., Amsterdam, 1976, pp. 33–55.
20. G.K. Batchelor, *J. Fluid Mech.*, 74: 1–29, 1978.
21. G.K. Batchelor, *J. Fluid Mech.*, 83: 97–117, 1977.
22. W.B. Russel, *J. Fluid Mech.*, 92: 401–419, 1979.
23. R. Herczynski, I. Pienkowska, *Annu. Rev. Fluid Mech.*, 12: 237–269, 1980.
24. W.B. Russel, *Annu. Rev. Fluid Mech.*, 13: 425–455, 1981.
25. W.B. Russel, A.P. Gast, *J. Chem. Phys.*, 84: 1815–1826, 1986.
26. G. Bossis, J.F. Brady, *J. Chem. Phys.*, 91: 1866–1874, 1989.
27. N.J. Wagner, W.B. Russel, *Physica A*, 155: 475–518, 1989.

28.   J.F. Brady, *J. Chem. Phys.*, 99: 567–581, 1993.
29.   N.J. Wagner, *J. Colloid Interface Sci.*, 161: 169–181.
30.   R.A. Lionberger, W.B. Russel, *J. Rheol.*, 38: 1885–1908, 1994.
31.   R.A. Lionberger, *J. Rheol.*, 42: 843–863, 1998.
32.   J.F. Morris, B. Katyol, *Phys. Fluids*, 14: 1920–1937, 2002.
33.   S. Dagreou, A. Allal, G. Marin, B. Mendiboure, *Rheol. Acta*, 41: 500–513, 2002.

Part II

Rheology of Dispersions of Rigid Particles

Part II consists of Chapter 4 through Chapter 6, and deals with the rheology of dispersions of *rigid* spherical and nonspherical particles.

Chapter 4 focuses on dispersions of rigid *spherical* particles. The steps involved in the derivation of the expression for dipole strength of a single (rigid and spherical) particle located in an infinite matrix fluid are discussed. Using the expression for the dipole strength of a single particle, the Einstein equation for the viscosity of dilute suspension of rigid spherical particles is derived. The rheology of dispersions of *porous* (rigid and spherical) particles is discussed next. From the dipole strength of a single porous (rigid and spherical) particle located in an infinite matrix fluid, the expression for the intrinsic viscosity of a dispersion of porous particles is derived. The remaining portion of the chapter is devoted to the rheology of dispersions of *electrically charged* solid particles. The fundamental equations governing the flow field around a charged particle are discussed. The expression for the bulk stress in dispersion of charged particles is derived. The rheological behavior of dispersions of charged particles predicted from the constitutive equation under different conditions is discussed.

Chapter 5 deals with the rheology of dispersions of rigid *nonspherical* particles. After a brief introduction, the motion of a single isolated spheroidal particle in shear flow is described. The Jeffery solution of the angular velocity of a non-Brownian spheroidal particle is presented. The orientation distribution of particles is discussed in terms of the Fokker–Planck equation. The constitutive equation of a dilute dispersion of axisymmetric non-Brownian particles is described in terms of the second and fourth moments of unit orientation vector of a particle. The rheological properties of a dispersion of non-Brownian axisymmetric particles under different conditions are discussed. The effects of rotary Brownian motion on the orientation distribution function and rheological properties of suspensions of nonspherical particles are presented next.

The rheology of dispersions of rigid, spherical and nonspherical, *magnetic* particles is dealt with in Chapter 6. The suspensions of *spherical* dipolar particles are discussed first. The motion of a single isolated dipolar (spherical) particle in shear flow, in the presence of an external magnetic field, is described. The equation for the bulk stress tensor of a suspension of dipolar particles in the presence of an external magnetic field is derived. The rheological properties predicted from the constitutive equation under different conditions are explained. The effect of rotary Brownian motion on the bulk stress of suspensions of dipolar (spherical) particles is also discussed. The remainder of the chapter is devoted to suspensions of *nonspherical* dipolar particles. The equations governing the angular velocity of a particle, the orientation state of a suspension, and the bulk stress of a suspension are described. The rheological properties of a dilute suspension of nonspherical, axisymmetric, magnetic particles subjected to rotary Brownian motion and an external magnetic field predicted from the constitutive equation are discussed.

4 Dispersions of Rigid Spherical Particles

4.1 INTRODUCTION

Dispersions of rigid particles in liquids, also referred to as *suspensions*, form a large group of materials of industrial importance. Both aqueous and nonaqueous dispersions are encountered in industrial applications [1–5].

In this chapter, the rheology of dispersions of the following three different types of *spherical* rigid particles is discussed: (1) solid spherical particles, (2) porous spherical particles, and (3) electrically charged solid spherical particles.

4.2 DISPERSIONS OF SOLID SPHERICAL PARTICLES

The constitutive equation for a dilute dispersion of identical spherical particles is given by (see Chapter 3):

$$\langle \bar{\bar{\sigma}} \rangle = -\langle P \rangle \bar{\bar{\delta}} + 2\eta_c \langle \bar{\bar{E}} \rangle + \frac{3\phi}{4\pi R^3} \bar{\bar{S}}^0 \qquad (4.1)$$

where $\langle \bar{\bar{\sigma}} \rangle$ is the bulk (average) stress tensor, $\langle P \rangle$ is the average pressure, $\bar{\bar{\delta}}$ is a unit tensor, η_c is matrix viscosity, $\langle \bar{\bar{E}} \rangle$ is the bulk rate of strain tensor, ϕ is the volume fraction of particles, R is the particle radius, and $\bar{\bar{S}}^0$ is the dipole strength of a single spherical particle located in an infinite matrix fluid. $\bar{\bar{S}}^0$ is given as:

$$\bar{\bar{S}}^0 = \int_{A_0} \left[\bar{\bar{\sigma}} \bullet \hat{n}\, \vec{x} - \frac{1}{3}\left(\vec{x} \bullet \bar{\bar{\sigma}} \bullet \hat{n} \right) \bar{\bar{\delta}} - \eta_c \left(\hat{n}\,\vec{u} + \vec{u}\,\hat{n} \right) \right] dA \qquad (4.2)$$

where A_0 is the surface area of the particle, $\hat{n}$ is a unit outward normal to the particle surface, $\vec{u}$ is fluid velocity at the particle surface, and $\bar{\bar{\sigma}}$ is the fluid stress tensor at the particle surface. To evaluate $\bar{\bar{S}}^0$, the velocity and stress distributions around the particle are needed.

Consider a spherical solid particle of radius R, suspended in a continuous-phase (matrix) liquid of viscosity η_c. Choose a coordinate system that has its origin fixed at the center of the particle. Let $r = |\vec{r}|$ represent the radial distance from the origin, where $\vec{r}$ is the position vector measured from the origin.

The system is subjected to linear shear flow such that far away from the particle, the imposed flow field varies linearly with position, that is,

$$\vec{u}_\infty = \bar{\bar{\Gamma}}_\infty \bullet \vec{r} \tag{4.3}$$

where $\bar{\bar{\Gamma}}_\infty$ is the spatially uniform (homogeneous) velocity gradient tensor $(\nabla \vec{u})^T$ far away from the particle. It is convenient to divide $\bar{\bar{\Gamma}}_\infty$ into its symmetric and antisymmetric parts:

$$\bar{\bar{E}}_\infty = \frac{1}{2}\left(\bar{\bar{\Gamma}}_\infty + \bar{\bar{\Gamma}}_\infty^T\right) \tag{4.4}$$

$$\bar{\bar{\Omega}}_\infty = \frac{1}{2}\left(\bar{\bar{\Gamma}}_\infty - \bar{\bar{\Gamma}}_\infty^T\right) \tag{4.5}$$

where $\bar{\bar{E}}_\infty$ is the rate of strain tensor and $\bar{\bar{\Omega}}_\infty$ is the vorticity tensor of the undisturbed flow far away from the particle. Thus, Equation 4.3 can be expressed as:

$$\vec{u}_\infty = \bar{\bar{E}}_\infty \bullet \vec{r} + \bar{\bar{\Omega}}_\infty \bullet \vec{r} \tag{4.6}$$

or

$$\vec{u}_\infty = \bar{\bar{E}}_\infty \bullet \vec{r} + \vec{\omega}_\infty \times \vec{r} \tag{4.7}$$

where $\vec{\omega}_\infty$ is the local angular velocity of the undisturbed flow.

In the absence of inertial effects, the equations governing the fluid flow around the particle are the Stokes equations:

$$\nabla^2 \vec{u} = \frac{1}{\eta_c} \nabla P \tag{4.8}$$

$$\nabla \bullet \vec{u} = 0 \tag{4.9}$$

where η_c, P, and $\vec{u}$ are the fluid viscosity, pressure, and velocity, respectively. The boundary conditions are:

1. Far away from the particle $(r \rightarrow \infty)$

$$\vec{u} \rightarrow \vec{u}_\infty \tag{4.10}$$

2. At the surface of the particle ($r = R$)

$$\vec{u} = \vec{\omega} \times \vec{r} \tag{4.11}$$

where $\vec{\omega}$ is the angular velocity of the particle.

The solution of the Stokes equations (Equation 4.8 and Equation 4.9) subject to the boundary conditions Equation 4.10 and Equation 4.11 is as follows [6,7]:

$$\vec{u}(\vec{r}) = \bar{\bar{E}}_\infty \bullet \vec{r} + \vec{\omega} \times \vec{r} - \left(\frac{R}{r}\right)^5 (\bar{\bar{E}}_\infty \bullet \vec{r}) + \frac{\vec{r}(\vec{r} \bullet \bar{\bar{E}}_\infty \bullet \vec{r})}{r^2}\left[\frac{-5}{2}\left(\frac{R}{r}\right)^3 + \frac{5}{2}\left(\frac{R}{r}\right)^5\right] \tag{4.12}$$

$$P = P_\infty - 5\left(\frac{R}{r}\right)^3 \frac{(\vec{r} \bullet \bar{\bar{E}}_\infty \bullet \vec{r})\eta_c}{r^2} \tag{4.13}$$

The local rate of strain tensor $\bar{\bar{E}}$ in the matrix fluid is as follows [6]:

$$\bar{\bar{E}}(\vec{r}) = \bar{\bar{E}}_\infty\left[1 - \left(\frac{R}{r}\right)^5\right] + \left[\frac{\vec{r}(\bar{\bar{E}}_\infty \bullet \vec{r}) + (\vec{r} \bullet \bar{\bar{E}}_\infty)\vec{r}}{r^2} - \frac{2}{3}\left(\frac{\vec{r}\bar{\bar{E}}_\infty \bullet \vec{r}}{r^2}\right)\bar{\bar{\delta}}\right]$$

$$\times\left[\frac{-5}{2}\left(\frac{R}{r}\right)^3 + 5\left(\frac{R}{r}\right)^5\right] + \left[\left(\frac{\vec{r} \bullet \bar{\bar{E}}_\infty \bullet \vec{r}}{r^2}\right)\left(\frac{\vec{r}\vec{r}}{r^2} - \frac{1}{3}\bar{\bar{\delta}}\right)\right]\left[\frac{25}{2}\left(\frac{R}{r}\right)^3 - \frac{35}{2}\left(\frac{R}{r}\right)^5\right] \tag{4.14}$$

The stress field in the matrix fluid can be determined from Equation 4.13 and Equation 4.14 using

$$\bar{\bar{\sigma}} = -P\bar{\bar{\delta}} + 2\eta_c\,\bar{\bar{E}} \tag{4.15}$$

Thus,

$$\bar{\bar{\sigma}} = -P_\infty\bar{\bar{\delta}} + 2\eta_c\,\bar{\bar{E}}_\infty\left[1 - \left(\frac{R}{r}\right)^5\right] + 2\eta_c\left[\frac{\vec{r}(\bar{\bar{E}}_\infty \bullet \vec{r}) + (\vec{r} \bullet \bar{\bar{E}}_\infty)\vec{r}}{r^2}\right]$$

$$\times\left[\frac{-5}{2}\left(\frac{R}{r}\right)^3 + 5\left(\frac{R}{r}\right)^5\right] + 5\eta_c\left(\frac{\vec{r} \bullet \bar{\bar{E}}_\infty \bullet \vec{r}}{r^2}\right)\left(\frac{R}{r}\right)^5\bar{\bar{\delta}}$$

$$-35\eta_c\left(\frac{R}{r}\right)^5\left(\frac{\vec{r}\vec{r}}{r^2}\right)\left(\frac{\vec{r} \bullet \bar{\bar{E}}_\infty \bullet \vec{r}}{r^2}\right) + 25\eta_c\left(\frac{R}{r}\right)^3\left(\frac{\vec{r}\vec{r}}{r^2}\right)\left(\frac{\vec{r} \bullet \bar{\bar{E}}_\infty \bullet \vec{r}}{r^2}\right) \tag{4.16}$$

Using the expression for local stress in the ambient fluid (Equation 4.16), the dipole strength $\bar{\bar{S}}^0$ can be evaluated from Equation 4.2. Note that for a solid rigid particle, the integral of $\vec{u}\,\hat{n}+\hat{n}\,\vec{u}$ over the surface A_0 of the particle is zero:

$$\int_{A_0} \left(\vec{u}\,\hat{n}+\hat{n}\,\vec{u}\right) dA = 0 \tag{4.17}$$

Equation 4.2 gives the following result for dipole strength:

$$\bar{\bar{S}}^0 = \frac{20}{3}\,\pi R^3\,\eta_c\,\bar{\bar{E}}_\infty \tag{4.18}$$

For a dilute dispersion, the rate of strain tensor $\bar{\bar{E}}_\infty$ far away from the particles can be equated to the bulk or imposed rate of strain tensor $\langle\bar{\bar{E}}\rangle$. Hence, for a dilute dispersion of solid spherical particles, the following constitutive equation can be obtained from Equation 4.1 and Equation 4.18:

$$\langle\bar{\bar{\sigma}}\rangle = -\langle P\rangle\bar{\bar{\delta}}+2\,\eta_c\,\langle\bar{\bar{E}}\rangle+5\,\eta_c\,\phi\langle\bar{\bar{E}}\rangle \tag{4.19}$$

To simplify the notation, the angular brackets < > can be dropped from Equation 4.19. Thus,

$$\bar{\bar{\sigma}} = -P\,\bar{\bar{\delta}}+2\,\eta_c\left(1+\frac{5}{2}\phi\right)\bar{\bar{E}} \tag{4.20}$$

where $\bar{\bar{\sigma}}$ is the bulk stress tensor of the dispersion and $\bar{\bar{E}}$ is the imposed or bulk rate of strain tensor on the dispersion.

The constitutive equation for a homogeneous incompressible Newtonian fluid possessing a shear viscosity η is given as:

$$\bar{\bar{\sigma}} = -P\,\bar{\bar{\delta}}+2\,\eta\,\bar{\bar{E}} \tag{4.21}$$

Clearly, Equation 4.20 is similar to Equation 4.21, indicating that a dilute dispersion of solid spherical particles is a Newtonian fluid of viscosity η given by:

$$\eta = \eta_c\left(1+\frac{5}{2}\phi\right) \tag{4.22}$$

Equation 4.22 is the celebrated Einstein relation [8,9] for the viscosity of a dilute suspension of spherical solid particles. Interestingly, Einstein [8,9] derived Equation 4.22 using a different approach, namely, the energy dissipation approach.

The energy dissipation approach, as described by Happel and Brenner [10], involves comparison of the rates of mechanical energy dissipation in dispersion and matrix fluid alone without particles when they are subjected to identical boundary conditions. If $\dot{Q}$ is the rate of energy dissipation in a dispersion and $\dot{Q}_c$ is the rate of energy dissipation in matrix fluid alone (without particles), under identical boundary conditions, then:

$$\frac{\eta}{\eta_c} = \frac{\dot{Q}}{\dot{Q}_c} = \frac{\int\limits_{A'} \left(\hat{n} \bullet \bar{\bar{\sigma}} \bullet \vec{u} \right) dA}{\int\limits_{A'} \left(\hat{n} \bullet \bar{\bar{\sigma}}_c \bullet \vec{u}_c \right) dA} \tag{4.23}$$

where $\hat{n}$ is a unit outward normal to surface A', $\bar{\bar{\sigma}}$ and $\bar{\bar{\sigma}}_c$ are stress tensors of dispersion and matrix fluid, respectively, $\vec{u}$ and $\vec{u}_c$ are velocities of dispersion and matrix fluid, respectively, and A' is the bounding surface of the region where fluid motion occurs. The velocities of dispersion and matrix fluid at the boundary A' must be the same in order to compare the energy dissipation rates, that is,

$$\vec{u} = \vec{u}_c \quad \text{on} \quad A' \tag{4.24}$$

The energy dissipation approach has the inherent limitation of yielding only the viscosity of the dispersion and not the complete constitutive equation.

4.3 DISPERSION OF POROUS RIGID PARTICLES

Porous particles are often used to model cross-linked polymers and drug delivery microspheres [11–13]. The rheology of dispersions of porous particles is of interest from both practical and fundamental points of view.

Consider a porous spherical particle of radius R suspended in a matrix liquid having viscosity η_c. The porous material is taken to be homogeneous and isotropic, characterized by a constant permeability K. For convenience, we assume the coordinate system to be fixed at the center of the particle. The external liquid is subjected to linear shear flow. Assuming that the inertia effects are negligible, the equations governing the fluid flow outside the porous particle are the Stokes equation and continuity equation:

$$\nabla^2 \vec{u} = \frac{1}{\eta_c} \nabla P \quad \text{for} \quad r > R \tag{4.25}$$

$$\nabla \bullet \vec{u} = 0 \quad \text{for} \quad r > R \tag{4.26}$$

In the interior of the porous particle, the fluid velocity and pressure are described by the Brinkman equation and continuity equation [14–16]:

$$\nabla^2 \vec{u}^* - \frac{1}{K} \vec{u}^* = \frac{1}{\eta_c} \nabla P^* \quad \text{for} \quad r < R \tag{4.27}$$

$$\nabla \bullet \vec{u}^* = 0 \quad \text{for} \quad r < R \tag{4.28}$$

The Brinkman equation (Equation 4.27) is a modification of the well-known Darcy law:

$$\vec{u}^* = -\frac{K}{\eta_c} \nabla P^* \tag{4.29}$$

The Darcy law is valid only in the interior of a porous body (particle) far from the boundary region. The $\nabla^2 \vec{u}^*$ term in the Brinkman equation is needed to satisfy the boundary condition at the surface of the particle.

As the particle is rigid and spherical, it is not necessary to consider the rotational part of the shear flow. The vorticity of the shear flow rotates the particle with the fluid at an equal angular velocity and, therefore, no disturbance in the velocity field occurs because of the presence of the particle.

Equation 4.25 to Equation 4.28 have been solved for pure straining flow with the following boundary conditions [16]:

1. Far away from the particle ($r \rightarrow \infty$)

$$\vec{u}_{\infty} = \overline{\overline{E}}_{\infty} \bullet \vec{r} \tag{4.30}$$

where $\overline{\overline{E}}_{\infty}$ is the rate-of-strain tensor (which is symmetric and traceless) far away from the particle.

2. On the surface of the particle ($r = R$)

$$\vec{u} = \vec{u}^* \tag{4.31}$$

$$\hat{n} \bullet \overline{\overline{\sigma}} = \hat{n} \bullet \overline{\overline{\sigma}}^* \tag{4.32}$$

where $\overline{\overline{\sigma}}$ is the stress tensor outside the porous particle, and $\overline{\overline{\sigma}}^*$ is the stress tensor inside the porous particle.

The pressure and velocity fields outside the particle ($r > R$) are as follows [16]:

$$P = P_\infty + \eta_c \left[A(r) \right] \vec{r} \bullet \bar{\bar{E}}_\infty \bullet \vec{r} \tag{4.33}$$

$$\vec{u} = \left[1 + B(r) \right] \left(\bar{\bar{E}}_\infty \bullet \vec{r} \right) + \left[C(r) \right] \vec{r} \left(\vec{r} \bullet \bar{\bar{E}}_\infty \bullet \vec{r} \right) \tag{4.34}$$

where $A(r)$, $B(r)$, and $C(r)$ are functions of r given by:

$$A(r) = C_1 r^{-5} \tag{4.35}$$

$$B(r) = \frac{2}{35} C_2 r^{-5} \tag{4.36}$$

$$C(r) = \frac{1}{2} C_1 r^{-5} - \frac{1}{7} C_2 r^{-7} \tag{4.37}$$

The constants C_1 and C_2 are given as follows:

$$C_1 = -5 R^3 \left[\frac{H(\beta)}{1 + 10 \beta^{-2} H(\beta)} \right] \tag{4.38}$$

$$C_2 = R^5 \left[\frac{-35}{2} + \frac{175}{J\beta^2} \left[\left(\frac{-3}{2} \beta^5 \right) \sinh(-\beta) - \left(\frac{3}{2} \beta^6 \right) \cosh(-\beta) \right] \right] \tag{4.39}$$

where

$$\beta = R / \sqrt{K} \tag{4.40}$$

$$H(\beta) = 1 + 3 \beta^{-2} - 3 \beta^{-1} \coth(\beta) \tag{4.41}$$

$$J = 3 \left(30 \beta + \beta^5 + 10 \beta^3 \right) \sinh(-\beta) + 90 \beta^2 \cosh(-\beta) \tag{4.42}$$

From the pressure and velocity fields given by Equation 4.33 and Equation 4.34, one can evaluate the stress tensor $\bar{\bar{\sigma}}$ in the ambient fluid from:

$$\bar{\bar{\sigma}} = -P \bar{\bar{\delta}} + \eta_c \left[\nabla \vec{u} + \left(\nabla \vec{u} \right)^T \right] \tag{4.43}$$

Using the stress tensor of the ambient fluid at the surface of the particle $(r = R)$, one can obtain the dipole strength $\bar{\bar{S}}^0$ defined in Equation 4.2. Zackrisson and Bergenholtz [16] have shown that $\bar{\bar{S}}^0$ is:

$$\bar{\bar{S}}^0 = -\frac{4}{3}\pi\eta_c\, C_1\, \bar{\bar{E}}_\infty \qquad (4.44)$$

where C_1 is given by Equation 4.38. In the limit $K \to 0$ (i.e., the particle is completely impermeable), $C_1 = -5\,R^3$ and hence $\bar{\bar{S}}^0$ of the particle becomes equal to the $\bar{\bar{S}}^0$ of solid spherical particle (see Equation 4.18). In the limit $K \to \infty$ (i.e., the particle is completely permeable), $C_1 = 0$ and thus $\bar{\bar{S}}^0$ becomes zero, as expected.

The full constitutive equation of a dilute dispersion of rigid porous spheres can now be written as (see Equation 4.1):

$$\left\langle \bar{\bar{\sigma}} \right\rangle = -\left\langle P \right\rangle \bar{\bar{\delta}} + 2\eta_c \left\langle \bar{\bar{E}} \right\rangle + \left(-C_1\,R^{-3}\right)\eta_c\,\phi\left\langle \bar{\bar{E}} \right\rangle \qquad (4.45)$$

In arriving at Equation 4.45, $\bar{\bar{E}}_\infty$ is taken to be equal to the bulk or imposed rate of strain tensor $\left\langle \bar{\bar{E}} \right\rangle$. Dropping the angular brackets $\langle\ \rangle$ to simplify the notation, the constitutive equation becomes:

$$\bar{\bar{\sigma}} = -P\,\bar{\bar{\delta}} + 2\eta_c\left[1 + \left(\frac{-C_1\,R^{-3}}{2}\right)\phi\right]\bar{\bar{E}} \qquad (4.46)$$

The form of Equation 4.46 indicates that the dispersion is a Newtonian fluid of viscosity η given by:

$$\eta = \eta_c\left[1 + \left(\frac{-C_1 R^{-3}}{2}\right)\phi\right] \qquad (4.47)$$

The intrinsic viscosity $[\,\eta\,]$ of the dispersion, defined as:

$$[\eta] = \lim_{\phi \to 0}\left[\frac{\eta/\eta_c - 1}{\phi}\right] \qquad (4.48)$$

is equal to $(-C_1\,R^{-3}/2)$, that is,

$$[\eta] = -C_1\,R^{-3}/2 = \frac{5}{2}\left[\frac{H(\beta)}{1 + 10\beta^{-2}\,H(\beta)}\right] \qquad (4.49)$$

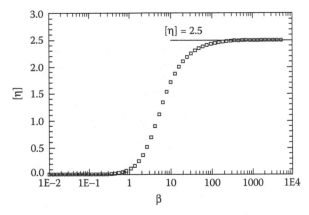

FIGURE 4.1 Variation of intrinsic viscosity of a dispersion of porous spheres with β. (*Note*: $\beta = R/\sqrt{K}$ where R is particle radius and K is the permeability of the particle material.)

For a completely impermeable particle, $K \rightarrow 0$, $\beta \rightarrow \infty$, and $H(\beta) = 1.0$. Therefore, the intrinsic viscosity is 5/2, as expected from the Einstein equation (Equation 4.22). For a completely permeable particle, $K \rightarrow \infty$, $\beta \rightarrow 0$, and $[\eta]$ reduces to zero.

Figure 4.1 shows the plot of intrinsic viscosity $[\eta]$ of a dispersion of porous spheres as a function of β. With the increase in β (i.e., a decrease in permeability K), the intrinsic viscosity increases.

4.4 DISPERSIONS OF ELECTRICALLY CHARGED SOLID PARTICLES

The particles of colloidal dispersions, particularly the aqueous dispersions, are often electrically charged. The surface of a colloidal particle can develop an electric charge through a number of possible mechanisms, such as (1) selective adsorption of ions present in the dispersion medium (matrix fluid), (2) ionization of acid or base groups present on the surface of the particle, and (3) selective dissolution of ions from the particle to the ambient fluid [17,18].

Figure 4.2 shows an electrically charged particle. The charge on the particle, shown as negative in the figure, attracts ions of opposite charge (the counterions) from the dispersion medium. Thus, an ionic atmosphere is formed around the particle. Adjacent to the particle surface, a monolayer of immobile counterions is formed. This layer is called the *Stern layer*. Outside the Stern layer, the ionic atmosphere consists of mobile co-ions and counterions and the concentration of counterions is much higher than that of co-ions. These mobile ions make up a diffuse outer layer referred to as *Gouy* or *Debye–Huckel layer*. The concentration of counterions in the diffuse layer gradually decreases toward the bulk concentration on moving away from the particle. The surface charge on the particle together with the surrounding charge cloud form an electrical double layer.

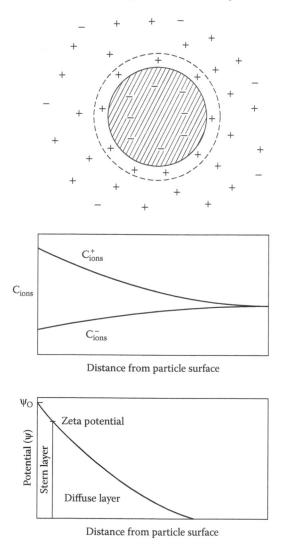

FIGURE 4.2 Negatively charged particle surrounded by an ionic atmosphere. The figure also shows variations in ionic concentration and electric potential with the distance from the particle surface.

Dispersions of electrically charged particles are generally more viscous compared to similar dispersions of uncharged particles. Furthermore, they can exhibit non-Newtonian behavior even at very low dispersed-phase concentrations [19].

The force dipole strength $\overline{\overline{S}}^0$ of charged particles is different from that of similar uncharged particles. When the fluid surrounding the charged particle is set in motion, the electrostatic attraction between the particle's surface charge and the surrounding charge cloud of counterions hinders the fluid flow inside the double layer and generates additional stresses (Maxwell stresses). This effect, whereby additional stresses are produced in a dispersion on account of distortion of an electrical double layer under shear, is called the *primary electroviscous effect*. The primary electroviscous effect is present even in a dilute dispersion of charged particles in which the interparticle interactions are absent.

4.4.1 FUNDAMENTAL EQUATIONS

Consider an isolated electrically charged spherical particle suspended in an infinite matrix fluid (electrolyte solution) with the origin of the coordinate system fixed at the center of the particle. Assume that: (1) the inertia effects are negligible, (2) the ionic species behave as point charges, and (3) the fluid properties such as viscosity η_c and permittivity ϵ are constants independent of ion concentration and electric field strength [19,20]. Based on these assumptions, the equations governing the motion of a fluid around a charged particle are:

$$\eta_c \nabla^2 \vec{u} = \nabla P + \rho \Delta \Psi \qquad (4.50)$$

$$\nabla \bullet \vec{u} = 0 \qquad (4.51)$$

$$\nabla^2 \Psi = -(\rho/\epsilon) \qquad (4.52)$$

$$\frac{\partial n_i}{\partial t} + \nabla \bullet \left[n_i \vec{V}_i \right] = 0 \qquad (4.53)$$

where $\vec{u}$ is fluid velocity, P is pressure, Ψ is electric potential, ρ is local charge density, n_i is the number density of ionic species i, and $\vec{V}_i$ is velocity of ionic species i. Equation 4.50 is the Stokes equation with an electrostatic body force $-\rho \nabla \Psi$. Equation 4.51 is the continuity equation. Equation 4.52 is the Poisson equation, and Equation 4.53 is the conservation equation for ions. The local charge density ρ is given by

$$\rho = \sum_{i=1}^{N} Z_i e \, n_i \qquad (4.54)$$

where Z_i is the valence of ionic species i, e is the elementary electronic charge, and N is the total number of ionic species present in the fluid. The velocity $\vec{V}_i$ of ionic species i can be expressed as:

$$\vec{V}_i = \vec{u} + (\mu_i/e) [-eZ_i \nabla \Psi - kT \nabla \ell n \, n_i] \qquad (4.55)$$

where μ_i is the mobility of ionic species i, k is the Boltzmann constant, and T is the absolute temperature. According to Equation 4.55, the ions slip (drift) relative to the flow velocity $\vec{u}$ because of the electrical force $-eZ_i \nabla \Psi$ and the entropic force (which causes diffusion) $-kT \nabla \ell n\, n_i$ [19]. Equation 4.53 and Equation 4.55 can be combined to give:

$$\frac{\partial n_i}{\partial t} + \vec{u} \bullet \nabla n_i = \left(\mu_i/e\right) kT \left[\nabla^2 n_i + \left(\frac{eZ_i}{kT}\right) \nabla \bullet \left(n_i \nabla \Psi\right) \right] \qquad (4.56)$$

The boundary conditions are:

1. Far away from the particle $(r \to \infty)$

$$\vec{u} \to \vec{u}_\infty \qquad (4.57)$$

$$n_i \to n_i^\infty \ \text{ with } \ \sum_{i=1}^{N} Z_i n_i^\infty = 0 \qquad (4.58)$$

$$\Psi \to 0 \qquad (4.59)$$

2. At the surface of the particle $(r = R)$

$$\vec{u} = \vec{\omega} \times \vec{r} \qquad (4.60)$$

where $\vec{\omega}$ is the angular velocity of the particle.

$$\Psi = \Psi_0 \qquad (4.61)$$

where Ψ_0 is the zeta potential.

$$\vec{n} \bullet \vec{V}_i = 0 \qquad (4.62)$$

where $\vec{n}$ is the unit outward normal to the particle surface.

The boundary condition specified by Equation 4.62 indicates that there is no flux of ions into the particle. Also, note that the boundary conditions at the surface of the particle are complicated by the presence of the immobile Stern layer of adsorbed ions. The adsorbed ions of the Stern layer are usually considered to be part of the surface charge and, therefore, the electrostatic potential at the surface of the particle is taken to be the zeta potential, which is the potential at the slipping plane.

In principle, the governing equations subject to the given boundary conditions can be solved for $\vec{u}$, P, Ψ, and n_i. However, this is a complicated problem. Therefore, it is helpful to write the governing equations in dimensionless form and to identify the important dimensionless groups.

Equation 4.50 to Equation 4.52 and Equation 4.56 (under steady state) in dimensionless form are [20,21]:

$$\tilde{\nabla}^2 \tilde{u} = \tilde{\nabla} \tilde{P} + N_{Ha}(\tilde{\rho} \tilde{\nabla} \tilde{\Psi}) \qquad (4.63)$$

$$\tilde{\nabla} \bullet \tilde{u} = 0 \qquad (4.64)$$

$$\tilde{\nabla}^2 \tilde{\Psi} = - \tilde{\rho} \, (R\kappa)^2 \qquad (4.65)$$

$$N_{Pe} \, \tilde{u} \bullet \tilde{\nabla} \tilde{n}_i = \tilde{\mu}_i \left[\tilde{\nabla}^2 \tilde{n}_i + Z_i \, \tilde{\psi}_0 \, \tilde{\nabla} \bullet (\tilde{n}_i \, \tilde{\nabla} \tilde{\Psi}) \right] \qquad (4.66)$$

In these equations, the scaled variables are:

$$\tilde{\nabla} = R \nabla \qquad (4.67)$$

$$\tilde{\nabla}^2 = R^2 \nabla^2 \qquad (4.68)$$

$$\tilde{u} = \vec{u} / u_0 \qquad (4.69)$$

$$\tilde{P} = RP / \eta_c u_0 \qquad (4.70)$$

$$\tilde{\psi} = \psi / \psi_0 \qquad (4.71)$$

$$\tilde{\rho} = \frac{\rho}{\left(e^2 \Psi_0 \, I \, / \, kT \right)} \qquad (4.72)$$

$$\tilde{n}_i = \frac{n_i}{(e \Psi_0 \, I \, / \, kT)} \qquad (4.73)$$

$$\tilde{\mu}_i = \mu_i / \mu_0 \qquad (4.74)$$

where u_0 is the characteristic fluid velocity, ψ_0 is surface potential (zeta potential), I is the bulk ionic strength defined as $\sum_i z_i^2 n_i^\infty$, and μ_0 is characteristic ion mobility.

Equation 4.63 to Equation 4.66 involve four dimensionless groups:

$$N_{Ha} = \left(\frac{\in \Psi_0^2}{R \eta_c u_0} \right) (R\kappa)^2 \qquad (4.75)$$

$$N_{Pe} = \frac{R u_0 e}{\mu_0 kT} \qquad (4.76)$$

$$R\kappa = R\sqrt{e^2 I / \in kT} \qquad (4.77)$$

$$\tilde{\psi}_0 = \frac{e \psi_0}{kT} \qquad (4.78)$$

N_{Ha} is called the electric Hartmann number and is the ratio of electrical forces to viscous forces. The modification of the flow field around the particles caused by the electrostatic body force is small when $N_{Ha} \ll 1$. N_{Pe} is the ion Peclet number, which measures the extent to which the motion of the fluid relative to the particle disturbs the charge cloud; for small N_{Pe} ($N_{Pe} \ll 1$), the ionic atmosphere is distorted only slightly from its equilibrium shape because the diffusion of ions is strong as compared to the convection of ions (N_{Pe} could also be defined as the ratio of convection to Brownian diffusion of ions). $R\kappa$ is the ratio of particle size to Debye length (κ^{-1}), where the Debye length is defined as:

$$\kappa^{-1} = \sqrt{\in kT \Big/ \left(e^2 \sum_i Z_i^2 n_i^\infty \right)} \qquad (4.79)$$

As the double layer extends a distance of κ^{-1} from the surface of the particle, $1/R\kappa$ is a measure of the relative thickness of the electric double layer. If $R\kappa \gg 1$, the electric double layer is thin; if $R\kappa \ll 1$, the electric double layer is considered to be thick relative to the particle radius. The fourth dimensionless group in the governing equations is the dimensionless surface potential $\tilde{\psi}_0$ defined in Equation 4.78. When the surface potential (zeta potential) is small, $\tilde{\psi}_0 \ll 1$.

It should be noted that for thin double layers ($R\kappa \gg 1$), the appropriate length scale is κ^{-1} rather than the particle radius "R" utilized in the preceding dimensional analysis [19]. If κ^{-1} is used as a length scale, the dimensional analysis gives the following expressions for the Hartmann and Peclet numbers:

$$N_{Ha} = \left(\frac{\epsilon \Psi_0^2}{R \eta_c u_0} \right) \qquad (4.80)$$

$$N_{Pe} = \left(\frac{u_0 \kappa^{-1} e}{\mu_0 kT} \right) \qquad (4.81)$$

4.4.2 THE BULK STRESS

The bulk or average stress for the dispersion can be written as:

$$\langle \bar{\bar{\sigma}} \rangle = - \langle P \rangle \bar{\bar{\delta}} + 2\eta_c \langle \bar{\bar{E}} \rangle + \langle \bar{\bar{\sigma}}^P \rangle \qquad (4.82)$$

where $\langle \bar{\bar{\sigma}}^P \rangle$ is the particle contribution to the bulk stress.

For electroviscous problems, the average stress $\langle \bar{\bar{\sigma}} \rangle$ can be defined as [19]:

$$\langle \bar{\bar{\sigma}} \rangle = \frac{1}{V} \int_V \left(\bar{\bar{\sigma}}^H + \bar{\bar{\sigma}}^M \right) dV \qquad (4.83)$$

where the volume V is chosen to be large enough to contain many particles but much smaller than the flow scale, $\bar{\bar{\sigma}}^H$ is the hydrodynamic stress, and $\bar{\bar{\sigma}}^M$ is the Maxwell stress defined as:

$$\bar{\bar{\sigma}}^M = \epsilon \left(\nabla \Psi \nabla \Psi - \frac{1}{2} \bar{\bar{\delta}} \nabla \Psi \bullet \nabla \Psi \right) \qquad (4.84)$$

Using Equation 4.82 and Equation 4.83, the particle stress $\langle \bar{\bar{\sigma}}^P \rangle$ can be expressed as:

$$\langle \bar{\bar{\sigma}}^P \rangle = \frac{1}{V} \int_V \left[\bar{\bar{\sigma}}^H + \bar{\bar{\sigma}}^M + P \bar{\bar{\delta}} - \eta_c \left(\nabla \vec{u} + \nabla \vec{u}^T \right) \right] dV \qquad (4.85)$$

or

$$\langle \bar{\bar{\sigma}}^P \rangle = \frac{1}{V} \int_{V_c} \bar{\bar{\sigma}}^M dV + \frac{1}{V} \int_{V_d} \left[\bar{\bar{\sigma}}^H + \bar{\bar{\sigma}}^M + P \bar{\bar{\delta}} - \eta_c \left(\nabla \vec{u} + \nabla \vec{u}^T \right) \right] dV \qquad (4.86)$$

where V_c is the volume occupied by the continuous-phase (matrix) liquid and V_d is the volume occupied by the particles. Note that in the continuous-phase liquid:

$$\bar{\bar{\sigma}}^H + P\bar{\bar{\delta}} - \eta_c\left(\nabla\vec{u} + \nabla\vec{u}^T\right) = 0 \tag{4.87}$$

Using the identities:

$$\bar{\bar{\sigma}}^H + \bar{\bar{\sigma}}^M = \left[\nabla\bullet\left\{\left(\bar{\bar{\sigma}}^H + \bar{\bar{\sigma}}^M\right)\vec{x}\right\}\right]^T - \vec{x}\,\nabla\bullet\left[\bar{\bar{\sigma}}^H + \bar{\bar{\sigma}}^M\right] \tag{4.88}$$

$$\bar{\bar{\sigma}}^M = \left[\nabla\bullet\left(\bar{\bar{\sigma}}^M\vec{x}\right)\right]^T - \vec{x}\,\nabla\bullet\bar{\bar{\sigma}}^M \tag{4.89}$$

and the momentum balance equation:

$$\nabla\bullet\left(\bar{\bar{\sigma}}^H + \bar{\bar{\sigma}}^M\right) = 0 \tag{4.90}$$

the particle stress (Equation 4.86) can be expressed as:

$$\langle\bar{\bar{\sigma}}^P\rangle = \frac{1}{V}\int_{V_c}\left[\left\{\nabla\bullet\left(\bar{\bar{\sigma}}^M\vec{x}\right)\right\}^T - \vec{x}\,\nabla\bullet\bar{\bar{\sigma}}^M\right]dV + \frac{1}{V}\int_{V_d}\left[\nabla\bullet\left\{\left(\bar{\bar{\sigma}}^H + \bar{\bar{\sigma}}^M\right)\vec{x}\right\}\right]^T dV$$
$$+\frac{1}{V}\int_{V_d}\left[P\bar{\bar{\delta}} - \eta_c\left(\nabla\vec{u} + \nabla\vec{u}^T\right)\right]dV \tag{4.91}$$

Using the Gauss divergence theorem and:

$$\vec{x}\,\nabla\bullet\bar{\bar{\sigma}}^M = -\vec{x}\rho\nabla\Psi \tag{4.92}$$

Equation 4.91 can be rewritten as:

$$\langle\bar{\bar{\sigma}}^P\rangle = \frac{1}{V}\sum_{i=1}^{N}\int_{A_i}\left[\vec{x}\bar{\bar{\sigma}}^H\bullet\hat{n} - \eta_c(\vec{u}\hat{n} + \hat{n}\vec{u})\right]dA$$
$$+\frac{1}{V}\int_{V_c}\left(\rho\vec{x}\nabla\Psi\right)dV + \frac{1}{V}\int_{V}\left\{\nabla\bullet\left(\bar{\bar{\sigma}}^M\vec{x}\right)\right\}^T dV \tag{4.93}$$

where A_i is the surface area of the ith particle, and N is the total number of particles in volume V. The isotropic pressure term has not been included as it is irrelevant to the dynamics of incompressible fluids.

The last volume integral in Equation 4.93 can be neglected for dilute dispersions [19]. Thus, for a dilute dispersion of identical electrically charged particles of surface area A_0:

$$\left\langle \overline{\overline{\sigma}}^P \right\rangle = \frac{N}{V} \int_{A_0} \left[\vec{x} \, \overline{\overline{\sigma}}^H \bullet \hat{n} - \eta_c \left(\vec{u}\hat{n} + \hat{n}\vec{u} \right) \right] dA + \frac{1}{V} \int_{V_c} \left(\rho \vec{x} \nabla \Psi \right) dV \quad (4.94)$$

Because for rigid particles,

$$\int_{A_0} \left(\vec{u}\hat{n} + \hat{n}\vec{u} \right) dA = 0 \quad (4.95)$$

Equation 4.94 reduces to:

$$\left\langle \overline{\overline{\sigma}}^P \right\rangle = \frac{N}{V} \int_{A_0} \left(\vec{x} \, \overline{\overline{\sigma}}^H \bullet \hat{n} \right) dA + \frac{1}{V} \int_{V_c} \left(\rho \vec{x} \nabla \Psi \right) dV \quad (4.96)$$

4.4.3 CONSTITUTIVE EQUATION

The constitutive equation for a dilute dispersion of electrically charged solid particles, expressed in the form of Equation 4.82, involves stress contributions from the matrix fluid, as well as from the particle.

The particle stress $\langle \overline{\overline{\sigma}}^P \rangle$ given by Equation 4.96 can be rewritten as:

$$\left\langle \overline{\overline{\sigma}}^P \right\rangle = \frac{N}{V} \left[\int_{A_0} \left(\vec{x} \, \overline{\overline{\sigma}}^H \bullet \hat{n} \right) dA + \int_{V'-V_0} \left(\rho \vec{x} \nabla \Psi \right) dV \right] \quad (4.97)$$

where V_0 is the volume of a particle and $V' - V_0$ is an arbitrary volume of matrix fluid surrounding the particle. The volume V' satisfies the following conditions: $V' \gg V_0, V' \gg \kappa^{-3}$, and $V' \ll V/N$ ($V' - V_0$ is basically the volume of an electric double layer surrounding the particle). Thus, the average particle stress $\langle \overline{\overline{\sigma}}^P \rangle$ over a volume V containing N identical electrically charged spheres is N/V times dipole strength $\overline{\overline{S}}^0$ of a single particle where $\overline{\overline{S}}^0$ is given as [11]:

$$\overline{\overline{S}}^o = \int_{A_0} \left(\vec{x} \, \overline{\overline{\sigma}}^H \bullet \hat{n} \right) dA + \int_{V'-V_0} \left(\rho \vec{x} \nabla \Psi \right) dV \quad (4.98)$$

To evaluate $\bar{\bar{s}}^0$, the fundamental equations described in Section 4.4.1 need to be solved for velocity, pressure, electric potential, and number density of ionic species or charge density distributions around a particle.

Several researchers have solved this problem under different limiting conditions [19–27]. Booth [23] solved the problem under the conditions of small Peclet number ($N_{Pe} \ll 1$), small Hartmann number ($N_{Ha} \ll 1$), and small surface potential ($\tilde{\Psi}_0 \ll 1$). However, no restriction was placed on the thickness of the electric double layer (arbitrary $R\kappa$). Sherwood [25] solved the governing equations numerically for a small Peclet number ($N_{Pe} \ll 1$) and obtained solutions for the following cases: (1) arbitrary surface potential $\tilde{\Psi}_0$ and arbitrary electric double layer thickness ($R\kappa$) but small Hartmann number ($N_{Ha} \ll 1$), and (2) arbitrary Hartmann number and double-layer thickness but small surface potential $\tilde{\Psi}_0$. Watterson and White [26] obtained numerical solutions for arbitrary N_{Ha}, $\tilde{\Psi}_0$, and $R\kappa$ under the condition of low Peclet number ($N_{Pe} \ll 1$).

For weak flows characterized by $N_{Pe} \ll 1$, a dilute dispersion of electrically charged spherical particles exhibits Newtonian behavior and the constitutive equation is expressed as:

$$\bar{\bar{\sigma}} = -P\,\bar{\bar{\delta}} + 2\eta_c\,\bar{\bar{E}} + 5\left(1+q\right)\eta_c\,\phi\,\bar{\bar{E}} \tag{4.99}$$

where q is the primary electroviscous coefficient which is, in general, a function of surface potential $\tilde{\Psi}_0$, Hartmann number N_{Ha}, and double-layer thickness ($R\kappa$). The dispersion viscosity can be obtained from Equation 4.99 as:

$$\eta = \eta_c\left[1+\frac{5}{2}\left(1+q\right)\phi\right] \tag{4.100}$$

The primary electroviscous coefficient q at low surface potential ($\tilde{\Psi}_0 \ll 1$) and low Hartmann number ($N_{Ha} \ll 1$) is given as [26]:

$$q = \frac{4\pi \in kT}{\eta_c\, e}\left[\frac{\displaystyle\sum_{i=1}^{N} n_i^{\infty}\, Z_i^2\, \mu_i^{-1}}{\displaystyle\sum_{i=1}^{N} n_i^{\infty}\, Z_i^2}\right]\left(\frac{e\,\Psi_0}{kT}\right)^2\left(1+R\kappa\right)^2\, f\left(R\kappa\right) \tag{4.101}$$

where Ψ_0 is taken to be the zeta potential and $f(R\kappa)$ is a power series function of $R\kappa$ with the two limiting forms as follows:

$$f\left(R\kappa\right) = \frac{1}{200\,\pi\left(R\kappa\right)} + \frac{11\left(R\kappa\right)}{3200\,\pi} \tag{4.102}$$

$$f(R\kappa) = \frac{3}{2\pi(R\kappa)^4} \qquad (4.103)$$

The form of $f(R\kappa)$ given in Equation 4.102 is valid for small $R\kappa$ ($R\kappa \ll 1$), that is, thick double layers. The form of $f(R\kappa)$ expressed in Equation 4.103 is valid for large $R\kappa$ ($R\kappa \gg 1$), that is, thin double layers. Figure 4.3 shows the plot of intrinsic viscosity $[\eta]$, defined in Equation 4.48, as a function $R\kappa$. The data are generated from Equation 4.100 to Equation 4.102 using the following information: electrolyte is NaCl with $\mu_{Na^+} = 5.192 \times 10^{-8}$ m²/V.s and $\mu_{Cl^-} = 7.913 \times 10^{-8}$ m²/V.s; zeta potential $\psi_o = -10$ mV; T $= 298.15$ K; k $= 1.381 \times 10^{-23}$ J/K; $\epsilon = \epsilon_r \epsilon_o$ with $\epsilon_o = 8.854 \times 10^{-12}$ C²/(N.m²) and $\epsilon_r = 79$; e $= 1.602 \times 10^{-19}$ C; and $\eta_c = 0.89$ mPa.s. Clearly, for small values of $R\kappa$ (i.e., thick double layers), the intrinsic viscosity $[\eta]$ is much larger than the Einstein value of 2.5.

Dilute dispersions of electrically charged particles exhibit non-Newtonian behavior when N_{Pe} is not small. Russel [19] developed a constitutive equation for dilute dispersions of charged spheres under conditions of arbitrary flow strength (arbitrary N_{Pe}) and thin double layers ($R\kappa \gg 1$). The particle stress $\bar{\bar{\sigma}}^P$ was expressed as:

$$\bar{\bar{\sigma}}^P = \frac{-2\,\epsilon\,\Psi_0^2}{R^2}\phi\left(\frac{3\,\bar{\bar{C}}}{R\kappa}\right) \qquad (4.104)$$

where the tensor $\bar{\bar{C}}$ is expressed through the equation:

$$\left(\frac{D}{Dt} + \kappa^2\frac{\mu}{e}kT\right)\bar{\bar{C}} = -5(R\kappa)\bar{\bar{E}} \qquad (4.105)$$

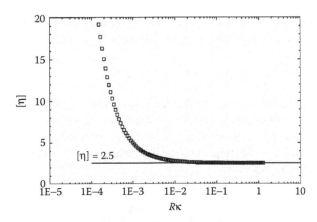

FIGURE 4.3 Variation of intrinsic viscosity of a dispersion of charged particles with $R\kappa$ (small $R\kappa$ implies thick double layer).

In Equation 4.105, D/Dt is the co-rotational (Jaumann) derivative defined as:

$$\frac{D}{Dt} = \frac{\partial}{\partial t} + \vec{u} \bullet \nabla + (\nabla \times \vec{u} \times) + (\nabla \times \vec{u} \times)^T \qquad (4.106)$$

The full constitutive equation for the dispersion can be obtained from Equation 4.82, Equation 4.104, and Equation 4.105, as:

$$\left[1 + \frac{e}{\kappa^2 \, \mu kT} \frac{D}{Dt}\right] \overline{\overline{\sigma}}$$

$$= -P \overline{\overline{\delta}} + 2\eta_c \left[1 + \frac{5}{2} \phi \left(1 + 6 \frac{e \, \epsilon \, \Psi_0^2}{\eta_c \, \mu kT \, (R\kappa)^2}\right)\right] \overline{\overline{E}} \qquad (4.107)$$

$$+ \frac{2\eta_c \left(1 + \frac{5}{2} \phi\right)}{(\kappa^2 \, \mu kT / e)} \frac{D}{Dt} \overline{\overline{E}}$$

This constitutive equation gives the following expression for dispersion viscosity:

$$\eta = \eta_c \left[1 + \frac{5}{2} \phi \left\{1 + 6 \frac{e \, \epsilon \, \Psi_0^2}{\eta_c \, \mu kT} \frac{1}{(R\kappa)^2}\left(\frac{1}{1 + N_{Pe}^2}\right)\right\}\right] \qquad (4.108)$$

In Equation 4.108, μ is the ionic mobility (assumed to be the same for all ions), and N_{Pe} is the ion Peclet number defined as:

$$N_{Pe} = \dot{\gamma} e / \kappa^2 \, \mu kT \qquad (4.109)$$

where $\dot{\gamma}$ is the shear rate. Clearly, the dispersion exhibits shear-thinning behavior in that the viscosity decreases with the increase in Peclet number N_{Pe}.

The constitutive equation, Equation 4.107, also predicts nonzero normal-stress differences, indicating viscoelastic behavior of dilute dispersions of electrically charged spheres:

$$N_1 = \sigma_{11} - \sigma_{22} = 30 \frac{\epsilon \, \Psi_0^2}{R^2} \phi \left(\frac{N_{Pe}^2}{1 + N_{Pe}^2}\right) \qquad (4.110)$$

$$N_2 = \sigma_{33} - \sigma_{22} = -15 \frac{\in \Psi_0^2}{R^2} \phi \left(\frac{N_{Pe}^2}{1+N_{Pe}^2} \right) \qquad (4.111)$$

Figure 4.4 shows the plots of intrinsic viscosity $[\eta]$ and normal stresses N_1 and $-N_2$ calculated from Equation 4.108, Equation 4.110, and Equation 4.111, respectively. The intrinsic viscosity decreases with the increase in Peclet number, indicating shear-thinning behavior. The normal stresses N_1 and N_2 increase initially with the increase in Peclet number and then level off.

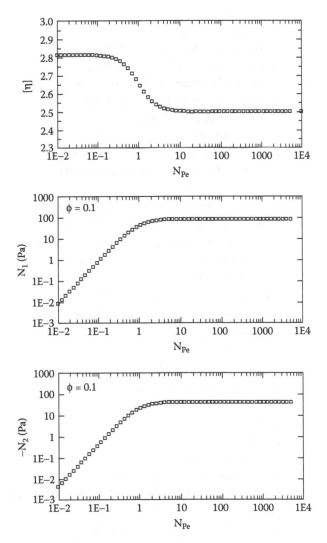

FIGURE 4.4 Variations of intrinsic viscosity, first normal stress difference, and second normal stress difference with Peclet number.

Rheology of Particulate Dispersions and Composites

In summary, the rheological behavior of dilute dispersions of electrically charged particles can be complex (shear-thinning, viscoelastic). The full constitutive equation (Equation 4.107) presented here is valid under the conditions: $R\kappa \gg 1$, thin double layer, $N_{Ha} \ll 1$, and $\tilde{\Psi}_0 \ll 1$. Lever [21] has presented a similar analysis of the primary electroviscous effect under the conditions $R\kappa \ll 1$ (thick double layer), $N_{Ha} \ll 1$, and $\tilde{\Psi}_0 \ll 1$.

NOTATION

A	surface area
A_0	surface area of a particle
A'	bounding surface of the region where fluid motion occurs
$A(r)$	function of r, given by Equation 4.35
$B(r)$	function of r, given by Equation 4.36
$C(r)$	function of r, given by Equation 4.37
C_1	constant defined by Equation 4.38
C_2	constant defined by Equation 4.39
$\overline{\overline{C}}$	second-order tensor expressed through Equation 4.105
D/Dt	co-rotational (Jaumann) derivative, defined by Equation 4.106
$\overline{\overline{E}}$	rate of strain tensor
$\overline{\overline{E}}_\infty$	rate of strain tensor of undisturbed flow far away from the particle
e	elementary electronic charge, 1.602×10^{-19} C
$f(R\kappa)$	function of dimensionless double-layer thickness $R\kappa$, given by Equation 4.102 and Equation 4.103
$H(\beta)$	function of β, given by Equation 4.41
I	bulk ionic strength, defined as $\sum_i z_i^2 n_i^\infty$
J	function of β, given by Equation 4.42
k	Boltzmann constant
K	permeability
N	total number of particles or ionic species
N_1	first normal stress difference
N_2	second normal stress difference
N_{Ha}	Hartmann number
N_{Pe}	Peclet number
$\hat{n}$	outer unit normal vector
n_i	number density of ionic species i
$n_{i\infty}$	number density of ionic species i far away from the particle
P	pressure
P_∞	pressure far away from the particle
$\dot{Q}$	rate of energy dissipation in dispersion
$\dot{Q}_c$	rate of energy dissipation in matrix fluid alone without particles

q	primary electroviscous coefficient; see Equation 4.101
R	particle radius
r	radial distance
$\vec{r}$	position vector
$\overline{\overline{S}}{}^{o}$	dipole strength of a single particle located in an infinite matrix fluid
T	temperature (as superscript, T refers to transpose)
t	time
u_o	characteristic fluid velocity
$\vec{u}$	fluid velocity vector
$\vec{u}_c$	velocity vector of matrix fluid
$\vec{u}_\infty$	fluid velocity vector far away from the particle
V	volume
V_c	volume occupied by the continuous-phase (matrix) fluid
V_d	volume occupied by dispersed phase (particles)
V_o	volume of a particle
$V' - V_o$	volume of an electric double layer surrounding the particle
$\vec{V}_i$	velocity of ionic species i
$\vec{x}$	position vector
Z_i	valence of ionic species i

GREEK

β	defined by Equation 4.40
$\dot{\gamma}$	shear rate
$\overline{\overline{\Gamma}}_\infty$	velocity gradient tensor far away from the particle
$\overline{\overline{\delta}}$	unit tensor
∇	del operator
$\in$	permittivity
$\in_o$	permittivity of vacuum, $8.854 \times 10^{-12}\ \mathrm{C^2\,N^{-1}m^{-2}}$
$\in_r$	relative permittivity or dielectric constant
η	viscosity
η_c	viscosity of continuous-phase (matrix) fluid
$[\eta]$	intrinsic viscosity
κ	Debye–Huckel constant (inverse of Debye length), see Equation 4.79
μ_i	mobility of ionic species i
μ_o	characteristic ion mobility
ρ	local charge density
$\overline{\overline{\sigma}}$	stress tensor
$\overline{\overline{\sigma}}_c$	stress tensor in continuous phase
$\overline{\overline{\sigma}}^{H}$	hydrodynamic stress tensor

$\bar{\bar{\sigma}}^{M}$ Maxwell stress tensor

$\bar{\bar{\sigma}}^{P}$ particle contribution to stress tensor

ϕ volume fraction of particles

ψ electric potential

ψ_{o} surface potential (zeta potential)

$\bar{\bar{\Omega}}_{\infty}$ vorticity tensor of the undisturbed flow far away from the particle

$\vec{\omega}$ angular velocity of the particle

$\vec{\omega}_{\infty}$ local angular velocity of the undisturbed flow far away from the particle

$\langle X \rangle$ volume average of quantity X

$\sim$ indicates dimensionless variable

REFERENCES

1.   E. Kissa, *Dispersions*, Marcel Dekker, New York, 1999.
2.   R.F. Conley, *Practical Dispersion*, VCH Publishers, New York, 1996.
3.   R.B. McKay, *Technological Applications of Dispersions*, Marcel Dekker, New York, 1994.
4.   I.D. Morrison, S. Ross, *Colloidal Dispersions*, Wiley, New York, 2002.
5.   D. Wasan, A. Nikolov, D. Henderson, *AIChE J.*, 49:550–556, 2003.
6.   G.K. Batchelor, J.T. Green, *J. Fluid Mech.*, 56:401–427, 1972.
7.   H.A. Stone, L.G. Leal, *J. Fluid Mech.*, 211:123–156, 1990.
8.   A. Einstein, *Ann. Physik*, 19:289–306, 1906.
9.   A. Einstein, *Ann. Physik*, 34:591–592, 1911.
10. J. Happel, H. Brenner, *Low Reynolds Number Hydrodynamics*, Noordhoff International Publishing, Leyden, the Netherlands, 1973.
11. V. Natraj, S.B. Chen, *J. Colloid Interface Sci*, 251:200–207, 2002.
12. B.U. Felderhof, J.M. Deutch, *J. Chem. Phys.*, 62:2391–2397, 1975.
13. P. Debye, A.B. Bueche, *J. Chem. Phys.*, 16:573–579, 1948.
14. J.J.L. Higdon, M. Kojina, *Int. J. Multiphase Flow*, 7:719–727, 1981.
15. S.B. Chen, X. Ye, *J. Colloid Interface Sci.*, 221:50–57, 2000.
16. M. Zackrisson, J. Bergenholtz, *Colloids & Surfaces A: Physicochem. Eng. Aspects*, 225:119–127, 2003.
17. M. Daoud, C.E. Williams, Eds., *Soft Matter Physics*, Springer, Berlin, 1999.
18. D.J. Shaw, *Introduction to Colloid and Surface Chemistry*, Butterworth-Heinemann, Oxford, 1992.
19. W.B. Russel, *J. Fluid Mech.*, 85:673–683, 1978.
20. W.B. Russel, *J. Colloid Interface Sci.*, 55:590–604, 1976.
21. D.A. Lever, *J. Fluid Mech.*, 92:421–433, 1979.
22. M. von Smoluchowski, *Kolloid-Z*, 18:190, 1916.
23. F. Booth, *Proc. R. Soc. London Ser. A.*, 203:533–551, 1950.
24. W.B. Russel, *J. Fluid Mech.*, 85:209–232, 1978.
25. J. D. Sherwood, *J. Fluid Mech.*, 101: 609–629, 1980.
26. I.G. Watterson, L.R. White, *J. Chem. Soc., Faraday Trans. 2*, 77:1115–1128, 1981.
27. E.J. Hinch, J.D. Sherwood, *J. Fluid Mech.*, 132:337–347, 1983.

5 Dispersions of Rigid Nonspherical Particles

5.1 INTRODUCTION

Suspensions of nonspherical (anisometric) rigid particles in liquids are ubiquitous in many industries, such as printing and papermaking, petroleum, polymer, aerospace, bioengineering, construction, ceramics, coal, magnetic media, pharmaceutical, and food [1–10]. Slurries of nonspherical particles (such as calcium carbonate, kaolin clay, titanium dioxide, talc) and water are widely used in the paper-coating process. Clay particles used in drilling muds are nonspherical in shape; the particles of bentonite and hectorite clays are platelike, whereas the particles of attapulgite and sepiolite clays are rod shaped. Platelike particles of mica, aluminum, magnesium hydroxide, and talc are commonly used as fillers/reinforcing agents in plastics to enhance stiffness [5]. Many polymeric composites are manufactured from suspensions of rodlike fibers in polymer melts. Fibers of rayon, vinylon, glass, nylon, and carbon materials are often utilized [8]. The solid propellants commonly used in aerospace propulsion are prepared from dispersion of nonspherical solid particles (oxidizer particles) in a rubbery matrix. Blood is a suspension of red blood cells (erythrocytes), white blood cells (leukocytes), and platelets in plasma [11]. Red blood cells are biconcave disks about 8 μm in diameter and 1 μm in thickness at the center. White blood cells are generally spherical with a mean diameter of about 7 μm. Platelets are irregularly oval or spherical bodies without any nucleus. Suspensions of acicular (needle or cigar shaped) magnetic particles are commonly used in the manufacture of digital tapes and magnetic recording media [6]. Molten chocolate is a suspension of solid nonspherical particles (mostly sugar granules and crushed cocoa beans) in cocoa butter [4]. Other examples of suspensions of nonspherical particles include suspensions of mineral pigments, cement, concrete, coal, starch, and insoluble drugs.

In light of the practical applications just mentioned, a good understanding of the rheological behavior of particulate suspensions of nonspherical particles is important. A vast amount of published literature exists on the rheology of dilute suspensions of nonspherical particles [12–62]. However, the published literature is mostly restricted to suspensions of spheroidal particles, that is, particles with the shape of an ellipse of revolution. Spheroidal particles have received a lot of attention mainly because the spheroid can be used as an approximation for a variety of shapes ranging from disks to rods. Plate- or disk-type particles can be modeled as oblate spheroids (spheroids generated by revolving an ellipse about its minor axis). Rod- or needle-type particles can be modeled as prolate spheroids (spheroids generated by revolving an ellipse about its major axis).

5.2 MOTION OF A SINGLE ISOLATED SPHEROIDAL PARTICLE IN SHEAR FLOW

Consider a single rigid particle of spheroidal shape suspended in an unbounded incompressible Newtonian liquid of viscosity of η_c. Let the coordinate system be fixed at the center of the particle. The liquid is subjected to a homogeneous shear at infinity. Assuming that the inertia effects are negligible, the equations governing the local flow field around the particle are:

$$\nabla^2 \vec{u} = \frac{1}{\eta_c} \nabla P \qquad (5.1)$$

$$\nabla \bullet \vec{u} = 0 \qquad (5.2)$$

Equation 5.1 and Equation 5.2 need to be solved subject to the following boundary conditions:

1. Far away from the particle

$$\vec{u} = \vec{u}_\infty \qquad (5.3)$$

2. On the surface of the particle

$$\vec{u} = \vec{\omega} \times \vec{x} \qquad (5.4)$$

where $\vec{\omega}$ is the angular velocity of the particle and $\vec{x}$ is the position vector measured from the center of the particle.

Jeffery [12] was the first to solve the Stokes equations (Equation 5.1 and Equation 5.2) subject to the boundary conditions (Equation 5.3 and Equation 5.4) for spheroidal particles. Jeffery's solution for the flow field around the particle contains the unknown angular velocity components $(\omega_1, \omega_2, \omega_3)$ of the particle. He obtained the angular velocity of the particle by assuming that the particles are non-Brownian and that the hydrodynamic torque on the particle is zero.

According to Jeffery's analysis, the rotation of non-Brownian spheroidal particles can be described by [18]:

$$\frac{d\vec{p}}{dt} = \overline{\overline{\Omega}} \bullet \vec{p} + \left(\frac{r^2 - 1}{r^2 + 1}\right)\left[\overline{\overline{E}} \bullet \vec{p} - \vec{p}\left(\vec{p} \bullet \overline{\overline{E}} \bullet \vec{p}\right)\right] \qquad (5.5)$$

where $\vec{p}$ is a unit vector parallel to the axis of symmetry of particle (and locked into the particle), $\overline{\overline{\Omega}}$ is the vorticity tensor $(\overline{\overline{\Omega}} = (-1/2)[\nabla\vec{u} - (\nabla\vec{u})^T])$, $\overline{\overline{E}}$ is the rate of strain tensor $(\overline{\overline{E}} = (1/2)[\nabla u + (\nabla u)^T])$, and r is the aspect ratio of the particle (length of the particle along its axis of symmetry to its length perpendicular to this axis). The aspect ratio r is larger than unity for prolate spheroids; $r < 1$ for oblate spheroids. Note that $d\vec{p}/dt$ is equal to $\vec{\omega} \times \vec{p}$ where $\vec{\omega}$ is the angular velocity of the particle. Also, Equation 5.5 can be rewritten as:

$$\frac{d\vec{p}}{dt} = -\vec{p} \bullet \overline{\overline{\Omega}} + \left(\frac{r^2 - 1}{r^2 + 1}\right)\left[\vec{p} \bullet \overline{\overline{E}} - \vec{p}\vec{p}\vec{p} : \overline{\overline{E}}\right] \qquad (5.6)$$

Equation 5.5 (or Equation 5.6) applies to nearly all axially symmetrical particles [26,53,64] provided that r is replaced by an effective aspect ratio. For example, cylindrical rods have an effective aspect ratio of 0.7 (L/d), where L/d is the true aspect ratio [53].

In a simple shear flow with

$$u_x = \dot{\gamma}\, y; \quad u_y = u_z = 0 \qquad (5.7)$$

the motion of the axis of revolution of the particle, described by Equation 5.5, can be expressed as [50]:

$$\frac{d\theta}{dt} = \frac{\dot{\gamma}\left(r^2 - 1\right)}{4\left(r^2 + 1\right)} \sin 2\theta \sin 2\phi \qquad (5.8)$$

$$\frac{d\phi}{dt} = -\frac{\dot{\gamma}}{\left(1 + r^2\right)}\left(r^2 \sin^2 \phi + \cos^2 \phi\right) \qquad (5.9)$$

where θ is the angle between $\vec{p}$ (unit vector parallel to the axis of symmetry of the particle) and the z-axis (see Figure 5.1), ϕ is the angle between the projection of $\vec{p}$ on the xy-plane and the x-axis, and $\dot{\gamma}$ is the shear rate. Note that some authors define ϕ to be the angle between the projection of $\vec{p}$ on the xy-plane and the y-axis (that is, their ϕ is equal to our $90° - \phi$).

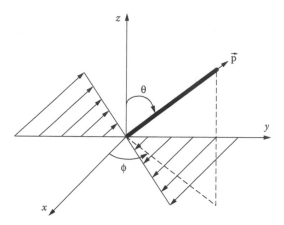

FIGURE 5.1 Coordinate system for a rotating particle.

Integration of Equation 5.8 and Equation 5.9 yield:

$$\tan \theta = \frac{C_o \, r}{\left(r^2 \sin^2 \phi + \cos^2 \phi\right)^{1/2}} \qquad (5.10)$$

$$\tan \phi = \frac{1}{r} \tan \left[\frac{-2\pi t}{T} + \tan^{-1}\left(r \tan \phi_0\right) \right] \qquad (5.11)$$

where C_o is a constant of integration referred to as the orbit constant, T is the period of rotation of the axis of revolution of a particle about the vorticity axis (z-axis), and ϕ_0 is the initial value of ϕ at time 0. The period T is given as:

$$T = \frac{2\pi}{\dot{\gamma}} \left(r + \frac{1}{r} \right) \qquad (5.12)$$

Thus, the motion of the particle is periodic in nature. The particle rotates indefinitely in a fixed orbit (referred to as *Jeffery orbit*) characterized by an orbit constant C_o. The orbit constant C_o varies from 0 to infinity, and the particle traverses any one of an infinite family of closed Jeffery orbits.

It should be noted that the angular velocity $d\phi/dt$ of the particle is not uniform. For example, in the case of prolate spheroids ($r > 1.0$), Equation 5.9 indicates that $-d\phi/dt$ is maximum when the particle is perpendicular to the flow direction ($\phi = 90°$) and $-d\phi/dt$ is minimum when it is aligned along the flow direction ($\phi = 0°$). That is, a prolate spheroid rotates slowly when its long axis is parallel to the flow direction, and rapidly when its long axis is perpendicular to the flow direction. The prolate spheroid spends the majority of its time in an orientation state close to the flow direction (near x-z or flow-vorticity plane). As the particle rotates away from the x-z plane, it flips rapidly until it becomes nearly aligned again. Oblate spheroids ($r < 1.0$), however, exhibit a different behavior. In this case, $-d\phi/dt$ is maximum when the axis of revolution is aligned with the flow direction ($\phi = 0°$), and $-d\phi/dt$ is minimum when the axis of revolution is perpendicular to the flow direction ($\phi = 90°$). In other words, oblate spheroids spend more time with their axis of revolution perpendicular to the flow.

Figure 5.2 shows the Jeffery orbits for a rod-shaped particle of effective aspect ratio of 35 (actual aspect ratio of 50) for several values of the modified orbit constant C_b, defined as $C_o/(C_o + 1)$. The particle is centered at the origin, and the curves shown represent the path of the end point of the particle [56]. The projection of the orbits onto the xy-plane is an ellipse. For $C_b = 0$ ($C_o = 0$), the axis of symmetry lies parallel to the z-axis. The particle merely rotates about its axis of symmetry and the rheological effect of this situation ($C_b = 0$) is equivalent to that of a sphere. When $C_b = 1.0$ ($C_o = \infty$), the axis of symmetry lies in the xy-plane and the particle rotates about the z-axis.

When $r \rightarrow 0$, the rotation period of the particle becomes infinite. The aspect ratio r is zero for a flat circular disk. The rotation period is also infinite when $r \rightarrow \infty$, that is, for long thin rods. In these situations when the rotation period is infinite, the particle orientation does not change at all and $d\vec{p}/dt = 0$. Consequently, $\vec{\omega} \times \vec{p} = 0$ and $\vec{\omega}$ is collinear with $\vec{p}$. Thus, the particle simply rotates about its symmetry axis [28] with an angular velocity $\vec{p}(\vec{\omega}_0 \bullet \vec{p})$, where $\vec{\omega}_0$ is fluid angular velocity. When $r \rightarrow \infty$, the rod takes up an orientation with its axis of symmetry lying in the x-z plane, and the angle θ depends on the initial

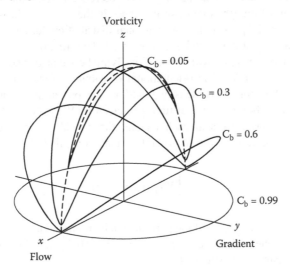

FIGURE 5.2 Jeffery orbits with different orbit constants (effective aspect ratio of particle = 35). (From M.P. Petrich, D.L. Koch, C. Cohen, An experimental determination of the stress-microstructure relationship in semi-concentrated fiber suspensions, *J. Non-Newtonian Fluid Mech.*, 95: 101–133, 2000. Reprinted with permission from Elsevier.)

orientation of the rod. When $r \rightarrow 0$, the disk takes up an orientation with its axis of symmetry in the y-z plane, and the angle θ depends on the initial orientation of the disk.

5.3 RHEOLOGY OF SUSPENSION OF NON-BROWNIAN AXISYMMETRIC PARTICLES

The constitutive equation for a dispersion of identical particles is given by:

$$\langle \overline{\overline{\sigma}} \rangle = -\langle P \rangle \overline{\overline{\delta}} + 2\eta_c \left\langle \overline{\overline{E}} \right\rangle + \frac{N}{V} \left\langle \overline{\overline{S}} \right\rangle \tag{5.13}$$

where $\langle \overline{\overline{\sigma}} \rangle$ is the bulk (average) stress tensor, $\langle P \rangle$ is the average pressure, $\overline{\overline{\delta}}$ is the unit tensor, η_c is the matrix viscosity, $\langle \overline{\overline{E}} \rangle$ is the bulk rate of strain tensor, N/V is the number density of particles in the dispersion, and $\langle \overline{\overline{S}} \rangle$ is the average value of dipole strength $\overline{\overline{S}}$ for a reference particle.

The mean dipole strength $\langle \overline{\overline{S}} \rangle$ for a dilute dispersion can be expressed as:

$$\left\langle \overline{\overline{S}}^o \right\rangle = \int \overline{\overline{S}}^o (\vec{p}) \Psi(\vec{p}, t) d\vec{p} \tag{5.14}$$

where $\overline{\overline{S}}^0 (\vec{p})$ is the value of the dipole strength of the reference particle for particle orientation $\vec{p}$, evaluated on the basis that the reference particle alone is present in an infinite matrix fluid (the rate of strain tensor in the matrix fluid is $\langle \overline{\overline{E}} \rangle$ far away the particle), and $\Psi\left(\vec{p}, t\right)$ is the orientational distribution function that specifies the orientation state of the particles at any time t. The fraction of the total number of particles in the dispersion with orientations lying between $\vec{p}$ and $\vec{p} + d\vec{p}$ at time t is given by $\Psi\left(\vec{p}, t\right) d\vec{p}$. The orientational distribution function Ψ can also be interpreted as the probability density function; the probability that a particle is oriented in a neighborhood of $\vec{p}$ at time t is given by $\Psi\left(\vec{p}, t\right) d\vec{p}$. As the particle must have some orientation at each instant, the probability density function is normalized by integrating over all possible orientations, that is:

$$\int \Psi(\vec{p}, t) d\vec{p} = 1.0 \tag{5.15}$$

The orientation vector $\vec{p}$ of the particle can be regarded as a unit radial vector in a system of spherical coordinates so that $\vec{p} = \vec{p}\left(\theta, \phi\right)$ and Equation 5.14 and Equation 5.15 can be rewritten as:

$$\left\langle \overline{\overline{S}}^o \right\rangle = \int_0^{2\pi} \int_0^{\pi} \overline{\overline{S}}^0 \left(\theta, \phi\right) \Psi \left(\theta, \phi, t\right) \sin \theta \; d\theta d\phi \tag{5.16}$$

$$\int_0^{2\pi} \int_0^{\pi} \Psi(\theta, \phi, t) \sin\theta d\theta \, d\phi = 1.0 \qquad (5.17)$$

The calculation of the mean dipole strength $\langle \bar{s}^o \rangle$, and hence the bulk stress $\langle \bar{\bar{\sigma}} \rangle$ of the dispersion, requires knowledge of $\bar{\bar{S}}^0 (\vec{p})$ and $\Psi(\vec{p}, t)$. The dipole strength of an isolated particle as a function of its instantaneous orientation can be determined easily from the knowledge of local flow fields around the particle. In the absence of rotary Brownian motion, the orientation distribution function is governed by the following Fokker–Planck equation in $\vec{p}$ space [20] derived from conservation of particle number:

$$\frac{\partial \Psi}{\partial t} = -\frac{\partial}{\partial \vec{p}} \bullet \left(\Psi \frac{d\vec{p}}{dt} \right) \qquad (5.18)$$

where $d\vec{p}/dt$ is given by Equation 5.5. The solution of Equation 5.18 gives the full information on the orientation state of the dispersion.

The expression for the bulk stress tensor for a dilute dispersion of axisymmetric non-Brownian particles is given as [18,53]:

$$\bar{\bar{\sigma}} = -P\bar{\bar{\delta}} + 2\eta_c \bar{\bar{E}} + 2\eta_c \phi \left\{ A \langle \vec{p}\vec{p}\vec{p}\vec{p} \rangle : \bar{\bar{E}} + B(\langle \vec{p}\vec{p} \rangle \bullet \bar{\bar{E}} + \bar{\bar{E}} \bullet \langle \vec{p}\vec{p} \rangle) + C\bar{\bar{E}} \right\} \qquad (5.19)$$

where A, B, and C are material coefficients that depend only on the particle aspect ratio ($\bar{\bar{\sigma}}, P, \bar{\bar{E}}$ are bulk or average quantities — the angular brackets have been dropped from these quantities to simplify the notation). The constitutive equation of the dispersion (Equation 5.19) consists of second and fourth moments of $\vec{p}$, defined as:

$$\langle \vec{p}\vec{p} \rangle = \int \vec{p}\vec{p}\Psi(\vec{p}, t) \, d\vec{p}$$

$$= \int_0^{2\pi} \int_0^{\pi} \vec{p}\,\vec{p}\,\Psi\left(\theta, \phi, t\right) \sin\theta \, d\theta \, d\phi \qquad (5.20)$$

$$\langle \vec{p}\vec{p}\vec{p}\vec{p} \rangle = \int \vec{p}\vec{p}\vec{p}\vec{p}\psi(\vec{p}, t) \, d\vec{p}$$

$$= \int_0^{2\pi} \int_0^{\pi} \left(\vec{p}\vec{p}\vec{p}\vec{p} \right) \Psi\left(\theta, \phi, t\right) \sin\theta \, d\theta \, d\phi \qquad (5.21)$$

Equation 5.5, Equation 5.18, and Equation 5.19 describe the complete rheological behavior of dilute suspensions of axisymmetric particles in the absence of Brownian motion and particle–particle interactions.

For near-sphere spheroidal particles, where the departure from sphericity parameter $\epsilon = r - 1 \ll 1$, the shape factors A, B, C in Equation 5.19 are given as [15,16,62]:

$$A = \frac{395}{147} \epsilon^2 \tag{5.22}$$

$$B = \frac{15}{14} \epsilon - \frac{895}{588} \epsilon^2 \tag{5.23}$$

$$C = \frac{5}{2}\left(1 - \frac{2}{7}\epsilon + \frac{47}{147}\epsilon^2\right) \tag{5.24}$$

For long thin rods ($r \to \infty$), the shape factors are [15]:

$$A = \frac{r^2}{2\left(\ell n\, 2r - \dfrac{3}{2}\right)} \tag{5.25}$$

$$B = \frac{6\,\ell n\, 2r - 11}{r^2} \tag{5.26}$$

$$C = 2 \tag{5.27}$$

As $C \ll A$ and $B \ll A$ for large aspect ratio particles (slender bodies), the constitutive equation (Equation 5.19) simplifies to:

$$\bar{\bar{\sigma}} = -P\,\bar{\bar{\delta}} + 2\eta_c\,\bar{\bar{E}} + 2\eta_c\,\phi\,A\,\langle \vec{p}\vec{p}\vec{p}\vec{p}\rangle : \bar{\bar{E}} \tag{5.28}$$

For a flat circular disk ($r \to 0$), the shape factors are [15]:

$$A = \frac{10}{3\pi r} + 2\left(\frac{104}{9\pi^2} - 1\right) \tag{5.29}$$

$$B = \frac{-8}{3\pi r} + \left(1 - \frac{128}{9\pi^2}\right) \tag{5.30}$$

$$C = \frac{8}{3\pi r} + \left(\frac{128}{9\pi^2}\right) \tag{5.31}$$

For a dilute suspension of identical axisymmetric particles (having the same aspect ratio) undergoing shear flow, Equation 5.5 predicts that each particle stays forever in the same orbit and that each particle possesses the same period of rotation regardless of its initial orientation and orbit. The distribution of particles in different orbits is determined by the orientation distribution function Ψ at time $t = 0$. As all particles have the same time period, the orientation distribution function Ψ (governed by Equation 5.18) is a periodic function of time. Consequently, the bulk stress given by Equation 5.19 is also time dependent even when the imposed shearing motion is steady. Interestingly, the bulk stress becomes indeterminate if the initial orientation state of the particles at time $t = 0$ is not known.

In simple shear flow described by Equation 5.7, the orientation distribution function for a suspension of rod-shaped particles is given by [30]:

$$\Psi(\theta,\phi,t) = \left[4\pi(1 - \dot{\gamma}t \sin^2\theta \sin 2\phi + \dot{\gamma}^2 t^2 \sin^2\theta \sin^2\phi)^{3/2}\right]^{-1} \tag{5.32}$$

provided the particles are initially randomly oriented, that is, $\Psi(t = 0) = \pi/4$. This assumption that the particles are randomly oriented before the onset of motion is often referred to as the *Eisenschitz hypothesis*. Note that a random distribution at time $t = 0$ is random again at $t = T/2$ when each particle has rotated through a ϕ angle of $180°$. However, the distribution is nonrandom at intermediate times [25].

As the orientation distribution function oscillates with a period of $T/2$ (frequency of $2/T$), the rheological properties of the suspension predicted by Equation 5.19 also exhibit oscillatory behavior. Both the intrinsic viscosity $[\eta]$ and normal stress differences oscillate with a period of $T/2$. As an example, Figure 5.3 shows the plot of intrinsic viscosity $[\eta]$ of a suspension with initially randomly oriented prolate spheroids ($r = 10$). As expected, the intrinsic viscosity exhibits oscillatory behavior. The intrinsic normal stress differences, defined as:

$$\left[N_{ij}\right] = \frac{\tau_{ii} - \tau_{jj}}{\eta_c\, \dot{\gamma}\, \phi} \tag{5.33}$$

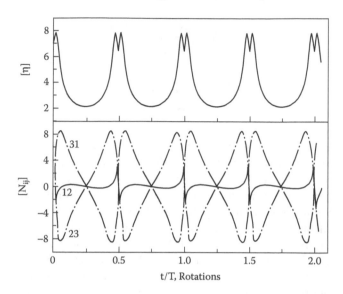

FIGURE 5.3 Plots of intrinsic viscosity [η] and normal stress difference [N_{ij}] for a suspension with initially randomly oriented prolate spheroids ($r = 10$). (From A. Okagawa, R.G. Cox, S.G. Mason, The kinetics of flowing dispersions VI. Transient orientation and rheological phenomena of rods and discs in shear flow, *J. Colloid Interface Sci.*, 45: 303–329, 1973. Reprinted with permission from Elsevier.)

also exhibit oscillatory behavior as shown in Figure 5.3 for initially randomly oriented prolate spheroids of aspect ratio $r = 10$.

If the particles have a uniform orientation, that is, all particles are oriented in the same direction $\vec{p}_0$, the orientation distribution function is given by a Dirac delta function [31]:

$$\Psi = \delta\,(\vec{p} - \vec{p}_0) \tag{5.34}$$

In this case:

$$\left\langle \vec{p}\vec{p}\vec{p}\vec{p} \right\rangle = \vec{p}_0\ \vec{p}_0\ \vec{p}_0\ \vec{p}_0 \tag{5.35}$$

$$\left\langle \vec{p}\vec{p} \right\rangle = \vec{p}_0\ \vec{p}_0 \tag{5.36}$$

and the constitutive equation (Equation 5.19) reduces to:

$$\bar{\bar{\sigma}} = -P\,\bar{\bar{\delta}} + 2\,\eta_c\,\bar{\bar{E}} + 2\eta_c\ \phi\,\Big\{ A\,(\vec{p}_0\ \vec{p}_0\ \vec{p}_0\ \vec{p}_0 : \bar{\bar{E}})$$
$$+ B\Big[\,\vec{p}_0\vec{p}_0 \bullet \bar{\bar{E}} + \bar{\bar{E}} \bullet \vec{p}_0\vec{p}_0\,\Big] + C\,\bar{\bar{E}}\Big\} \tag{5.37}$$

As the motion of the particles is periodic in shear flow, the rheological properties of the suspension predicted by Equation 5.37 are also expected to be periodic. Note that if at time $t = 0$ the particles have a uniform orientation with respect to each other (all particles are oriented in the same direction), then their orientation remains uniform with respect to each other at all times provided that all particles have the same period of rotation (same aspect ratio) and that there are no Brownian motion and particle–particle interactions.

Figure 5.4 shows the dependence of the intrinsic viscosity [η] of a suspension of rodlike particles ($r = 20$) on the orientation angle ϕ for the case where $C_b = 1$ [29]. Note that $C_b = 1$ corresponds to the situation where the axis of symmetry lies in the xy-plane and the particle rotates about the z-axis (see Figure 5.2). The values of [η] shown in Figure 5.4 were calculated by a particle simulation method [29]. The figure reveals the following information: (a) the intrinsic viscosity is periodic, (b) the intrinsic viscosity is minimum when the rods are either in the shearing plane ($\phi = 0$ or $180°$) or perpendicular to velocity field ($\phi = 90°$), and (c) [η] is maximum when the particles are oriented at $\phi = 45°$ and $135°$.

The effect of initial orientation of the particles (specified by orbit constant C_b) on the intrinsic viscosity is shown in Figure 5.5. The maximum value in intrinsic viscosity decreases with a decrease in the orbit constant C_b. When $C_b = 0$, the axis of symmetry of the particle lies parallel to the z-axis, and the particle merely rotates about its axis of symmetry (see Figure 5.2). Furthermore, the orientation angle ϕ at which a maximum in [η] occurs shifts to $90°$ from $45°$ and $135°$.

The time-averaged intrinsic viscosity, averaged over one period of rotation, as a function of orbit constant C_b is shown in Figure 5.6. The viscosity increases with the increase in C_b. The time-averaged viscosity also increases with the increase in the particle aspect ratio, as can be seen in Figure 5.7 for $C_b = 1$.

Thus, under the conditions that (1) there occurs no rotary Brownian motion of the particles, (2) there are no particle–particle interactions, and (3) the aspect ratio of all the particles is the same (identical particles); it can be concluded that the

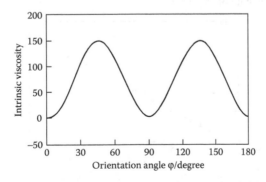

FIGURE 5.4 Dependence of intrinsic viscosity of a suspension of rodlike particles ($r = 20$) on the orientation angle ϕ for the case where orbit constant $C_b = 1$. (Adapted from S. Yamamoto, T. Matsuaka, Viscosity of dilute suspensions of rodlike particles: A numerical simulation method, *J. Chem. Phys.*, 100: 3317–3324, 1994. With permission from American Institute of Physics.)

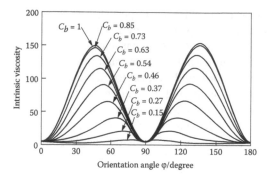

FIGURE 5.5 The effect of initial orientation of the particles (specified by orbit constant C_b) on the intrinsic viscosity of a suspension of rodlike particles ($r = 20$). (From S. Yamamoto, T. Matsuaka, Viscosity of dilute suspensions of rodlike particles: A numerical simulation method, *J. Chem. Phys.*, 100: 3317–3324, 1994. With permission from American Institute of Physics.)

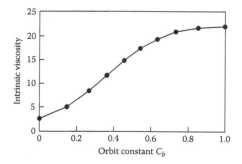

FIGURE 5.6 The time-averaged intrinsic viscosity, averaged over one period of rotation, as a function orbit constant C_b for a suspension of rodlike particles ($r = 20$). (Adapted from S. Yamamoto, T. Matsuaka, Viscosity of dilute suspensions of rodlike particles: A numerical simulation method, *J. Chem. Phys.*, 100: 3317–3324, 1994. With permission from American Institute of Physics.)

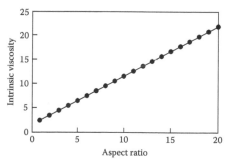

FIGURE 5.7 The time-averaged intrinsic viscosity, averaged over one period of rotation, as a function of particle aspect ratio for a suspension of rodlike particles ($C_b = 1$). (Adapted from S. Yamamoto, T. Matsuaka, Viscosity of dilute suspensions of rodlike particles: A numerical simulation method, *J. Chem. Phys.*, 100: 3317–3324, 1994. With permission from American Institute of Physics.)

rheological properties of a dilute suspension of axisymmetric particles do not exhibit any steady values in steady shear flow. The intrinsic viscosity and normal stresses exhibit oscillatory behavior. Furthermore, the rheological properties and the orientation distribution function depend on the initial orientation state of the particles.

However, in uniaxial elongational (extensional) flows where the vorticity tensor $\bar{\bar{\Omega}} = 0$ and the rate of deformation tensor $\bar{E}$ is given as:

$$\bar{\bar{E}} = \begin{pmatrix} E_{11} & 0 & 0 \\ 0 & -E_{11}/2 & 0 \\ 0 & 0 & -E_{11}/2 \end{pmatrix} \tag{5.38}$$

all the particles (non-Brownian) align themselves in the same direction and the orientation distribution function becomes the Dirac delta function regardless of the initial orientation state of the particles [20–22]. The prolate spheroids ($r > 1$) align themselves with their axis of symmetry parallel to the direction of tension (see Figure 5.8), whereas the oblate spheroids ($r < 1$) align themselves with their axis of symmetry perpendicular to the direction of tension (see Figure 5.8). The particles remain aligned as long as the flow field is maintained. Under these conditions, the suspension exhibits steady rheological properties provided that the imposed flow field is steady. Batchelor [22] has shown that for a dilute suspension of axisymmetric particles in uniaxial extensional flow:

$$\sigma_{11} = -P + 2\eta_c \, E_{11} + \left(\frac{2}{3}\right) \eta_c \, E_{11} \, \phi \, r^2 \, \varepsilon \, Q \, (\varepsilon) \tag{5.39}$$

$$\sigma_{22} = -P + 2\eta_c \, E_{22} = -P + 2\eta_c \left(\frac{-E_{11}}{2}\right) \tag{5.40}$$

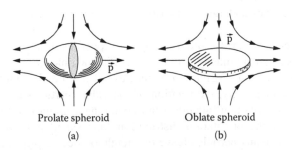

Prolate spheroid   Oblate spheroid

(a)   (b)

FIGURE 5.8 Orientation of (a) prolate and (b) oblate spheroids in uniaxial elongational flow. (Adapted from H. Brenner, Suspension rheology, *Prog. Heat Mass Transfer*, 5: 89–129, 1972. With permission from Elsevier.)

$$\sigma_{33} = -P + 2\eta_c \ E_{33} = -P + 2\eta_c \left(\frac{-E_{11}}{2} \right) \tag{5.41}$$

where $\varepsilon = 1 / \ell n(2r)$ and $Q(\varepsilon)$ is given as:

$$Q(\varepsilon) = \left(\frac{1 + 0.64 \ \varepsilon}{1 - 1.5 \ \varepsilon} \right) + 1.659 \ \varepsilon^2 \tag{5.42}$$

The extensional viscosity η_E, defined as:

$$\eta_E = \frac{\sigma_{11} - \sigma_{22}}{E_{11}} = \frac{\sigma_{11} - \sigma_{33}}{E_{11}} \tag{5.43}$$

is given by:

$$\eta_E = \eta_c \left[3 + \frac{2}{3} \phi \, r^2 \, \varepsilon \, Q(\varepsilon) \right] \tag{5.44}$$

Thus, the extensional viscosity is independent of the rate of extension, indicating Newtonian behavior.

In steady shear flows, the rheological properties of a suspension of axisymmetric nonspherical particles exhibit steady values if one or more of the following conditions are satisfied:

1. The suspension is polydisperse (the aspect ratio of all the particles is not the same) [33,34].
2. Particle–particle hydrodynamic interactions occur [55].
3. The continuous phase (matrix) of the suspension is viscoelastic in nature [41–44].
4. The particles undergo rotary Brownian motion [13–21].

The existence of steady-state rheological properties of a suspension implies that the suspension ultimately attains (if sufficient time is allowed) a unique steady-state orientation distribution, regardless of the initial orientation state of the particles. Interestingly, the rotation of individual particles does not prevent the orientation distribution function from becoming stationary in time. The time independence of the orientation distribution function Ψ implies that for any orientation $\vec{p}$ as many particles leave the neighborhood of $\vec{p}$ as enter it.

When a suspension consists of particles of different aspect ratios, the particles of the same suspension rotate with different periods of rotation, depending on the aspect ratio. As a consequence, the orientation distribution function is no longer periodic. If sufficient time is allowed, a unique steady-state orientation distribution is reached. As an example, Figure 5.9 shows the plots of η_{sp}/ϕ for polydisperse suspensions of non-Brownian rods as a function of t/T. The specific viscosity η_{sp} is defined as $(\eta - \eta_c)/\eta_c$, and the period T is given by Equation 5.12. From the figure, it is clear that the viscosity reaches a steady-state value after some oscillation. The damping of oscillations in the specific viscosity η_{sp} is mainly due to the spread in the particle's aspect ratio.

Several authors have studied the effects of particle–particle hydrodynamic interactions [55,57] and viscoelasticity of matrix phase [41–44] on the orientation distribution function and rheological properties of suspensions of non-Brownian axisymmetric particles. In the presence of hydrodynamic interactions between the particles or when the matrix of the suspension is viscoelastic, the particles attain a steady-state orbit distribution regardless of the initial orientation state of the particles. As a consequence of the steady-state orientation distribution, the rheological properties of the suspension under these conditions also exhibit steady behavior.

5.4 EFFECTS OF BROWNIAN MOTION ON THE RHEOLOGY OF AXISYMMETRIC PARTICLES

The effects of rotary Brownian motion on the orientation distribution function and rheological properties of suspensions of axisymmetric particles have been studied in detail [13–21]. Nonspherical particles of suspensions undergo rotary Brownian motion; that is, they suffer random fluctuations in their orientation because of thermal agitation if their dimensions are sufficiently small. Particles

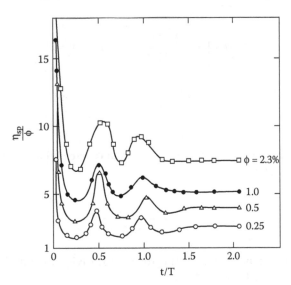

FIGURE 5.9 Variation of η_{sp}/ϕ with t/T for polydisperse suspensions of non-Brownian rods. (Adapted from Y. Ivanov, T.G.M. Van de Ven, S.G. Mason, Damped oscillations in the viscosity of suspensions of rigid rods. I. Monomodal suspensions, *J. Rheol.*, 26: 213–230, 1982. With permission from the Society of Rheology.)

having the greatest dimension, less than about 10 µm, are expected to undergo significant Brownian motion [53]. The rate at which a particle reorients by Brownian motion is determined by the rotary diffusivity of the particle D_r for rotation of the axisymmetric particle about a transverse axis (axis perpendicular to the symmetry axis). The rotary diffusivity is given by the Stokes–Einstein equation [13, 28]:

$$D_r = \frac{3\,k\,T}{16\,\pi\,\eta_c\,a_{11}\,a_\perp^2}\left(\frac{r^2\,\alpha_{11} + \alpha_\perp}{r^2 + 1}\right) \tag{5.45}$$

where k is the Boltzmann constant, T is the absolute temperature, a_{11} is half the length of the particle along its axis of symmetry, $a_\perp$ is half the length of the particle perpendicular to the axis of symmetry, r is the aspect ratio ($a_{11}/a_\perp$), and

$$\alpha_{11} = \frac{2}{r^2 - 1}\left(r^2\,\alpha - 1\right) \tag{5.46}$$

$$\alpha_\perp = \frac{r^2}{r^2 - 1}\left(1 - \alpha\right) \tag{5.47}$$

where α is given as:

$$\alpha = \left[\frac{\cosh^{-1} r}{r(r^2 - 1)^{1/2}}\right] \quad\quad \text{for} \quad r > 1$$

$$\left[\frac{\cos^{-1} r}{r(1 - r^2)^{1/2}}\right] \quad\quad \text{for} \quad r < 1 \tag{5.48}$$

For spherical particles where $r = 1$, α_{11} and $\alpha_\perp$ are equal to 2/3. Thus, D_r for spherical particles is

$$D_r = \frac{k\,T}{8\pi\eta_c R^3} \tag{5.49}$$

where R is the particle radius. The rotary diffusivity for infinitesimally thin circular disks ($r \ll 1$) is:

$$D_r = \frac{3\,k\,T}{32\,\eta_c R^3} \tag{5.50}$$

where R is the radius of the disk. For long, thin, prolate spheroids ($r \gg 1$), D_r is:

$$D_r = \frac{3 \, k \, T \, (\ell n \, 2r - 0.5)}{8\pi\eta_c a_{11}^3} \tag{5.51}$$

Rotary Brownian motion affects the rheology of suspensions of nonspherical particles in two ways. First, it contributes to an angular velocity of the particle. The angular velocity of the particle on account of Brownian torque alone is given by [21]:

$$\vec{\omega}^{Br} = -D_r \, \vec{p} \times \frac{\partial \, \ell n \, \Psi}{\partial \, \vec{p}} \tag{5.52}$$

Thus, the effective rotation rate of the particle is $\vec{\omega}^{Hyd} + \vec{\omega}^{Br}$, where $\vec{\omega}^{Hyd}$ is the angular velocity of the particle on account of hydrodynamic torque alone. Note that the sum of the hydrodynamic torque and Brownian torque acting on the suspended particle is zero:

$$\vec{T}^{Hyd} + \vec{T}^{Br} = 0 \tag{5.53}$$

The Brownian angular velocity $\vec{\omega}^{Br}$ is of the same order as the hydrodynamic angular velocity $\vec{\omega}^{Hyd}$ [21], and therefore it cannot be neglected. As the dipole strength $\overline{\overline{S}}^0$ (see Equation 5.14), and hence the bulk stress of the suspension, is sensitive to the angular velocity of the particle, a change in the angular velocity (owing to the rotary Brownian motion) produces a corresponding change in the bulk stress. This contribution of rotary Brownian motion to bulk stress of the suspension is referred to as *direct contribution* and is given by [15–17]:

$$\overline{\overline{\sigma}} = 2 \, \eta_c \, \phi \, F \, \langle \vec{p}\vec{p} \rangle \, D_r \tag{5.54}$$

where F is a material coefficient that depends on the particle aspect ratio.

The second effect of rotary Brownian motion on the rheology of suspensions of nonspherical particles stems from the fact that rotational Brownian motion affects the orientational distribution function [21]. This effect of Brownian motion, often referred to as *indirect contribution*, on suspension rheology is taken into account by adding a term $D_r \, \nabla^2 \, \Psi$ to the right-hand side of the Fokker–Planck equation (Equation 5.18); that is, the Fokker–Planck equation in $\vec{p}$ space is now given as [13–21]:

$$\frac{\partial\Psi}{\partial t} = -div\left(\Psi \, \frac{d\vec{p}}{dt}\right) + D_r \, \nabla^2 \, \Psi \tag{5.55}$$

The solution of Equation 5.55, using Jeffery's expression (Equation 5.5) for $d\vec{p}/dt$, can be expressed in the following general form for a specified Ψ_0 at time $t = 0$:

$$\Psi = \Psi\left(\vec{p}, t, r, N_{Pe}\right) \tag{5.56}$$

where N_{Pe} is the rotary Peclet number. To obtain a solution for Ψ, the following boundary conditions are applied: Ψ is positive, single valued, and finite, and it satisfies the normalization condition (Equation 5.15). It is important to note that if sufficient time is allowed, the orientation distribution function Ψ reaches a unique steady-state value that is independent of the initial value Ψ_0. This is true regardless of the magnitude of the rotary Brownian diffusion; that is, even an infinitesimal amount of rotary Brownian motion produces a unique steady-state Ψ independent of Ψ_0, if sufficient time is allowed.

In summary, the rheological behavior of suspensions of rigid axisymmetric particles in the presence of rotary Brownian motion is governed by Jeffery's equation (Equation 5.5), Fokker–Planck equation (Equation 5.55), and the following expression for the bulk stress tensor:

$$\bar{\bar{\sigma}} = -P\,\bar{\bar{\delta}} + 2\eta_c\,\bar{\bar{E}} + 2\eta_c\,\phi\left\{A\left\langle\vec{p}\vec{p}\vec{p}\vec{p}\right\rangle : \bar{\bar{E}}\right.$$
$$\left. + B\left[\left\langle\vec{p}\vec{p}\right\rangle \bullet \bar{\bar{E}} + \bar{\bar{E}} \bullet \left\langle\vec{p}\vec{p}\right\rangle\right] + C\bar{\bar{E}} + F\left\langle\vec{p}\vec{p}\right\rangle D_r\right\} \tag{5.57}$$

where the last term involving the material coefficient F is referred to as the diffusion-stress representing the direct contribution of the rotary Brownian motion arising from angular velocity changes. For near-sphere spheroidal particles, the material coefficient F is given as [15]:

$$F = 9 \in \tag{5.58}$$

where $\in = r - 1$. For long, thin rods ($r \to \infty$), F is [15]:

$$F = \frac{3\,r^2}{\left(\ell n\,2r - 0.5\right)} \tag{5.59}$$

In the case of flat circular disks ($r \to 0$), F is given by [15]:

$$F = \frac{-12}{\pi r} \tag{5.60}$$

To predict the rheological behavior of dilute suspensions of axisymmetric particles in the presence of rotary Brownian motion, Equation 5.5, Equation 5.55, and Equation 5.57 need to be solved. The solution of these equations is difficult.

In general, the equations are solved numerically. Analytical solutions can be obtained only for the two limiting cases of near-equilibrium and strong-flow. In the near-equilibrium situation, the Brownian forces dominate ($N_{Pe} \ll 1$) and the orientation distribution function is nearly isotropic. In the strong-flow limit, $N_{Pe} \gg 1$ and the Brownian motion is weak.

The rheological behavior of suspensions predicted by Equation 5.5, Equation 5.55, and Equation 5.57 is non-Newtonian. For example, in simple shear flow the equations predict shear-thinning behavior in that the intrinsic viscosity $[\eta]$ decreases with the increase in rotary Peclet number (defined as $\dot{\gamma}/D_r$, where $\dot{\gamma}$ is the shear rate). The shear-thinning behavior in suspensions of rigid nonspherical particles is not unexpected as viscous stresses tend to orient the particles with the flow in a manner such that they are least disruptive to flow, whereas rotary Brownian motion tends to randomize the particle orientations. When $N_{Pe} \ll 1$, $[\eta]$ is high as the Brownian forces dominate and the particles are randomly oriented. When $N_{Pe} \gg 1$, $[\eta]$ is low as the viscous stresses dominate and the particles are aligned with the flow. In addition to shear-thinning behavior, the suspensions of rigid nonspherical particles also exhibit normal stresses in simple shear flow.

In uniaxial elongational flows, the suspensions of nonspherical particles in the presence of rotary Brownian motion exhibit strain thickening (extensional thickening), that is, the extensional viscosity increases with increase in rotary Peclet number. When the Peclet number is small, the extensional viscosity is low as the particles are randomly oriented and cause little disturbance to elongational flow. When $N_{Pe} \gg 1$, the extensional viscosity is high as the particles are aligned in the direction of flow, resulting in an increase in the drag.

In Table 5.1, the expressions for intrinsic shear viscosity $[\eta]$, intrinsic normal stress differences $[N_1]$ and $[N_2]$, and intrinsic extensional viscosity $[\eta_E]$ in the limiting cases of strong and weak Brownian motion are given for long, thin prolate spheroids ($r \to \infty$) and infinitesimally thin circular disks ($r \to 0$). The intrinsic quantities are defined as follows:

$$[\eta] = \lim_{\phi \to 0} \left(\frac{\eta - \eta_c}{\phi \eta_c} \right) \qquad (5.61a)$$

$$[N_1] = \lim_{\phi \to 0} \left(\frac{\tau_{11} - \tau_{22}}{\phi \eta_c D_r} \right) \qquad (5.61b)$$

$$[N_2] = \lim_{\phi \to 0} \left(\frac{\tau_{22} - \tau_{33}}{\phi \eta_c D_r} \right) \qquad (5.61c)$$

$$[\eta_E] = \lim_{\phi \to 0} \left(\frac{\eta_E - 3\eta_c}{3\phi \eta_c} \right) \qquad (5.61d)$$

TABLE 5.1

Rheological Properties of a Dilute Suspension of Rigid Axisymmetric Spheroids with Extreme Values of Aspect Ratio r [13,15,24,26,62]

Particle Shape	$[\eta]$	$[N_1] = N_1/\eta_c\,\phi\,D_r$	$[N_2] = N_2/\eta_c\,\phi\,D_r$	$[\eta_E] = (\eta_E - 3\eta_c)/3\eta_c\,\phi$
$N_{Pe} \ll 1$, Strong Brownian Motion; Low Shear Rates				
Prolate spheroids $(r \to \infty)$ rod-shaped particles	$\dfrac{4r^2}{15\,\ell nr}$	$\dfrac{2r^2\,N_{Pe}^2}{35\,\ell nr}$	$\dfrac{-r^2\,N_{Pe}^2}{105\,\ell nr}$	$\dfrac{4r^2}{15\,\ell nr}$
Oblate spheroids $(r \to 0)$ disk-shaped particles	$\dfrac{32}{15\,\pi r}$	$\dfrac{4N_{Pe}^2}{21\pi r}$	$\dfrac{-8N_{Pe}^2}{105\pi r}$	$\dfrac{32}{15\,\pi r}$
$N_{Pe} \gg 1$, Weak Brownian Motion; High Shear Rates				
Prolate spheroids $(r \to \infty)$ rod-shaped particles	$\dfrac{0.315r}{\ell nr}$	$\dfrac{r^4}{4\ell nr}$	$-[N_2] \ll \dfrac{r^4}{\ell nr}$	$\dfrac{1}{3}\left(\dfrac{r^2}{\ell nr}\right)$
Oblate spheroids $(r \to 0)$ disk-shaped particles	3.13	$\dfrac{4}{3\,\pi r^3}$	$-\dfrac{1}{3\,\pi r^3}$	$\dfrac{7}{3\,\pi r}$

where τ_{11}, τ_{22}, and τ_{33} are normal stresses (here, the subscript 1 denotes the direction of flow, subscript 2 denotes the direction perpendicular to the flow — the direction pointing to the velocity gradient — and subscript 3 denotes neutral direction).

The intrinsic shear viscosity $[\eta]$ for a given shape (rod or disk) is larger at low N_{Pe} as compared to its value at high N_{Pe}, indicating shear thinning. In the case of rod-shaped particles, $[\eta]_0/[\eta]_\infty$ is 0.846 r and for disk-shaped particles, $[\eta]_0/[\eta]_\infty$ is 0.217/r. This is further illustrated in Figure 5.10. The magnitude of

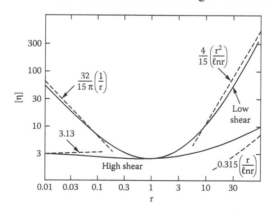

FIGURE 5.10 The low-shear and high-shear limiting intrinsic viscosities of a suspension of spheroidal particles. (Adapted from R.J. Hunter, *Foundations of Colloid Science*, Vol. 1, Clarendon Press, Oxford, 1987.)

the intrinsic normal stress differences $[N_1]$ or $[N_2]$ is much larger at high shear rates as compared to the values at low shear rates. The intrinsic extensional viscosity for a given particle shape is larger at high N_{Pe} compared to its value at low N_{Pe}, indicating extensional thickening; for example, in the case of rod-shaped particles, $[\eta_E]_\infty / [\eta_E]_0$ is 1.25 and for disk-shaped particles, $[\eta_E]_\infty / [\eta_E]_0$ is 1.094.

Figure 5.11 and Figure 5.12 show the complete plots of intrinsic shear viscosity $[\eta]$ for suspensions of prolate and oblate spheroids. The plots are generated from the data given in Reference 13. The suspensions clearly exhibit a shear-thinning behavior. The plots of intrinsic normal stress differences are shown in Figure 5.13 and Figure 5.14 for suspension of both prolate and oblate

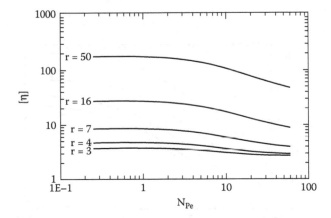

FIGURE 5.11 Intrinsic viscosity as a function of rotary Peclet number for prolate spheroids of different aspect ratios suspended in simple shear flow.

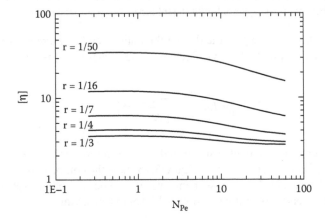

FIGURE 5.12 Intrinsic viscosity as a function of rotary Peclet number for oblate spheroids of different aspect ratios suspended in simple shear flow.

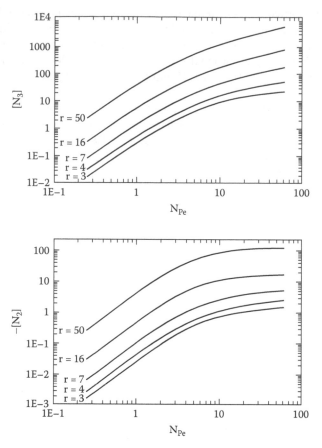

FIGURE 5.13 Variation of intrinsic normal stress differences $[N_3]$ and $-[N_2]$ with rotary Peclet number for prolate spheroids of different aspect ratios suspended in simple shear flow.

spheroids. The plots are generated from the data given in Reference 13. The data are plotted as $[N_3]$ and $-[N_2]$ vs. Peclet number, where $[N_3]$ is defined as:

$$[N_3] = \operatorname*{Lim}_{\phi \to 0} \left(\frac{\tau_{11} - \tau_{33}}{\phi \eta_c D_r} \right) = [N_1] + [N_2] \qquad (5.62)$$

The magnitude of the intrinsic normal stress difference increases with increase in Peclet number. Figure 5.15 and Figure 5.16 show the complete plots of intrinsic extensional viscosity $[\eta_E]$ for suspensions of prolate and oblate spheroids. The suspensions clearly exhibit extensional-thickening behavior.

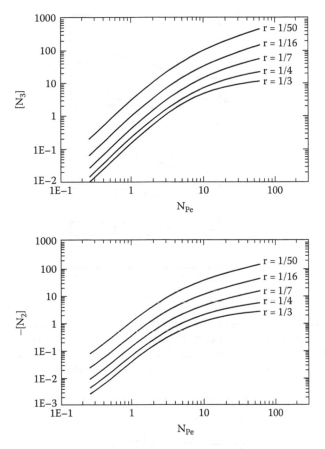

FIGURE 5.14 Variation of intrinsic normal stress differences [N_3] and –[N_2] with rotary Peclet number for oblate spheroids of different aspect ratios suspended in simple shear flow.

For dilute suspensions of nearly spherical particles (spheroids with $\in = r - 1$ $\ll 1$), the constitutive equation (Equation 5.57) reduces to [16]:

$$\overline{\overline{\sigma}} = -P\,\overline{\overline{\delta}} + 2\eta_c\,\overline{\overline{E}} + 2\eta_c\,\phi\left\{\frac{395}{147}\in^2\langle\vec{p}\vec{p}\vec{p}\vec{p}\rangle:\overline{\overline{E}}\right.$$

$$+\left(\frac{15}{14}\in-\frac{895}{588}\in^2\right)\left[\langle\vec{p}\vec{p}\rangle\bullet\overline{\overline{E}}+\overline{\overline{E}}\bullet\langle\vec{p}\vec{p}\rangle\right] \qquad (5.63)$$

$$\left.+\frac{5}{2}\left(1-\frac{2}{7}\in+\frac{1}{3}\in^2\right)\overline{\overline{E}}+9D_r\in\langle\vec{p}\vec{p}\rangle\right\}$$

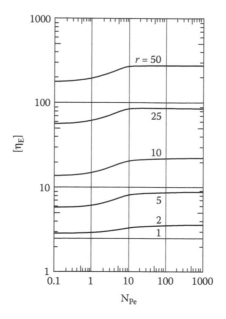

FIGURE 5.15 Intrinsic extensional viscosity $[\eta_E]$ as a function of rotary Peclet number for prolate spheroids of different aspect ratios suspended in axisymmetric uniaxial extensional flow. (Adapted from H. Benner, Rheology of a dilute suspension of axisymmetric Brownian particles, *Int. J. Multiphase Flow*, 1: 195–341, 1974. With permission from Elsevier.)

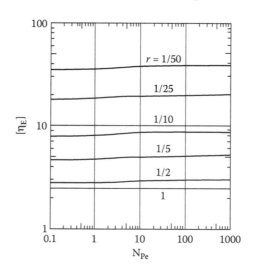

FIGURE 5.16 Intrinsic extensional viscosity $[\eta_E]$ as a function of rotary Peclet number for oblate spheroids of different aspect ratios suspended in axisymmetric uniaxial extensional flow. (Adapted from H. Benner, Rheology of a dilute suspension of axisymmetric Brownian particles, *Int. J. Multiphase Flow*, 1: 195–341, 1974. With permission from Elsevier.)

Leal and Hinch [16] have shown that Equation 5.63 along with Equation 5.5 and Equation 5.55 can be reduced to the following set of two equations:

$$\frac{D\overline{\overline{A}}}{Dt} + 6\,D_r\,\overline{\overline{A}} = 3\,\overline{\overline{E}} \tag{5.64}$$

$$\overline{\overline{\sigma}} = -P\,\overline{\overline{\delta}} + 2\,\eta_c\,\overline{\overline{E}} + 2\eta_c\,\phi\left\{\frac{5}{2}\,\overline{\overline{E}} + \epsilon^2\right.$$

$$\left.\times\left[\frac{26}{147}\,\overline{\overline{E}} + \frac{6}{5}\,D_r\,\overline{\overline{A}} + \frac{1}{7}\left(\overline{\overline{E}}\bullet\overline{\overline{A}} + \overline{\overline{A}}\bullet\overline{\overline{E}} - \frac{2}{3}\,\overline{\overline{\delta}}\,(\overline{\overline{A}}:\overline{\overline{E}})\right)\right]\right\} \tag{5.65}$$

where $\overline{\overline{A}}$ is the symmetric and traceless tensor and D/Dt is the Jaumann derivative.

In simple shear flow, Equation 5.64 and Equation 5.65 lead to the following results for intrinsic viscosity $[\eta]$ and normal stress differences:

$$[\eta] = \frac{5}{2} + \epsilon^2\left\{\frac{26}{147} + \frac{3}{5(1 + N_{Pe}^2/36)}\right\} \tag{5.66}$$

$$[N_1] = N_1/\eta_c D_r \phi = \frac{36}{35}\,\epsilon^2\left\{\frac{6}{1 + (6/N_{Pe})^2}\right\} \tag{5.67}$$

$$[N_2] = N_2/\eta_c D_r \phi = -\frac{6}{35}\,\epsilon^2\left\{\frac{6}{1 + (6/N_{Pe})^2}\right\} \tag{5.68}$$

For general flows that are weak and slowly varying, Equation 5.64 and Equation 5.65 lead to the following approximate solution for the bulk stress:

$$\overline{\overline{\sigma}} = -P\,\overline{\overline{\delta}} + 2\,\eta_c\,\overline{\overline{E}}$$

$$+ 2\eta_c\,\phi\left\{\frac{5}{2}\,\overline{\overline{E}} + \epsilon^2\left[\frac{571}{735}\,\overline{\overline{E}} - \frac{1}{10D_r}\frac{D\overline{\overline{E}}}{Dt} + \frac{1}{7D_r}\left(\overline{\overline{E}}\bullet\overline{\overline{E}} - \frac{1}{3}\,\overline{\overline{\delta}}(\overline{\overline{E}}:\overline{\overline{E}})\right)\right]\right\} \tag{5.69}$$

Rheology of Particulate Dispersions and Composites

NOTATION

A	material coefficient
$\bar{\bar{A}}$	symmetric and traceless tensor expressed through Equation 5.64
a_{11}	half the length of a particle along its axis of symmetry
$a_{\perp}$	half the length of a particle perpendicular to axis of symmetry
B	material coefficient
C	material coefficient
C_o	orbit constant; see Equation 5.10
C_b	modified orbit constant defined as $C_o/(C_o+1)$, where C_o is orbit constant
d	diameter of a particle
D_r	rotary diffusivity of a particle
D/Dt	Jaumann derivative
$\bar{\bar{E}}$	rate of strain tensor
E_{11}	rate of extension in direction "1"
E_{22}	rate of extension in direction "2"
E_{33}	rate of extension in direction "3"
F	material coefficient
k	Boltzmann constant
L	length of a particle
N	number of particles
$[N_1]$	intrinsic first normal stress difference
$[N_2]$	intrinsic second normal stress difference
$[N_3]$	intrinsic normal stress difference defined by Equation 5.62
$[N_{ij}]$	intrinsic normal stress difference defined by Equation 5.33
N_{Pe}	Peclet number defined as $\dot{\gamma}/D_r$
P	pressure
$\vec{p}$	unit vector parallel to the axis of symmetry of particle
$Q(\varepsilon)$	function of ε given by Equation 5.42
R	particle radius
r	aspect ratio of a particle
$\bar{\bar{S}}$	dipole strength of a particle
$\bar{\bar{S}}^o(\vec{p})$	dipole strength of a reference particle for particle orientation $\vec{p}$, evaluated on the basis that the reference particle alone is present in an infinite matrix
T	either temperature or period of rotation of the axis of revolution of a particle about the vorticity axis (as superscript, T refers to transpose)
$\vec{T}^{Hyd}$	hydrodynamic torque acting on the particle
$\vec{T}^{Br}$	torque acting on the particle on account of Brownian motion
t	time
$\vec{u}$	fluid velocity vector
$\vec{u}_\infty$	fluid velocity vector far away from the particle
u_x, u_y, u_z	velocity components
V	volume
$\vec{x}$	position vector

GREEK

α	function of aspect ratio r, defined by Equation 5.48
α_{11}	function of aspect ratio r, defined by Equation 5.46
$\alpha_{\perp}$	function of aspect ratio r, defined by Equation 5.47
$\dot{\gamma}$	shear rate
∇	del operator
δ	delta function
$\underline{\underline{\delta}}$	unit tensor
ε	defined as $1/\ln(2r)$, where r is the particle aspect ratio
$\in$	defined as $(r-1)$, where r is the particle aspect ratio
η_c	continuous-phase shear viscosity
η_E	extensional viscosity
$[\eta]_o$	intrinsic shear viscosity at low Peclet number
$[\eta]_\infty$	intrinsic shear viscosity at high Peclet number
$[\eta_E]$	intrinsic extensional viscosity
$[\eta_E]_o$	intrinsic extensional viscosity at low Peclet number
$[\eta_E]_\infty$	intrinsic extensional viscosity at high Peclet number
θ	angle between $\vec{p}$ (unit vector parallel to the axis of symmetry of a particle) and the z-axis
$\bar{\bar{\sigma}}$	stress tensor
σ_{11}	normal total stress
σ_{22}	normal total stress
σ_{33}	normal total stress
τ_{11}	normal viscous stress
τ_{33}	normal viscous stress
τ_{ii}	normal viscous stress
τ_{jj}	normal viscous stress
ϕ	either volume fraction of particles or the angle between the projection of $\vec{p}$ (unit vector parallel to the axis of symmetry of a particle) on the xy-plane and the x-axis
ψ	orientation distribution function
$\bar{\bar{\Omega}}$	vorticity tensor
$\vec{\omega}$	angular velocity of a particle
$\vec{\omega}_o$	fluid angular velocity
$\vec{\omega}^{Br}$	angular velocity of a particle due to Brownian torque
$\vec{\omega}^{Hyd}$	angular velocity of a particle due to hydrodynamic torque
$\langle X \rangle$	volume average of quantity X

124 Rheology of Particulate Dispersions and Composites

REFERENCES

1. L.J. Struble, C.F. Zukoski, G.C. Maitland, *Flow and Microstructure of Dense Suspensions*, Materials Research Society, Pittsburgh, PA, 1993.
2. R.F. Conley, *Practical Dispersion: A Guide to Understanding and Formulating Slurries*, VCH Publishers, New York, 1996.
3. E.S. Boek, P.V. Coveney, H.N.W. Lekkerkerker, P. van der Schoot, *Phys. Rev. E*, 55: 3124–3133, 1997.
4. P.C. Hiemenz, R. Rajagopalan, *Principles of Colloid and Surface Chemistry*, Marcel Dekker, New York, 1997.
5. P.R. Hornsby, *Adv. Polym. Sci.*, 139: 155–217, 1999.
6. S.J. Gason, D.V. Boger, D.E. Dunstan, *Langmuir*, 15: 7446–7453, 1999.
7. A. Luciani, Y. Leterrier, J.E. Manson, *Rheol. Acta*, 38: 437–442, 1999.
8. E. Kissa, *Dispersions:Characterization, Testing, and Measurement*, Marcel Dekker, New York, 1999.
9. D. Qi, L.S. Luo, *J. Fluid Mech.*, 477: 201–213, 2003.
10. R. Kutteh, *Phys. Rev. E*, 69: 011406-1–011406-16, 2004.
11. R. Pal, *J. Biomech.*, 36: 981–989, 2003.
12. G.B. Jeffery, *Proc. Roy. Soc. A,* 102: 161–179, 1922.
13. H. Brenner, *Int. J. Multiphase Flow*, 1: 195–341, 1974.
14. L.G. Leal, E.J. Hinch, *J. Fluid Mech.*, 46: 685–703, 1971.
15. E.J. Hinch, L.G. Leal, *J. Fluid Mech.*, 52: 683–712, 1972.
16. L.G. Leal, E.J. Hinch, *J. Fluid Mech.*, 55: 745–765, 1972.
17. E.J. Hinch, L.G. Leal, *J. Fluid Mech.*, 57: 753–767, 1973.
18. L.G. Leal, E.J. Hinch, *Rheol. Acta*, 12: 127–132, 1973.
19. E.J. Hinch, L.G. Leal, *J. Fluid Mech.*, 76: 187–208, 1976.
20. G.K. Batchelor, *Annu. Rev. Fluid Mech.*, 6: 227–255, 1974.
21. H. Brenner, *Prog. Heat Mass Transfer*, 5: 89–129, 1972.
22. G.K. Batchelor, *J. Fluid Mech.*, 46: 813–829, 1971.
23. G.K. Batchelor, in *Theoretical and Applied Mechanics*, W.T. Koiter, Ed., North-Holland Publishing Co., Amsterdam, 1976.
24. D. Barthes-Biesel, in *Rheological Measurement*, A.A. Collger, D.W. Clegg, Eds., Elsevier, London, 1988.
25. T.G.M. Van de Ven, *Colloidal Hydrodynamics*, Academic Press, London, 1989.
26. C.J.S. Petrie, *J. Non-Newtonian Fluid Mech.*, 87: 369–402, 1999.
27. J. Jezek, S. Saic, K. Segeth, *Appl. Maths*, 44: 469–479, 1999.
28. H. Brenner, *Prog. Heat Mass Transfer*, 6: 509–574, 1972.
29. S. Yamamoto, T. Matsuaka, *J. Chem. Phys.*, 100: 3317–3324, 1994.
30. S.E. Barbosa, M.A. Bibbo, *J. Polym. Sci. Part B: Polym. Phys.*, 38: 1788–1799, 2000.
31. M.A. Bibbo, S.M. Dinh, R.C. Armstrong, *J. Rheol.*, 29: 905–929, 1985.
32. S.M. Dinh, R.C. Armstrong, *J. Rheol.*, 28: 207–227, 1984.
33. Y. Ivanov, T.G.M. Van de Ven, S.G. Mason, *J. Rheol.*, 26: 213–230, 1982.
34. Y. Ivanov, T.G.M. Van de Ven, *J. Rheol.*, 26: 231–244, 1982.
35. E. Anczurowski, S.G. Mason, *J. Colloid Interface Sci.*, 23: 522–532, 1967.
36. E. Anczurowski, S.G. Mason, *J. Colloid Interface Sci.*, 23: 533–546, 1967.
37. E. Anczurowski, R.G. Cox, S.G. Mason, *J. Colloid Interface Sci.*, 23: 547–562, 1967.
38. A. Okagawa, R.G. Cox, S.G. Mason, *J. Colloid Interface Sci.*, 45: 303–329, 1973.

39. A. Okagawa, S.G. Mason, *J. Colloid Interface Sci.*, 45: 330–358, 1973.
40. H.A. Scheraga, *J. Chem. Phys.*, 23: 1526–1532, 1955.
41. A. Karnis, S.G. Mason, *Trans. Soc. Rheol.*, 10: 571–592, 1966.
42. P.G. Sooffman, *J. Fluid Mech.*, 1: 540–553, 1956.
43. L.G. Leal, *J. Fluid Mech.*, 69: 305–337, 1975.
44. R.K. Gupta, *Polymer and Composite Rheology*, Marcel Dekker, New York, 2000.
45. M.A. Zirnsak, D.U. Hurr, D.V. Boger, *J. Non-Newtonian Fluid Mech.*, 54: 153–193, 1994.
46. D.J. Jeffrey, A. Acrivos, *AIChE J.*, 22: 417–432, 1976.
47. R.J. Hunter, *Foundations of Colloid Science*, Vol. 1, Clarendon Press, Oxford, 1987.
48. G.G. Lipscombe II, M.M. Denn, D.U. Hurr, D.V. Boger, *J. Non-Newtonian Fluid Mech.*, 26: 297–325, 1988.
49. M. Manhart, *J. Non-Newtonian Fluid Mech.*, 112: 269–293, 2003.
50. K.B. Moses, S.G. Advani, A. Reinhardt, *Rheol. Acta*, 40: 296–306, 2001.
51. M.C. Altan, S.G. Advani, S.I. Guceri, R.S. Pipes, *J. Rheol.*, 33: 1129–1155, 1989.
52. C.A. Storer, D.L. Koch, C. Cohen, *J. Fluid Mech.*, 238: 277–296, 1992.
53. R.G. Larson, *The Structure and Rheology of Complex Fluids*, Oxford University Press, New York, 1999.
54. E.S.G. Shagfeh, G.H. Fredrickson, *Phys. Fluids*, 2: 7–24, 1990.
55. M. Rahnama, D.L. Koch, E.S.G. Shagfeh, *Phys. Fluids*, 7: 487–506, 1995.
56. M.P. Petrich, D.L. Koch, C. Cohen, *J. Non-Newtonian Fluid Mech.*, 95: 101–133, 2000.
57. D.L. Koch, *Phys. Fluids*, 7: 2086–2088, 1995.
58. A. Mongruel, M. Cloitre, *Rheol. Acta*, 38: 451–457, 1999.
59. A.M. Wierenga, A.P. Philipse, *Colloids Surf.*, 137: 355–372, 1998.
60. C.W. Macosko, *Rheology Principles, Measurements, and Applications*, VCH Publishers, New York, 1994.
61. F.P. Bretherton, *J. Fluid Mech.*, 14: 284–304, 1962.
62. S.R. Strand, S. Kim, *Rheol. Acta*, 31: 94–117, 1992.

6 Dispersions of Rigid (Spherical and Nonspherical) Magnetic Particles

6.1 INTRODUCTION

Suspensions of small (~10 nm) dipolar or magnetic nanoparticles, also referred to as *ferrofluids* or *magnetic fluids*, are of significant practical interest as they exhibit unique physical properties [1–35]. Owing to the small size of the magnetic particles, generally made up of magnetic iron oxide, the particles can be treated as single-domain permanent magnets unlike macroscopic objects of magnetic materials, which do not possess a permanent magnetic moment because of random orientation of many domains. Ferrofluids are "superparamagnetic" and, therefore, their flow and properties can be controlled using only moderate magnetic fields of the order of 10 to 100 mT. However, in the absence of an external magnetic field, ferrofluids have no net magnetization as Brownian motion orients the particles randomly. The magnetic nanoparticles of ferrofluids are often coated with a surfactant, made of long-chain organic molecules to prevent agglomeration of particles due to attractive Van der Waal's forces.

One important characteristic of ferrofluids is that they flow toward regions of strong magnetic fields and remain fixated there (while preserving their liquid character) as long as the magnetic field is present. This characteristic of ferrofluids is exploited in several practical applications. For example, ferrofluids are widely used as liquid seals in rotary shafts for vacuum systems and hard disk drives of personal computers. The rotating shaft is surrounded by a circular permanent magnet. A small amount of ferrofluid is placed in the gap between the rotating shaft and the magnet. Whereas an ordinary liquid would drip out, the ferrofluid remains in the gap due to the presence of a magnetic field. Ferrofluids are also used in the cooling and damping of loudspeakers. Their potential biomedical use is drug targeting. The drug-carrying ferrofluid can be injected into the blood vessel and can be concentrated at a desired body site by applying a strong magnetic field [19]. Ferrofluids are also being considered as magnetic inks for jet printing [24] and as contrasting agents for MRI scans.

The interaction of ferrofluids (undergoing shear flow) with an external magnetic field leads to some interesting rheological properties. As an example, consider the situation where the magnetic particles are spherical in shape and possess a permanent dipole moment locked into them so that the rotation of the particle also rotates the dipole moment fixed in the particle. In the absence of an external magnetic field, the suspended particle rotates freely with the local vorticity of the fluid motion. When an external magnetic field is applied, the particle experiences a magnetic torque whenever its dipole vector is not parallel with the external field. The existence of a magnetic torque results in hindered rotation, and the particle is no longer able to rotate freely with the fluid. Only in the special case where the field is aligned with the vorticity of the flow does the particle align with the field and rotate freely with the local vorticity of the fluid motion (see Figure 6.1).

Owing to hindered rotation of particles of a suspension of magnetic particles, the rate of mechanical energy dissipation, and hence the viscosity of the suspension, increases. This field-induced enhancement of the suspension viscosity, also referred to as *rotational viscosity*, was first observed experimentally by McTague [1].

It is interesting to note that the presence of external couples on the suspended particles gives rise to a state of antisymmetric stress within the flowing suspension,

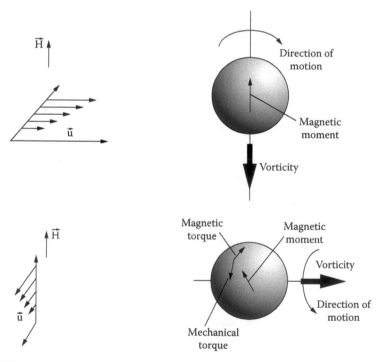

FIGURE 6.1 Interaction of external magnetic field and dipole moment of a magnetic particle suspended in simple shear flow. (Adapted from S. Odenbach, Magnetoviscous and viscoelastic effects in ferrofluids, *Int. J. Mod. Phys. B*, 14: 1615–1631, 2000. With permission from World Scientific.)

that is, the bulk stress tensor for suspensions of dipolar particles subjected to an external magnetic field is not symmetric [3,4,6,8].

6.2 SUSPENSIONS OF SPHERICAL DIPOLAR PARTICLES

6.2.1 MOTION OF A SINGLE ISOLATED DIPOLAR PARTICLE IN SHEAR FLOW

Consider a single dipolar particle, rigid and spherical, suspended in an unbounded incompressible Newtonian liquid of viscosity η_c. Let m be the strength of the dipole moment and $\vec{p}$ be the unit vector locked into the particle in the direction of the dipole. When an external magnetic field $\vec{H}$ is applied, the particle experiences a magnetic torque or couple given as:

$$\vec{T}^{ext} = m\,\vec{p} \times \vec{H} \tag{6.1}$$

The hydrodynamic torque exerted by the fluid on the particle is given as:

$$\vec{T}^{Hyd} = 8\pi\eta_c R^3 \left(\vec{\omega}_o - \vec{\omega}\right) \tag{6.2}$$

where R is the particle radius, $\vec{\omega}_o$ is half the vorticity of the undistributed shear flow, and $\vec{\omega}$ is the angular velocity of the particle.

Thus, the particle experiences two competing torques. The external torque tends to align the particle such that its $\vec{p}$ vector is collinear with the magnetic field $\vec{H}$, whereas the hydrodynamic torque tends to align the particle such that its $\vec{w}$ vector is collinear with the vorticity vector associated with the undisturbed flow. In the absence of rotary Brownian motion, the condition that the particle experiences no net couple can be expressed as:

$$\vec{T}^{ext} + \vec{T}^{Hyd} = 8\pi\eta_c R^3 (\vec{\omega}_o - \vec{\omega}) + m\,\vec{p} \times \vec{H} = 0 \tag{6.3}$$

Equation 6.3 leads to the following expression for the particle angular velocity $\vec{\omega}$:

$$\vec{\omega} = \vec{\omega}_o + \left[\lambda \left| \vec{\omega}_o \right| / \left| \vec{H} \right| \right] (\vec{p} \times \vec{H}) \tag{6.4}$$

where λ is a dimensionless measure of the magnetic field strength with respect to the strength of the hydrodynamic couple and is given as:

$$\lambda = \frac{m \left| \vec{H} \right|}{8\pi\eta_c R^3 \left| \vec{\omega}_o \right|} \tag{6.5}$$

The equation of motion of the particle is:

$$\frac{d\vec{p}}{dt} = \vec{\omega} \times \vec{p} \tag{6.6}$$

From Equation 6.4 and Equation 6.6, the time rate of change of the $\vec{p}$, $d\vec{p}/dt$, with respect to a space-fixed observer can be written as:

$$\frac{d\vec{p}}{dt} = \vec{\omega}_o \times \vec{p} + \frac{m}{8\pi\eta_c R^3}\left[\vec{H} - \vec{p}(\vec{p} \bullet \vec{H})\right] \tag{6.7}$$

The solution of the differential equation (Equation 6.7) gives the particle orientation $\vec{p}$ as a function of time. In general, the solution depends on the following factors: (1) initial orientation of the particle at time $t = 0$, (2) dimensionless field strength parameter λ, and (3) the angle γ between the directions of the magnetic field $\vec{H}$ and the undistributed vorticity vector $2\vec{\omega}_o$. Except for the special case when $\gamma = \pi/2$ and $0 < \lambda < 1$, Hall and Busenberg [2] and Brenner [3] have shown that in all other cases the particle dipole vector $\vec{p}$ ultimately achieves a certain fixed orientation regardless of the initial orientation, and thereafter the particle spins about this axis with an angular velocity of $\vec{p}(\vec{\omega}_o \bullet \vec{p})$. Note that $d\vec{p}/dt$ is zero at steady state and vectors $\vec{\omega}$ and $\vec{p}$ are collinear.

As an example, let us consider the motion of a spherical dipolar particle in simple shear where the external magnetic field is perpendicular to the vorticity vector, that is, angle γ is $\pi/2$. When $\lambda \geq 1$, the spherical particle does not rotate, that is, the particle angular velocity $\vec{\omega} = 0$ at steady state. The dipole vector $\vec{p}$ of the particle lies in a plane perpendicular to fluid vorticity vector $2\vec{\omega}_o$ and the angle α between $\vec{p}$ and magnetic field vector $\vec{H}$ is given as:

$$\alpha = \sin^{-1}(1/\lambda) \tag{6.8}$$

When λ is large ($>>1$), $\vec{p}$ and $\vec{H}$ are almost collinear. As λ is decreased, the sphere rotates by a fixed amount to a point where the external couple is exactly balanced by the hydrodynamic couple, that is,

$$m\left|\vec{H}\right|\sin\alpha = 8\pi\eta_c R^3 \left|\vec{\omega}_o\right| \tag{6.9}$$

Figure 6.2 shows the orientation of a dipolar sphere in simple shear flow when $\gamma = \pi/2$.

When $0 < \lambda < 1$ and $\gamma = \pi/2$, the particle dipole vector $\vec{p}$ does not align in any fixed direction, but instead traverses one of an infinite family of closed orbits. In each orbit, the dipole vector $\vec{p}$ rotates about a fixed central axis of rotation.

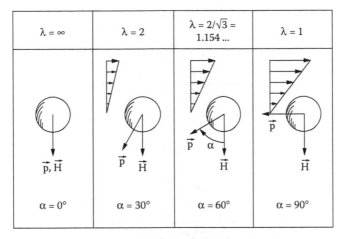

$\lambda = \infty$	$\lambda = 2$	$\lambda = 2/\sqrt{3} =$ 1.154 ...	$\lambda = 1$
$\alpha = 0°$	$\alpha = 30°$	$\alpha = 60°$	$\alpha = 90°$

FIGURE 6.2 Orientation of a dipolar particle in simple shear flow when the external magnetic field is perpendicular to vorticity vector ($\gamma = \pi/2$) and $\lambda \geq 1$ (λ is dimensionless field strength). (Adapted from H. Brenner, Dynamics of neutrally buoyant particles in low Reynold number flows, *Prog. Heat Mass Transfer*, 6: 509–574, 1972. With permission from Elsevier.)

The angular velocity of the particle is nonuniform and the period of rotation is given as:

$$T = \frac{4\pi}{\dot{\gamma}\sqrt{(1-\lambda^2)}} \tag{6.10}$$

Figure 6.3 shows the typical orbits for the following values of λ: 0, 0.5, and about 1. As the dipole axis rotates in a particular orbit, it sweeps over the surface of a circular cone. The orbit selected by the particle is determined by the value of λ and the initial orientation of the dipole vector $\vec{p}$. Note that when λ is zero (no magnetic field), the time period of rotation given by Equation 6.10 is $4\pi/\dot{\gamma}$ which is the time period of a spherical particle rotating in simple shear flow, with shear rate of $\dot{\gamma}$ and angular velocity of $\dot{\gamma}/2$, in the absence of an external couple.

6.2.2 RHEOLOGY OF SUSPENSIONS OF SPHERICAL DIPOLAR PARTICLES

The total deviatoric stress $\langle \bar{\bar{\tau}} \rangle$ in a suspension is the sum of contributions due to the matrix ($\langle \bar{\bar{\tau}}_c \rangle$) and the suspended particles ($\langle \bar{\bar{\tau}}_p \rangle$), that is:

$$\langle \bar{\bar{\tau}} \rangle = \langle \bar{\bar{\tau}}_c \rangle + \langle \bar{\bar{\tau}}_p \rangle \tag{6.11}$$

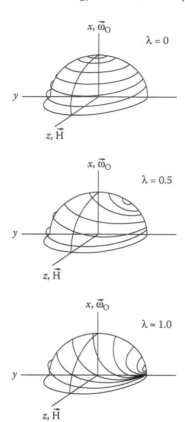

FIGURE 6.3 Typical orbits of rotation of dipolar particle when $\gamma = \pi/2$ and $0 < \lambda < 1$. (From E.J. Hinch, L.G. Leal, Note on the rheology of a dilute suspension of dipolar spheres with weak Brownian couples, *J. Fluid Mech.*, 56: 803–813, 1972. Reprinted with the permission of Cambridge University Press.)

The matrix stress tensor (assuming that the matrix is a Newtonian fluid) is:

$$\left\langle \overline{\overline{\tau}}_c \right\rangle = 2\eta_c \left\langle \overline{\overline{E}} \right\rangle \tag{6.12}$$

where $\left\langle \overline{\overline{E}} \right\rangle$ is the imposed macroscopic rate of deformation tensor on the suspension.

The particle deviatoric stress tensor $\left\langle \overline{\overline{\tau}}_p \right\rangle$, as pointed out earlier, is not symmetric if the particles are subjected to external couples. It is convenient to separate the particle stress tensor $\left\langle \overline{\overline{\tau}}_p \right\rangle$ into symmetric and antisymmetric parts. Thus,

$$\left\langle \overline{\overline{\tau}}_p \right\rangle = \left\langle \overline{\overline{\tau}}_p^S \right\rangle + \left\langle \overline{\overline{\tau}}_p^A \right\rangle \tag{6.13}$$

where $\left\langle \overline{\overline{\tau}}_p^S \right\rangle$ is the symmetric part and $\left\langle \overline{\overline{\tau}}_p^A \right\rangle$ is the antisymmetric part of the particle deviatoric stress.

Brenner and Weissman [4] have shown that for a dilute suspension of spherical dipolar particles, the symmetric portion of the particle deviatoric stress tensor is given as:

$$\left\langle \overline{\overline{\tau}}_p^S \right\rangle = \frac{5}{2}\phi\left(2\eta_c\left\langle \overline{\overline{E}} \right\rangle\right) \tag{6.14}$$

This is true whether or not rotary Brownian motion of particles is present.

The antisymmetric portion of the particle deviatoric stress tensor is related to the bulk external couple density $\langle \vec{N} \rangle$, defined as the couple per unit volume of the suspension, as follows:

$$\left\langle \overline{\overline{\tau}}_p^A \right\rangle = -\frac{1}{2}\overline{\overline{\overline{\varepsilon}}} \bullet \left\langle \vec{N} \right\rangle \tag{6.15}$$

where $\overline{\overline{\overline{\varepsilon}}}$ is the unit-alternating isotropic triadic (third-order tensor) and the external couple density is given as:

$$\left\langle \vec{N} \right\rangle = n\left\langle \vec{T}^{ext} \right\rangle \tag{6.16}$$

Here, n is the number density of particles (number of particles per unit volume of the suspension given by $3\phi/4\pi R^3$) and $\langle \vec{T}^{ext} \rangle$ is the orientation-averaged external couple, defined as:

$$\left\langle \vec{T}^{ext} \right\rangle = \int_{\vec{p}} \vec{T}^{ext}\Psi\left(\vec{p}\right)d\vec{p} \tag{6.17}$$

where $\Psi(\vec{p})$ is the orientation distribution function. The external couple $\vec{T}^{ext}$ on the particle is dependent on the orientation of the particle and is given by Equation 6.1. From Equation 6.1, Equation 6.15, and Equation 6.16, $\langle \overline{\overline{\tau}}_p^A \rangle$ can be expressed as:

$$\left\langle \overline{\overline{\tau}}_p^A \right\rangle = -\frac{1}{2}nm\,\overline{\overline{\overline{\varepsilon}}} \bullet \left(\left\langle \vec{p} \right\rangle \times \vec{H}\right) \tag{6.18}$$

where $\langle \vec{p} \rangle$ is the orientation average of the unit dipole vector. Thus, the particle deviatoric stress tensor becomes

$$\left\langle \overline{\overline{\tau}}_p \right\rangle = \frac{5}{2}\phi\left(2\eta_c\left\langle \overline{\overline{E}} \right\rangle\right) - \frac{3\phi m}{8\pi R^3}\overline{\overline{\overline{\varepsilon}}} \bullet \left(\left\langle \vec{p} \right\rangle \times \vec{H}\right) \tag{6.19}$$

Consequently, the bulk stress tensor of the suspension can be written as:

$$\langle \overline{\overline{\sigma}} \rangle = -\langle P \rangle \overline{\overline{\delta}} + 2\eta_c \left[1 + \frac{5}{2}\phi \right] \langle \overline{\overline{E}} \rangle - \frac{3m\phi}{8\pi R^3} \overline{\overline{\varepsilon}} \bullet \left(\langle \vec{p} \rangle \times \vec{H} \right) \qquad (6.20)$$

As $\overline{\overline{\varepsilon}} \bullet \left(\langle \vec{p} \rangle \times \vec{H} \right)$ is equal to $\left(\langle \vec{p} \rangle \vec{H} - \vec{H} \langle \vec{p} \rangle \right)$ Equation 6.20 can be rewritten as:

$$\langle \overline{\overline{\sigma}} \rangle = -\langle P \rangle \overline{\overline{\delta}} + 2\eta_c \left[1 + \frac{5}{2}\phi \right] \langle \overline{\overline{E}} \rangle - \frac{3m\phi}{8\pi R^3} \left[\langle \vec{p} \rangle \vec{H} - \vec{H} \langle \vec{p} \rangle \right] \qquad (6.21)$$

The mean of the unit dipole vector, $\langle \vec{p} \rangle$, present in Equation 6.21 can be calculated provided that the orientation distribution function Ψ is known:

$$\langle \vec{p} \rangle = \int_{\vec{p}} \vec{p} \Psi d\vec{p} \qquad (6.22)$$

Except for the special case where $\gamma = \pi/2$ and $0 < \lambda < 1$ (γ is the angle between the directions of the magnetic field $\vec{H}$ and the undisturbed vorticity vector; λ is the dimensionless field strength defined in Equation 6.5), the particle dipole vector $\vec{p}$ achieves a certain fixed orientation at steady state in all other situations. Thus, the orientation distribution function in all situations, other than the special case just mentioned, becomes the Dirac delta function δ at steady state, that is:

$$\Psi = \delta \left(\vec{p} - \vec{p}_s \right) \qquad (6.23)$$

where $\vec{p}_s$ is the dipole vector of the particles at steady state. Consequently, $\langle \vec{p} \rangle$ is the same as $\vec{p}_s$.

In the special case where $\gamma = \pi/2$ and $0 < \lambda < 1$, Equation 6.7 predicts that each particle rotates forever in the same closed orbit and that each particle possesses the same period of rotation. The distribution of particles in different orbits is determined by the initial orientation state of the suspension. As all particles (assumed identical) have the same time period, the orientation distribution function Ψ is a periodic function of time. Consequently, the bulk stress given by Equation 6.21 is also periodic, even when the imposed shearing motion is steady. If the initial orientation state of the suspension is not known, the bulk stress becomes indeterminate.

In simple shear flow given as:

$$u_y = \dot{\gamma} x, \quad u_x = 0, \quad u_z = 0 \qquad (6.24)$$

the deviatoric stress components predicted from Equation 6.21 are as follows [3]:

$$\tau_{xx} = \tau_{yy} = \tau_{zz} = 0 \tag{6.25}$$

$$\tau_{xy} = \eta_c \dot{\gamma} \left[1 + \frac{5}{2}\phi + \frac{3}{2}\phi \sin^2 \theta_s \right] \tag{6.26}$$

$$\tau_{yx} = \eta_c \dot{\gamma} \left[1 + \frac{5}{2}\phi - \frac{3}{2}\phi \sin^2 \theta_s \right] \tag{6.27}$$

$$\tau_{yz} = -\tau_{zy} \neq 0 \tag{6.28}$$

$$\tau_{zx} = -\tau_{xz} \neq 0 \tag{6.29}$$

where θ_s is the angle between the dipole vector $\vec{p}$ and the undisturbed vorticity vector $2\vec{\omega}_o$ at steady state, given by:

$$\sin^2 \theta_s = \frac{1}{2}\left(1 + \lambda^2\right)\left\{ 1 - \left[1 - \left(\frac{2\lambda}{1+\lambda^2} \right)^2 \sin^2 \gamma \right]^{1/2} \right\} \tag{6.30}$$

Although there are no normal stresses in a suspension, the stress is not symmetric in that $\tau_{ij} \neq \tau_{ji}$. This behavior is clearly different from Newtonian fluids, where the stress tensor is symmetric and $\tau_{ij} = \tau_{ji}$. Furthermore, the suspension of dipolar spherical particles exhibits shear thinning. The intrinsic viscosity of the suspension $[\eta]$, defined as

$$[\eta] = \lim_{\phi \to 0} \left[\frac{(\tau_{xy} / \dot{\gamma}) - \eta_c}{\eta_c \phi} \right] \tag{6.31}$$

is given by:

$$[\eta] = \frac{5}{2} + \frac{3}{2}\sin^2 \theta_s \tag{6.32}$$

With the increase in the dimensionless shear rate λ^{-1} given as:

$$\lambda^{-1} = \frac{4\pi \eta_c R^3 \dot{\gamma}}{m|\vec{H}|} \tag{6.33}$$

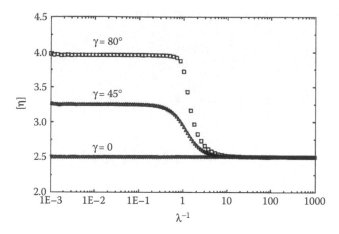

FIGURE 6.4 Variation of intrinsic viscosity with dimensionless shear rate λ^{-1} for different values of γ (γ is the angle between the external magnetic field and the undisturbed vorticity vector).

the intrinsic viscosity decreases (see Figure 6.4). At high values of λ^{-1}, [η] approaches the Einstein limit of 2.5. In the limit of $\lambda^{-1} \to 0$, the intrinsic viscosity is given as:

$$\left[\eta \right] = \frac{5}{2} + \frac{3}{2} \sin^2 \gamma \tag{6.34}$$

When the angle γ between magnetic field and undisturbed vorticity vector is $\pi/2$ Equation 6.34 gives [η] of 4.

In extensional flows (irrotational flows) where $\vec{\omega}_o$ is zero, the particles do not rotate and their angular velocity is zero; at steady state, the dipole vectors $\vec{p}$ of all the particles become parallel to the magnetic field vector $\vec{H}$. As no external magnetic torque is present on the particles ($\vec{p}$ and $\vec{H}$ are collinear), the stress tensor is now symmetric. The suspension behaves as a Newtonian fluid with viscosity given by the Einstein equation:

$$\eta = \eta_c \left(1 + \frac{5}{2} \phi \right) \tag{6.35}$$

Thus, the external magnetic field has no effect on the rheology of dilute suspension of spherical dipolar particles in extensional flows.

6.2.2.1 Effect of Rotary Brownian Motion

In the discussion thus far, the effect of rotary Brownian motion on the rheology of suspensions of spherical dipolar particles has been neglected. As most practical suspensions of magnetic particles consist of nanosize particles, the effect of rotary Brownian motion is generally important.

The symmetric part of the particle deviatoric stress tensor, given by Equation 6.14, is unaffected by the rotary Brownian motion. However, the antisymmetric part of the particle deviatoric stress tensor, given by Equation 6.18, is affected because the mean of the unit dipole vector $\vec{p}$ depends on the orientation distribution function Ψ, and the orientation distribution function in turn depends strongly on the rotary Brownian motion of the particles.

The orientation distribution function Ψ is governed by the following Fokker–Planck equation in $\vec{p}$ space:

$$\frac{\partial \Psi}{\partial t} = - div\left(\Psi \frac{d\vec{p}}{dt} \right) + D_r \nabla^2 \Psi \tag{6.36}$$

where $d\vec{p}/dt$ is given by Equation 6.7 and D_r is the rotary diffusion coefficient of a spherical particle given by the Stokes–Einstein equation:

$$D_r = \frac{kT}{8\pi \eta_c R^3} \tag{6.37}$$

The orientation distribution function Ψ, also referred to as the *probability density function*, is required to be nonnegative, continuous, and single valued, subject to the normalization condition:

$$\int_{\vec{p}} \Psi \, d\vec{p} = 1 \tag{6.38}$$

The integration in Equation 6.38 is carried out over all possible orientations.

The bulk stress tensor equation for the suspension, given by Equation 6.21, is still valid. Thus, the rheological behavior of a dilute suspension of spherical dipolar particles subjected to an external magnetic field and rotary Brownian motion is governed by Equation 6.7, Equation 6.21, and Equation 6.36. In simple shear flow given by $\vec{u} = (0, \dot{\gamma}x, 0)$, the solution of Equation 6.7, Equation 6.21, and Equation 6.36 has been obtained by Brenner and Weissman [4] as:

$$\tau_{xx} = \tau_{yy} = \tau_{zz} = 0 \tag{6.39}$$

$$\tau_{xy} = \eta_c \dot{\gamma} \left[1 + \frac{5}{2}\phi + \frac{3}{2}\phi K_z \right] \tag{6.40}$$

$$\tau_{yx} = \eta_c \dot{\gamma} \left[1 + \frac{5}{2}\phi - \frac{3}{2}\phi K_z \right] \tag{6.41}$$

$$\tau_{yz} = -\tau_{zy} = \frac{3}{2}\phi \eta_c \dot{\gamma} K_x \tag{6.42}$$

$$\tau_{zx} = -\tau_{xz} = \frac{3}{2}\phi \eta_c \dot{\gamma} K_y \tag{6.43}$$

where K_x, K_y, and K_z are functions of λ, γ, and the rotary Peclet number N_{Pe} defined as:

$$N_{Pe} = \frac{\omega_o}{D_r} = \frac{\dot{\gamma}}{2D_r} \tag{6.44}$$

From Equation 6.40, the intrinsic viscosity $[\eta]$ can be expressed as:

$$[\eta] = \frac{5}{2} + \frac{3}{2}K_z \tag{6.45}$$

Thus, the viscosity η or intrinsic viscosity $[\eta]$ of a dilute suspension is dependent on shear rate; it decreases with the increase in shear rate. There are no normal stresses but the stress is not symmetric. Under the condition that the angle γ between the magnetic field and undisturbed vorticity vector is $\pi/2$, the analytical expressions for K_x, K_y, and K_z as obtained by Brenner and Weissman [4] using a perturbation method are given in Table 6.1. Note that N_{pe} is defined by Equation 6.44 in Table 6.1.

It is interesting to note that the suspension stress in the presence of rotary Brownian motion is determinate under all conditions, regardless of the initial orientation state of the suspension. Inclusion of even a small degree of rotary Brownian motion is sufficient to remove the indeterminacy observed for the special case when $\gamma = \pi/2$ and $0 < \lambda < 1$ [5].

TABLE 6.1
Rheological Properties of a Dilute Suspension of Spherical Diopolar Particles Subjected to External Magnetic Field and Rotary Brownian Motion, under the Condition $\gamma = \pi/2$ [4]

K_x	K_y	K_z	$[\eta]$
		Weak External Field ($\lambda \ll 1$) and $N_{Pe} = $ Order(1)	
0	0	$\dfrac{2\lambda^2 N_{Pe}^2}{3(4 + N_{Pe}^2)}$	$\dfrac{5}{2} + \left(\dfrac{\lambda^2 N_{Pe}^2}{4 + N_{Pe}^2}\right)$
		Dominant Brownian Motion ($N_{Pe} \ll 1$) and $\lambda = $ Order(1)	
0	0	$\dfrac{\lambda^2 N_{Pe}^2}{6}$	$\dfrac{5}{2} + \dfrac{\lambda^2 N_{Pe}^2}{4}$
		Dominant Shear ($N_{Pe} \gg 1$, $\lambda N_{Pe} = $ Order(1))	
0	0	$\dfrac{2\lambda^2}{3}\left[1 + \dfrac{7}{30}\lambda^2 - \dfrac{4}{N_{Pe}^2}\right]$	$\dfrac{5}{2} + \lambda^2\left[1 + \dfrac{7}{30}\lambda^2 - \dfrac{4}{N_{Pe}^2}\right]$

6.3 SUSPENSIONS OF NONSPHERICAL DIPOLAR PARTICLES

Particulate suspensions utilized in the manufacture of magnetic tapes and disks generally consist of nonspherical magnetic particles [31–34] such as rodlike particles of γ-Fe_2O_3 and CrO_2 and platelike particles of Ba-ferrite. Thus, a good understanding of the rheology of suspensions of nonspherical dipolar particles is important from a practical point of view.

The symmetric part of the particle deviatoric stress tensor is independent of the external magnetic field and rotary Brownian motion in the case of spherical dipolar particles. However, in the case of suspensions of nonspherical dipolar particles, the external couple (magnetic torque) as well as the Brownian couple both contribute directly and indirectly to the symmetric part of the particle stress. The direct contribution to the particle stress results from the modification of the particle angular velocity by Brownian and magnetic torques. The indirect contribution is due to the effects of Brownian motion and external torque on the orientation distribution function.

The torques acting on a nonspherical dipolar particle subjected to rotary Brownian motion and external magnetic field, in a homogeneous shear flow, are:

$$\vec{T}^{Hyd} = \zeta\left[\vec{\omega}_o - \vec{\omega} + C(\vec{p} \times (\bar{\bar{E}} \bullet \vec{p}))\right] \tag{6.46}$$

Rheology of Particulate Dispersions and Composites

$$\vec{T}^{Br} = -\vec{p} \times \frac{\partial}{\partial \vec{p}} (kT \, \ell n \psi) \qquad (6.47)$$

$$\vec{T}^{ext} = m \left(\vec{p} \times \vec{H} \right) \qquad (6.48)$$

where $\vec{T}^{Hyd}$, $\vec{T}^{Br}$, and $\vec{T}^{ext}$ refer to hydrodynamic, Brownian, and external magnetic torques, respectively; ζ is hydrodynamic resistance in the direction perpendicular to the particle symmetry axis [8,36]; $\vec{\omega}_o$ is the undisturbed angular velocity of the fluid; $\vec{\omega}$ is angular velocity of the particle; and C is a shape factor defined in terms of the particle aspect ratio r as $(r^2 - 1)/(r^2 + 1)$. The dipole moment of the particle is assumed to be oriented parallel to the particle symmetry axis.

The net rotation rate of a particle can be determined by equating the sum of all torques to zero:

$$\vec{T}^{Hyd} + \vec{T}^{Br} + \vec{T}^{ext} = 0 \qquad (6.49)$$

Substituting the expressions for $\vec{T}^{Hyd}$, $\vec{T}^{Br}$, and $\vec{T}^{ext}$ into Equation 6.49 gives the following expression for the net angular velocity of a particle:

$$\vec{\omega} = \vec{\omega}_o + C\left[\vec{p} \times (\bar{\bar{E}} \bullet \vec{p}) \right] - \frac{kT}{\zeta} \vec{p} \times \frac{\partial (\ell n \psi)}{\partial \vec{p}} + \frac{m}{\zeta} (\vec{p} \times \vec{H}) \qquad (6.50)$$

The rate of change of orientation vector $\vec{p}$ in the absence of rotary Brownian motion can be written as:

$$\frac{d\vec{p}}{dt} = \vec{\omega} \times \vec{p}$$

$$= \vec{\omega}_o \times \vec{p} + C(\bar{\bar{\delta}} - \vec{p}\vec{p}) \bullet (\bar{\bar{E}} \bullet \vec{p}) + \frac{m}{\zeta} (\bar{\bar{\delta}} - \vec{p}\vec{p}) \bullet \vec{H} \qquad (6.51)$$

The motion of the orientation vector $\vec{p}$ described by Equation 6.51 has been studied by Almog and Frankel [25] in homogeneous shear flow. The motion is governed by two competing factors: (1) the shear field that tends to orient the particle in the direction of flow, and (2) the magnetic torque that tends to align the particle dipole with the external field. When the polar angle (γ) of the external field orientation (Figure 6.5) is $\pi/2$ (external field is perpendicular to the fluid vorticity axis), the motion of the particle depends on the azimuthal angle (β) of the field orientation and the field parameter λ. For small values of λ and when $\beta = 0$ or $\pi/2$ (field parallel or perpendicular to flow), the particle moves in one of a family of closed orbits;

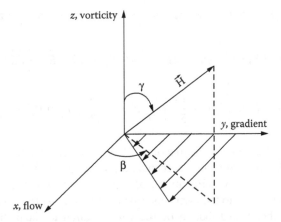

FIGURE 6.5 Coordinate system.

when β is different from 0 or $\pi/2$ and λ is small, the particle either reaches a stable equilibrium orientation or approaches a limit cycle on the equator [25]. For large values of λ, the particle ultimately reaches a stable equilibrium orientation. When the external field is not oriented at $\gamma = \pi/2$ and $\cos \gamma \approx \text{order}(1)$, the particle approaches a stable equilibrium orientation regardless of λ.

The equation governing the orientation state of a suspension, in the presence of rotary Brownian motion and external field, is the Fokker–Planck equation in $\vec{p}$ space:

$$\frac{\partial \Psi}{\partial t} = -\,div\left(\Psi \frac{d\vec{p}}{dt} \right) + D_r \nabla^2 \psi \tag{6.52}$$

where $d\vec{p}/dt$ is given by Equation 6.51 and D_r (rotary diffusion coefficient for rotation of particle about a tranverse axis) is given by the Stokes–Einstein equation:

$$D_r = \frac{kT}{\zeta} \tag{6.53}$$

For thin circular disks of radius R and thin prolate spheroids, D_r is given as:

$$D_r = \frac{3kT}{32\eta_c R^3} \quad \text{for thin circular disks } (r \ll 1) \tag{6.54a}$$

$$D_r = \frac{3kT\,(\ell n\, 2r - 0.5)}{8\pi\eta_c a_{11}^3} \quad \text{for thin prolate spheroids } (r \gg 1) \tag{6.54b}$$

where a_{11} is half the length of the particle along its axis of symmetry.

The symmetric part of the particle deviatoric stress tensor is given as [8]:

$$\langle \overline{\overline{\tau}}_p^S \rangle = 2\eta_c \phi \left\{ 2A_H \overline{\overline{E}} : \langle \vec{p}\vec{p}\vec{p}\vec{p} \rangle + 2B_H \left(\overline{\overline{E}} \bullet \langle \vec{p}\vec{p} \rangle + \langle \vec{p}\vec{p} \rangle \bullet \overline{\overline{E}} - \frac{2}{3} \overline{\overline{\delta}} \overline{\overline{E}} : \langle \vec{p}\vec{p} \rangle \right) \right.$$

$$\left. + C_H \overline{\overline{E}} + F_H D_r \left(\langle \vec{p}\vec{p} \rangle - \frac{1}{3}\overline{\overline{\delta}} \right) - \frac{1}{6} F_H D_r \left(\frac{m}{kT} \right) \langle \vec{H}_\perp \vec{p} + \vec{p}\vec{H}_\perp \rangle \right\} \tag{6.55}$$

where A_H, B_H, C_H, and F_H are functions of particle aspect ratio, and $\vec{H}_\perp$ is the projection of $\vec{H}$ in the direction perpendicular to $\vec{p}$ given as $\vec{H} \bullet (\overline{\overline{\delta}} - \vec{p}\vec{p})$. It should be noted that the last term in Equation 6.55 represents the direct contribution of the external magnetic field to the particle stress; the second-last term $F_H D_r (\langle \vec{p}\vec{p} \rangle - \overline{\overline{\delta}}/3)$ represents the direct contribution of Brownian motion to the particle stress; the remaining terms represent hydrodynamic contribution to the particle stress.

The antisymmetric portion of the particle deviatoric stress tensor is given as [8]:

$$\langle \overline{\overline{\tau}}_p^A \rangle = 3\eta_c \phi D_o \left(\frac{m}{kT} \right) \langle \vec{p}\vec{H}_\perp - \vec{H}_\perp \vec{p} \rangle \tag{6.56}$$

where D_o is the rotary diffusivity of a sphere of volume equal to that of the spheroidal particle under consideration.

The total stress tensor of the suspension is:

$$\langle \overline{\overline{\sigma}} \rangle = -\langle P \rangle \overline{\overline{\delta}} + 2\eta_c \langle \overline{\overline{E}} \rangle + \langle \overline{\overline{\tau}}_p^S \rangle + \langle \overline{\overline{\tau}}_p^A \rangle \tag{6.57}$$

On substituting the expressions for $\langle \overline{\overline{\tau}}_p^S \rangle$ and $\langle \overline{\overline{\tau}}_p^A \rangle$ into Equation 6.57, the following equation is obtained for the bulk stress tensor of the suspension of spheroidal dipolar particles subjected to rotary Brownian motion and an external magnetic field ($\langle \ \rangle$ is dropped from P and $\overline{\overline{E}}$ for simplicity):

$$\langle \overline{\overline{\sigma}} \rangle = -P\overline{\overline{\delta}} + 2\eta_c \overline{\overline{E}} + 2\eta_c \phi \left\{ 2A_H \overline{\overline{E}} : \langle \vec{p}\vec{p}\vec{p}\vec{p} \rangle \right.$$

$$+ 2B_H \left(\overline{\overline{E}} \bullet \langle \vec{p}\vec{p} \rangle + \langle \vec{p}\vec{p} \rangle \bullet \overline{\overline{E}} - \frac{2}{3}\overline{\overline{\delta}}\,\overline{\overline{E}} : \langle \vec{p}\vec{p} \rangle \right) + C_H \overline{\overline{E}}$$

$$+ F_H D_r \left(\langle \vec{p}\vec{p} \rangle - \frac{\overline{\overline{\delta}}}{3} \right) + 3D_o \frac{m}{kT} \left[\frac{(1-C)}{2} \langle \vec{p}\vec{H}_\perp \rangle - \frac{(1+C)}{2} \langle \vec{H}_\perp \vec{p} \rangle \right] \right\} \tag{6.58}$$

Table 6.2 gives the expressions for the coefficients A_H, B_H, C_H, and F_H, in terms of the particle aspect ratio r for the following limiting cases: particle is nearly spherical [$\epsilon = (r-1) \ll 1$], particle has the shape of a flat circular disk ($r \rightarrow 0$), and particle is a thin, long, prolate spheroid ($r \rightarrow \infty$).

TABLE 6.2
Stress Coefficients A_H, B_H, C_H, and F_H for Different Limiting Cases [8]

Coefficient	Particles As Near Spheres $[\in = (r-1) \to 0]$	Particles As Flat Circular Disks $(r \to 0)$	Particles As Thin and Long Prolate Spheroids $(r \to \infty)$
A_H	$\dfrac{395}{294}\in^2$	$\dfrac{5}{3\pi r} + \left(\dfrac{104}{9\pi^2} - 1\right)$	$\dfrac{r^2}{4\left[(\ell n 2r) - \dfrac{3}{2}\right]}$
B_H	$\dfrac{15}{28}\in - \dfrac{895}{1176}\in^2$	$-\dfrac{4}{3\pi r} + \left(\dfrac{1}{2} - \dfrac{64}{9\pi^2}\right)$	$\dfrac{3\ell n(2r) - (11/2)}{r^2}$
C_H	$\dfrac{5}{2} - \dfrac{5}{7}\in + \dfrac{235}{294}\in^2$	$\dfrac{8}{3\pi r} + \dfrac{128}{9\pi^2}$	2
F_H	$9\in$	$-\dfrac{12}{\pi r}$	$\dfrac{3r^2}{\ell n(2r) - 0.5}$

Thus, the rheological behavior of a dilute suspension of nonspherical axisymmetric magnetic particles, subjected to rotary Brownian motion and an external magnetic field, is governed by Equation 6.51, Equation 6.52, and Equation 6.58. These suspensions exhibit non-Newtonian characteristics, such as shear-rate dependent intrinsic viscosity, nonzero normal stress differences (unlike a suspension of dipolar spheres, which exhibits zero normal stress differences), and nonsymmetric state of stress.

The determination of the rheological properties of a suspension requires the solution of Equation 6.51, Equation 6.52, and Equation 6.58. In general, the equations are solved numerically. Strand and Kim [8] have solved this problem numerically in simple shear flow, given as $\vec{u} = (\dot{\gamma}y, 0, 0)$. The steady-state rheological property of a suspension in simple shear flow depends on five parameters: polar angle (γ) and azimuthal angle (β) of the external field orientation (Figure 6.5), aspect ratio of the particle, Peclet number (defined as $\dot{\gamma}/D_r$), and dimensionless external field strength $\tilde{H}$ defined as $m|\vec{H}|/(kT)$. Note that in the case of suspensions of spherical dipolar particles, the stress was independent of the azimuthal angle (β) of the external field orientation and, therefore, only one angle (polar angle, γ) was important.

The intrinsic viscosity $[\eta]$, defined as

$$[\eta] = \lim_{\phi \to 0}\left[\dfrac{\tau_{xy} - \eta_c\dot{\gamma}}{\phi\eta_c\dot{\gamma}}\right] \tag{6.59}$$

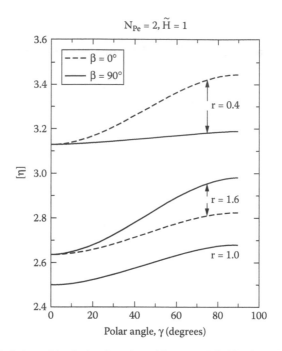

FIGURE 6.6 Variation of intrinsic viscosity with external field orientation for different particle aspect ratios. (Adapted from S.R. Strand, S. Kim, Dynamics and rheology of a dilute suspension of dipolar nonspherical particles in an external field: Part 1. Steady shear flows, *Rheol. Acta*, 31: 94–117, 1992. With kind permission of Springer Science and Business Media.)

is plotted as a function of the external field orientation in Figure 6.6 for aspect ratios of 0.4, 1.0, and 1.6 [8]. The Peclet number value is fixed at 2 and the dimensionless external field strength $(\tilde{H})$ value is fixed at 1. Note that the polar angle (γ) value of 0° corresponds to the situation where the external magnetic field is parallel to the vorticity axis. When $\gamma = 90°$ and $\beta = 0$, the external magnetic field is parallel to the flow direction. Figure 6.6 reveals the following interesting information: (1) for spherical particles ($r = 1$), the azimuthal angle β has no affect on [η]. When the polar angle (γ) value is zero (magnetic field is parallel to vorticity axis), [η] is equal to 2.5, corresponding to Einstein's value. As γ increases, [η] increases and reaches a maximum value when $\gamma = \pi/2$ (magnetic field is perpendicular to vorticity axis); (2) for nonspherical particles ($r = 0.4$ or 1.6), [η] not only increases with the increase in polar angle γ, it also depends on the azimuthal angle β; and (3) for prolate spheroids ($r = 1.6$), [η] increases when the external field orientation is changed from $\gamma = \pi/2$ and $\beta = 0$ (field parallel to flow direction) to $\gamma = \pi/2$ and $\beta = \pi/2$ (field perpendicular to flow direction). Opposite behavior is seen for the oblate spheroid ($r = 0.4$); [η] decreases when the external field orientation is changed from parallel to perpendicular to the flow direction.

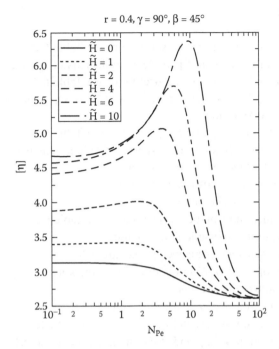

r = 0.4, γ = 90°, β = 45°

FIGURE 6.7 Variation of intrinsic viscosity of a suspension of oblate spheroids with the Peclet number for different values of magnetic field strength. (Adapted from S.R. Strand, S. Kim, Dynamics and rheology of a dilute suspension of dipolar nonspherical particles in an external field: Part 1. Steady shear flows, *Rheol. Acta*, 31: 94–117, 1992. With kind permission of Springer Science and Business Media.)

Figure 6.7 shows the plots of $[\eta]$ for a suspension of oblate spheroids as a function of the Peclet number (N_{Pe}) for different values of the magnetic field strength $\tilde{H}$ [8]. The magnetic field is parallel to (1,1,0), that is, the polar angle γ is held constant at 90° and the azimuthal angle β is 45°. At low values of $\tilde{H}$, $[\eta]$ decreases with the increase in N_{Pe}, indicating shear-thinning behavior. Interestingly, at high values of the field strength $\tilde{H}$, a maximum in $[\eta]$ occurs at some intermediate value of N_{Pe}. For example, at $\tilde{H}=10$ the maximum in $[\eta]$ occurs at $N_{Pe} \approx 10$; the low shear-limiting value of $[\eta]$ at $\tilde{H}=10$ is about 4.6, the high shear-limiting value of $[\eta]$ is about 2.6, and the maximum value of $[\eta]$ (at $N_{Pe} \approx 10$) is about 6.4. The peak in $[\eta]$ observed at some intermediate Peclet number is referred to as the *resonance phenomenon* [8]. The resonance effect seems to occur in strong external fields at selected field orientations when the aspect ratio of a particle differs from unity to a considerable extent. In a strong external field, the particles tend to align with the external field when N_{Pe} is low; at high N_{Pe}, the particles tend to spend most of their time aligned with the flow. It appears that at intermediate N_{Pe}, where the resonance effect is observed, the particles tend to spend a significant amount of their time oriented in the direction of maximum strain.

NOTATION

A_H	function of particle aspect ratio; see Table 6.2
a_{11}	half the length of particle along its axis of symmetry
B_H	function of particle aspect ratio; see Table 6.2
C	shape factor defined as $(r^2 - 1)/(r^2 + 1)$, where r is particle aspect ratio
C_H	function of particle aspect ratio; see Table 6.2
D_o	rotary diffusivity of a sphere of volume equal to that of the particle under consideration
$\underline{\underline{D}}_r$	rotational diffusion coefficient
$\underline{\underline{E}}$	rate of strain tensor
F_H	function of particle aspect ratio; see Table 6.2
$\vec{H}$	external magnetic field strength
$\tilde{H}$	dimensionless magnetic field strength
$\vec{H}_\perp$	projection of $\vec{H}$ in the direction perpendicular to $\vec{p}$
k	Boltzmann constant
K_x	function of λ, γ, and Peclet number; see Table 6.1
K_y	function of λ, γ, and Peclet number; see Table 6.1
K_z	function of λ, γ, and Peclet number; see Table 6.1
m	strength of dipole moment
$\vec{N}$	external couple density (couple per unit volume of the suspension)
N_{Pe}	Peclet number (N_{pe} is $\dot{\gamma}/2D_r$ in Table 6.1; N_{pe} is $\dot{\gamma}/D_r$ in Figure 6.7)
n	number density of particles (number of particles per unit volume of suspension)
P	pressure
$\vec{p}$	unit dipole vector (unit vector locked into the particle in the direction of the dipole)
$\vec{p}_s$	unit dipole vector at steady state
R	particle radius
r	particle aspect ratio
T	temperature or period of rotation of particle
$\vec{T}^{Br}$	Brownian torque on a particle
$\vec{T}^{ext}$	external torque on a particle
$\vec{T}^{Hyd}$	hydrodynamic torque on a particle
$\vec{u}$	fluid velocity vector
u_x, u_y, u_z	fluid velocity components

GREEK

α	angle between dipole unit vector $\vec{p}$ and magnetic field vector $\vec{H}$
β	azimuthal angle of the external magnetic field orientation; see Figure 6.5
γ	angle between the magnetic field strength vector $\vec{H}$ and undisturbed fluid vorticity vector
$\dot{\gamma}$	shear rate
δ	Dirac delta function
$\bar{\bar{\delta}}$	unit tensor
ϵ	equals $(r-1)$, where r is the particle aspect ratio
$\bar{\bar{\varepsilon}}$	unit alternating isotropic triadic, defined as $\sum_i\sum_j\sum_k \hat{\delta}_i \hat{\delta}_j \hat{\delta}_k \varepsilon_{ijk}$ where $\hat{\delta}_i, \hat{\delta}_j, \hat{\delta}_k$, are unit vectors and ε_{ijk} is permutation symbol
ζ	hydrodynamic resistance in the direction perpendicular to the particle symmetry axis
η	shear viscosity
η_c	continuous-phase viscosity
$[\eta]$	intrinsic viscosity
θ_s	angle between dipole vector $\vec{p}$ and undisturbed fluid vorticity vector at steady state
λ	dimensionless magnetic field strength given by Equation 6.5
$\bar{\bar{\sigma}}$	total stress tensor
$\bar{\bar{\tau}}$	deviatoric stress tensor
$\bar{\bar{\tau}}_c$	deviatoric stress contribution due to matrix
$\bar{\bar{\tau}}_p$	deviatoric stress contribution due to suspended particles
$\bar{\bar{\tau}}_p^S$	symmetric part of particle deviatoric stress
$\bar{\bar{\tau}}_p^A$	antisymmetric part of particle deviatoric stress
$\tau_{xx}, \tau_{yy}, \tau_{zz}$	normal stress components
$\tau_{xy}, \tau_{yx}, \tau_{xz}$	shear stress components
ϕ	volume fraction of particles
$\psi(\vec{p})$	orientation distribution function
$\vec{\omega}$	angular velocity of particle
$\vec{\omega}_o$	half the vorticity of undisturbed flow
$\langle X \rangle$	volume average of quantity X

REFERENCES

1. J.P. McTague, *J. Chem. Phys.*, 51: 133–136, 1969.
2. W.F. Hall, S.N. Busenberg, *J. Chem. Phys.*, 51: 137–144, 1969.
3. H. Brenner, *J. Colloid Int. Sci.*, 32: 141–158, 1970.
4. H. Brenner, M.H. Weissman, *J. Colloid Int. Sci.*, 41: 499–531, 1972.
5. E.J. Hinch, L.G. Leal, *J. Fluid Mech.*, 56: 803–813, 1972.
6. H. Brenner, *Prog. Heat Mass Transfer*, 5: 89–129, 1972.
7. R.E. Rosenzweig, *Ferrohydrodynamics*, Cambridge University Press, Cambridge, 1985.
8. S.R. Strand, S. Kim, *Rheol. Acta*, 31: 94–117, 1992.
9. K. Raj, B. Moskowitz, R. Casciari, *J. Magn. Magn. Mater.*, 149: 174–180, 1995.
10. Y. Almog, I. Frankel, *J. Fluid Mech.*, 366: 289–310, 1998.
11. S. Odenbach, H. Stork, *J. Magn. Magn. Mater.*, 183: 188–194, 1998.
12. I. Puyesky, I. Frankel, *J. Fluid Mech.*, 369: 191–216, 1998.
13. R.G. Larson, *The Structure and Rheology of Complex Fluids*, Oxford University Press, New York, 1999.
14. S. Odenbach, T. Rylewicz, M. Heyen, *J. Magn. Magn. Mater.*, 201: 155–158, 1999.
15. S. Odenbach, *Int. J. Mod. Phys. B*, 14: 1615–1631, 2000.
16. S. Odenbach, K. Raj, *Magnetohydrodynamics*, 36: 312–319, 2000.
17. B.U. Felderhof, *Magnetohydrodynamics*, 36: 329–334, 2000.
18. A.Y. Zubarev, L.Y. Iskakova, *Colloid J.*, 63: 706–713, 2001.
19. S. Odenbach, *Proc. Appl. Math. Mech.*, 1: 28–32, 2002.
20. A.Y. Zubarev, S. Odenbach, J. Fleischer, *J. Magn. Magn. Mater.*, 252: 241–243, 2002.
21. K. Melzner, S. Odenbach, *J. Magn. Magn. Mater.*, 252: 250–252, 2002.
22. S. Odenbach, *Colloids Surf. A.*, 217: 171–178, 2003.
23. R. Patel, R.V. Upadhyay, R.V. Mehta, *J. Colloid Int. Sci.*, 263: 661–664, 2003.
24. S.W. Charles, J. Popplewell, *Endeavour*, New Series 6: 153, 1982.
25. Y. Almog, I. Frankel, *J. Fluid Mech.*, 289: 243–261, 1995.
26. K.M. Jansons, *J. Fluid Mech.*, 137: 187–216, 1983.
27. S. Kim, C.J. Lawrence, *J. Non-Newtonian Fluid Mech.*, 24: 297–310, 1987.
28. H. Brenner, *Annu. Rev. Fluid Mech.*, 2: 137–175, 1970.
29. H. Brenner, *Prog. Heat Mass Transfer*, 6: 509–574, 1972.
30. T.E. Karis, M.S. Jhon, *Colloids Surf. A*, 80: vii–viii, 1993.
31. T.M. Kwon, M.S. Jhon, H.J. Choi, T.E. Karis, *Colloids Surf. A*, 80: 39–46, 1993.
32. T.M. Kwon, P.L. Frattini, L.N. Sadani, M.S. Jhon, *Colloids Surf. A*, 80: 47–61, 1993.
33. T.M. Kwon, M.S. Jhon, H.J. Choi, *Mat. Chem. Phys.*, 49: 225–228, 1997.
34. T.M. Kwon, M.S. Jhon, H.J. Choi, *J. Mol. Liq.*, 75: 115–126, 1998.
35. L.G. Leal, *J. Fluid Mech.*, 46: 395–416, 1971.
36. H. Brenner, *Int. J. Multiphase Flow*, 1:195–341,1974.

Part III

Rheology of Dispersions of Nonrigid Particles

Part III, consisting of Chapter 7 to Chapter 11, deals with the rheology of dispersions of a variety of deformable particles, such as soft solid particles, liquid droplets, bubbles, deformable capsules, and core–shell particles.

Chapter 7 covers the rheology of dispersions of soft (deformable) solid particles. After a brief introduction, the fundamental equations governing the flow fields inside and outside a single soft-solid particle suspended in a Newtonian fluid are given. The expression for the bulk stress tensor in a dilute dispersion of soft, purely elastic, Hookean-solid particles is discussed next. The rheological properties predicted from the constitutive equation under different conditions are described. The chapter concludes with a section on the rheology of dispersions of solid viscoelastic particles.

Chapter 8 describes the rheology of emulsions, that is, dispersions of liquid droplets. After a brief discussion of the industrial applications of emulsions, the deformation and orientation of droplets caused by the flow field is described. The equations governing the flow fields both inside and outside a single droplet suspended in a Newtonian fluid are given along with the boundary conditions. The steps involved in the derivation of the expression for dipole strength of a single droplet located in an infinite matrix fluid (Newtonian) are discussed. The constitutive equations for a dilute emulsion with zero-order, first-order, and second-order deformation of droplets are described. The rheological properties predicted from the constitutive equations under different conditions are discussed. The chapter concludes with a section on the effects of surfactants on emulsion rheology.

Chapter 9 deals with the rheology of dispersions of bubbles. The initial part of the chapter deals with the dilational properties of a bubbly suspension. The expression for the dilational viscosity of a bubbly suspension is derived. The full constitutive equations for a dilute bubbly suspension with zero-order and first-order deformation of bubbles are given. The rheological properties predicted from the constitutive equations are discussed. The final section of the chapter is devoted to the effects of surfactants on the rheology of bubbly suspensions.

Chapter 10 focuses on the rheology of dispersions of capsules (particles consisting of a drop of liquid surrounded by a thin deformable elastic-solid membrane). Capsules are thought to be models of red blood cells. The fundamental equations governing the flow fields inside and outside a single capsule suspended in a Newtonian matrix fluid, along with the boundary conditions, are given. As this problem requires knowledge of membrane mechanics, a section is devoted to membrane mechanics. Two types of membranes, namely Mooney–Rivlin (MR) membrane and red blood cell (RBC) membrane, are discussed. The equations describing the time evolution of the shape of initially spherical capsules are described. The constitutive equation for a dilute dispersion of capsules with first-order deformation is given. The rheological properties predicted from the constitutive equation for capsules with MR and RBC membranes are discussed.

The rheology of dispersions of core–shell particles is described in Chapter 11. Three types of core–shell particles are considered, namely *solid core–hairy shell*, *solid core–liquid shell*, and *liquid core–liquid shell (double-emulsion droplets)*. The equations governing the flow fields outside and within a single core–shell particle suspended in a Newtonian fluid are discussed along with the appropriate boundary conditions. Only zero-order deformation of particles is generally considered. The expressions for the dipole strength of the three types of core–shell particles are given. The viscous properties of dispersions of core–shell particles are described.

7 Dispersions of Soft (Deformable) Solid Particles

7.1 INTRODUCTION

The need for understanding the rheological behavior of dispersions of soft (deformable) solid particles has increased in recent years, mainly because of their wide range of applications. Many dispersions of practical interest can be represented by a simplified physical model in which the particles are approximated as soft solid-like particles. Examples of such systems are (1) suspensions of swollen starch granules [1–13], (2) suspensions of microgel particles [14–23], and (3) suspensions of rubbery particles in thermoplastic melts [24–26].

Starch is commonly used to thicken liquid foods such as sauces, soups, and custards. Uncooked starch granules have diameters in the range of 1 to 100 μm. However, when the starch suspension is heated above a certain temperature (gelatinization temperature), the crystalline order of the granules is lost and the granules imbibe many times their weight of water, swelling to much larger sizes compared to the uncooked granules. The swollen starch granules have been treated as soft elastic spheres [4,5] with a shear modulus value of about 10^4 Pa.

A microgel particle is a cross-linked polymer particle swollen by a good solvent. Microgel suspensions have a wide range of potential applications in industries involved in coatings, printing, and pharmaceuticals. The extent of swelling of microgel particles depends on factors such as degree of cross-linking, nature of the solvent, and temperature. In case of a thermosensitive microgel dispersion, the particle diameter and, hence, effective volume fraction of dispersed phase, can be varied simply by changing the temperature. Thus, the rheological properties of a thermosensitive microgel dispersion can be controlled by varying the temperature. The swollen microgel particles have been treated as soft elastic particles [21]. The shear modulus value of the Sephadex microgel particles [21] varies from 0.3×10^4 to 12×10^4 Pa, depending on the degree of cross-linking of particles.

The problem of motion and deformation of soft solid-like particles of viscoelastic material has been studied by Goddard and Miller [27] and Roscoe [28]. In simple shear flow, the initially spherical particle deforms into an ellipsoid of fixed dimensions and orientation at a given shear rate. The material inside the ellipsoidal particle undergoes continuous deformation and rotation. It is important to note that identical spheres deform in the same fashion and have the same

151

orientation. Thus, the orientation distribution of the particles is uniform, and the orientation of the particles is uniquely determined by the hydrodynamics of the problem [29,30].

According to the rheological constitutive models developed by Goddard and Miller [27] and Roscoe [28], dilute suspensions of deformable solid particles exhibit non-Newtonian rheological behavior. The suspension exhibits shear-thinning and normal-stress effects in shear flows. In uniaxial elongational flow, the suspension exhibits a strain-thickening effect.

7.2 FUNDAMENTAL EQUATIONS

Consider a particle composed of an isotropic and incompressible solid-like vis-coelastic material. In the unstressed state, the particle is spherical in shape. The particle is freely suspended in a Newtonian incompressible liquid of viscosity η_c. Buoyancy effects are assumed to be negligible. The system is subjected to linear shear flow such that far away from the particle, the velocity field is $\vec{u}_\infty = \overline{\overline{\Gamma}}_\infty \bullet \vec{r}$, where $\overline{\overline{\Gamma}}_\infty$ is a spatially uniform (homogeneous) velocity gradient tensor that could be time dependent. The fundamental equations governing the motion and deformation of the particle are expressed with respect to a reference frame centered on the particle center of mass and translating with it.

The instantaneous external shape of the particle can be expressed as:

$$r = \left(x_1^2 + x_2^2 + x_3^2\right)^{1/2} = f(x_1, x_2, x_3) \tag{7.1}$$

where r is the distance from the particle center of mass to the surface of the particle. The function f is *a priori* unknown and must be determined as part of the problem solution. Assuming that the particle Reynolds number is small (inertia effects are negligible), the internal and external velocity and stress fields are described by the following equations:

$$\nabla \bullet \overline{\overline{\sigma}}^* = 0, \qquad \nabla \bullet \vec{u}^* = 0 \quad \text{for } r \leq f \tag{7.2}$$

$$\nabla \bullet \overline{\overline{\sigma}} = 0, \qquad \nabla \bullet \vec{u} = 0 \quad \text{for } r \geq f \tag{7.3}$$

where $\vec{u}^*$ and $\overline{\overline{\sigma}}^*$ denote velocity and stress tensor inside the particle, and $\vec{u}$ and $\overline{\overline{\sigma}}$ denote velocity and stress tensor external to the particle.

Because the external phase is a Newtonian liquid (viscosity η_c), the stress tensor in the external phase is given by:

$$\overline{\overline{\sigma}} = -P\overline{\overline{\delta}} + \eta_c \left[\nabla \vec{u} + \left(\nabla \vec{u}\right)^T\right] \tag{7.4}$$

Thus Equation 7.3 can be rewritten as:

$$\eta_c \nabla^2 \vec{u} - \nabla P = 0, \quad \nabla \bullet \vec{u} = 0 \quad \text{for } r \geq f \qquad (7.5)$$

The associated boundary conditions are:

1. Far away from the particle ($r \to \infty$), the disturbance due to its presence disappears

$$\vec{u} \to \vec{u}_\infty \text{ as } r \to \infty \qquad (7.6)$$

2. On the surface of the particle ($r = \text{f}$), there is continuity of velocities and stresses:

$$\vec{u} = \vec{u}^* = \frac{\partial \vec{x}}{\partial t} \quad \text{at } r = f \qquad (7.7)$$

where $\partial \vec{x}/\partial t$ is the velocity of the surface, given as a time derivative of the position $\vec{x}$ of the surface material points. Also,

$$\bar{\bar{\sigma}} \bullet \hat{n} = \bar{\bar{\sigma}}^* \bullet \hat{n} \quad \text{at } r = f \qquad (7.8)$$

where $\hat{n}$ is an outward unit normal at the particle surface $r = f$.

To complete the formulation of the problem, the rheological constitutive equation of the solid particle material is also required.

7.2.1 Constitutive Equation of the Particle Material

The rheological behavior of the particle material can be described by a constitutive equation of the form:

$$\bar{\bar{e}}^* = f\left(\bar{\bar{\tau}}^*\right) \qquad (7.9)$$

where $\bar{\bar{e}}^*$ is the finite strain tensor, defined as (see Chapter 2):

$$\bar{\bar{e}}^* = \frac{1}{2}\left[\bar{\bar{\delta}} - \left(\frac{\partial \vec{X}}{\partial \vec{x}}\right)^T \bullet \left(\frac{\partial \vec{X}}{\partial \vec{x}}\right)\right] \qquad (7.10)$$

In Equation 7.10, $\vec{X}$ refers to the initial position of the material point in the reference (unstressed) configuration of the particle, and $\vec{x}$ refers to the current position of the material point in the deformed configuration of the particle.

The deviatoric or "extra" stress tensor $\bar{\bar{\tau}}^*$ present in Equation 7.9 can be expressed as:

$$\bar{\bar{\tau}}^* = \bar{\bar{\sigma}}^* + P^* \bar{\bar{\delta}} \tag{7.11}$$

where P^* is the mean pressure given by:

$$P^* = -\frac{1}{3} tr\,(\bar{\bar{\sigma}}^*) = -\frac{1}{3}\bar{\bar{\sigma}}^* : \bar{\bar{\delta}} \tag{7.12}$$

For particles of Hookean solids, Equation 7.9 can be written as:

$$\bar{\bar{e}}^* = \bar{\bar{\tau}}^*/2G_p \tag{7.13}$$

where G_p is the elastic modulus of the particle material.

For particles of viscoelastic material, Roscoe [28] assumed that the stress inside the particle is a summation of two terms: elastic stress and viscous stress. The elastic stress is one that would exist in a purely elastic material under the same strain as the particle material. The viscous stress is a Newtonian fluid type stress that is proportional to the rate of strain. Thus, the constitutive equation for viscoelastic particle material can be written as:

$$\bar{\bar{\tau}}^* = 2G_p\,\bar{\bar{e}}^* + 2\eta_p\bar{\bar{E}}^* \tag{7.14}$$

where η_p is the viscosity of the particle material and $\bar{\bar{E}}^*$ is the rate of strain inside the particle.

In their analysis, Goddard and Miller [27] and Roscoe [28] further assumed that the material inside the particle experiences homogeneous deformation; that is, the velocity gradient and stress field are homogeneous (uniform, independent of position) inside the particle.

7.3 RHEOLOGY OF DISPERSIONS OF SOFT, SOLID-LIKE ELASTIC PARTICLES

Goddard and Miller [27,31] have developed the following expression for the bulk stress $\bar{\bar{\sigma}}$ of a dilute suspension of soft purely elastic particles obeying Hooke's law (Equation 7.13):

$$\bar{\bar{\sigma}} = -P\bar{\bar{\delta}} + 2\eta_c \bar{\bar{E}} + 5\eta_c \phi \left\{ \begin{array}{c} \bar{\bar{E}} - \left(\dfrac{D\bar{\bar{e}}^*}{Dt}\right) + \dfrac{6}{7} S_d (\bar{\bar{E}} \bullet \bar{\bar{e}}^*) \\[4mm] -\dfrac{20}{7} S_d \left(\dfrac{D\bar{\bar{e}}^*}{Dt} \bullet \bar{\bar{e}}^*\right) \end{array} \right\} \tag{7.15}$$

where $\bar{\bar{E}}$ is the macroscopic (bulk) rate of strain tensor and the particle strain tensor $\bar{\bar{e}}^*$ satisfies the differential equation:

$$\frac{D\bar{\bar{e}}^*}{Dt} = \frac{5}{3}\bar{\bar{E}} - \left(\frac{2G_p}{3\eta_c}\right)\bar{\bar{e}}^* + \frac{10}{7}S_d(\bar{\bar{E}} \bullet \bar{\bar{e}}^*)$$

$$-\frac{24}{7}S_d\left(\frac{D\bar{\bar{e}}^*}{Dt} \bullet \bar{\bar{e}}^*\right) \tag{7.16}$$

The symbol S_d refers to the symmetric and traceless part of the indicated tensor, and is defined, for an arbitrary tensor $\bar{\bar{B}}$, as:

$$S_d\left(\bar{\bar{B}}\right) = \frac{1}{2}\left[\bar{\bar{B}} + \left(\bar{\bar{B}}\right)^T - \frac{2}{3}\left(tr\,\bar{\bar{B}}\right)\bar{\bar{\delta}}\right] \tag{7.17}$$

The time derivative, $D\bar{\bar{e}}^*/Dt$, is the Jaumann derivative of $\bar{\bar{e}}^*$ defined as:

$$\frac{D\bar{\bar{e}}^*}{Dt} = \frac{\partial \bar{\bar{e}}^*}{\partial t} + \bar{u} \bullet \nabla\bar{\bar{e}}^* + \bar{\bar{e}}^* \bullet \bar{\bar{\Omega}} - \bar{\bar{\Omega}} \bullet \bar{\bar{e}}^* \tag{7.18}$$

where $\bar{\bar{\Omega}}$ is the vorticity tensor for the undisturbed flow.

Using a method of successive approximations, Goddard and Miller [27,31] solved Equation 7.16 to eliminate the product between $\bar{\bar{e}}^*$ and $D\bar{\bar{e}}^*/Dt$, and obtained:

$$\frac{D\bar{\bar{e}}^*}{Dt} = \frac{5}{3}\bar{\bar{E}} - \left(\frac{2G_p}{3\eta_c}\right)\bar{\bar{e}}^* - \frac{30}{7}S_d(\bar{\bar{E}} \bullet \bar{\bar{e}}^*)$$

$$+\frac{16}{7}\left(\frac{G_p}{\eta_c}\right)S_d(\bar{\bar{e}}^* \bullet \bar{\bar{e}}^*) \tag{7.19}$$

The constitutive equation (Equation 7.15) can now be recast into:

$$\bar{\bar{\sigma}} = -P\bar{\bar{\delta}} + 2\eta_c\left(1 - \frac{5}{3}\phi\right)\bar{\bar{E}} + 5\eta_c\phi\left[\begin{array}{c}\left[\left(\dfrac{2G_p}{3\eta_c}\right)\bar{\bar{e}}^* + \left(\dfrac{8}{21}\right)S_d(\bar{\bar{E}} \bullet \bar{\bar{e}}^*)\right] \\[2ex] -\dfrac{8}{21}\left(\dfrac{G_p}{\eta_c}\right)S_d(\bar{\bar{e}}^* \bullet \bar{\bar{e}}^*)\end{array}\right] \tag{7.20}$$

For weakly time-dependent flows and provided that $(3\,\eta_c / 2\,G_p)$ is small compared to unity, Equation 7.20 leads to the following constitutive equation [31]:

$$\bar{\bar{\sigma}} = -P\bar{\bar{\delta}} + 2\eta_c\left(1 + \frac{5}{2}\phi\right)\bar{\bar{E}}$$

$$+ 25\eta_c\phi\left[-\left(\frac{\eta_c}{2G_p}\right)\frac{D\bar{\bar{E}}}{Dt} + \left(\frac{3\eta_c}{7G_p}\right)S_d(\bar{\bar{E}} \bullet \bar{\bar{E}})\right] \tag{7.21}$$

This equation could also be rewritten in the form:

$$\left[1 + \left(\frac{3\,\eta_c}{2\,G_p}\right)\frac{D}{Dt}\right]\bar{\bar{\tau}} = 2\,\eta_c\left[\begin{array}{c}\left[\left(1 + \dfrac{5}{2}\phi\right)\bar{\bar{E}} + \left(1 - \dfrac{5}{3}\phi\right)\left(\dfrac{3\eta_c}{2G_p}\right)\dfrac{D\bar{\bar{E}}}{Dt} + \right. \\[2ex] \left. \dfrac{25}{7}\phi\left(\dfrac{3\eta_c}{2G_p}\right)\left\{(\bar{\bar{E}} \bullet \bar{\bar{E}}) - \dfrac{1}{3}tr(\bar{\bar{E}} \bullet \bar{\bar{E}})\bar{\bar{\delta}}\right\}\right]\end{array}\right] \tag{7.22}$$

Note that Equation 7.22 is an approximate version of Equation 7.21, and is obtained by operating Equation 7.21 with $\left(1 + \dfrac{3\,\eta_c}{2\,G_p}\dfrac{D}{Dt}\right)$ and neglecting certain terms.

When a dispersion of soft elastic particles is subjected to steady shearing flow, Equation 7.22 predicts the presence of both shear and normal stresses. It can be shown that for steady shearing flow, Equation 7.22 gives the following expressions for the shear viscosity and normal stresses:

$$\eta = \frac{\eta_c}{1 + \tau_1^2\dot{\gamma}^2}\left[\left(1 + \frac{5}{2}\phi\right) + \tau_1^2\dot{\gamma}^2\left(1 - \frac{5}{3}\phi\right)\right] \tag{7.23}$$

$$N_1 = \frac{(25/3)\eta_c\tau_1\dot{\gamma}^2}{1 + \tau_1^2\dot{\gamma}^2}\phi \tag{7.24}$$

$$N_2 = -\frac{\eta_c \tau_1 \dot{\gamma}^2}{1+\tau_1^2 \dot{\gamma}^2} \phi \left[\frac{50}{21} - \left(\frac{25}{14}\right) \tau_1^2 \dot{\gamma}^2 \right] \qquad (7.25)$$

where η is the shear viscosity, $\dot{\gamma}$ is the shear rate, N_1 is the first normal stress difference, N_2 is the second normal stress difference, and τ_1 is the characteristic time constant for the suspension given as $3\eta_c/2G_p$. Equation 7.23 to Equation 7.25 can be rewritten in terms of relative or reduced quantities as:

$$\eta_r = \frac{\eta}{\eta_c} = 1 + 2.5\phi \left[\frac{1-(3/2)N_{se}^2}{1+(3N_{se}/2)^2} \right] \qquad (7.26)$$

$$N_{1r} = \frac{N_1}{\eta_c \dot{\gamma}} = \frac{25}{2} \phi \left[\frac{N_{se}}{1+(3N_{se}/2)^2} \right] \qquad (7.27)$$

$$N_{2r} = \frac{N_2}{\eta_c \dot{\gamma}} = -\frac{25}{7} \phi N_{se} \left[\frac{1-(27/16)N_{se}^2}{1+(3N_{se}/2)^2} \right] \qquad (7.28)$$

where N_{se} is a dimensionless group defined as:

$$N_{se} = \frac{\eta_c \dot{\gamma}}{G_p} \qquad (7.29)$$

Figure 7.1 shows the plot of intrinsic viscosity $[\eta]$, defined as $(\eta_r - 1)/\phi$, as a function of N_{se}. Clearly the suspension of soft elastic particles exhibits shear thinning. Figure 7.2 shows the plots of reduced normal stress differences (N_{1r} and N_{2r}) as functions of N_{se}. The reduced first normal stress difference N_{1r} rises linearly (on a log-log scale) with N_{se} up to a certain value of N_{se}. With further increase in N_{se}, N_{1r} decreases. The reduced second normal stress difference $(-N_{2r})$ exhibits a similar behavior except that it undergoes an abrupt decrease after a certain value of N_{se}.

In uniaxial elongational flow of a suspension of soft elastic particles, the constitutive equation (Equation 7.22) predicts strain-thickening behavior in that the elongation viscosity increases with the increase in elongation rate. The elongation viscosity η_E predicted from Equation 7.22 is as follows:

$$\eta_E = 3\eta_c \left[1 + \frac{5}{2}\phi + \frac{25}{14}\tau_1 \dot{\gamma}_E \phi \right] \qquad (7.30)$$

where $\dot{\gamma}_E$ is the rate of elongation.

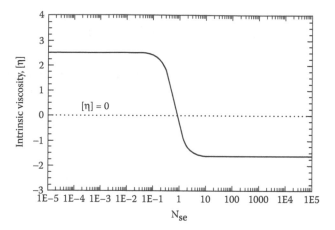

FIGURE 7.1 Variation of intrinsic viscosity with dimensionless group N_{se}.

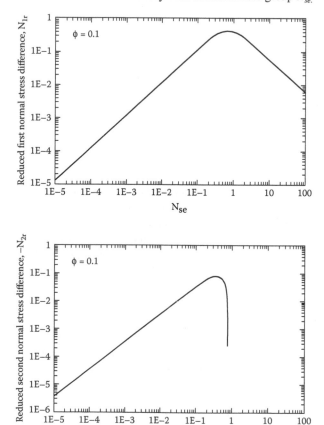

FIGURE 7.2 Variations of reduced normal stress differences (N_{1r} and $-N_{2r}$) with the dimensionless group N_{se}.

7.4 RHEOLOGY OF DISPERSIONS OF SOFT SOLID-LIKE VISCOELASTIC PARTICLES

In simple shear flow, Roscoe's analysis [28] gives the following expressions for the intrinsic viscosity and normal stress differences, provided that the deformation of the initially spherical particles of viscoelastic material is small:

$$[\eta] = 2.5 \left[1 - \frac{5}{3} \frac{\tau_1 \tau_2 \dot{\gamma}^2}{1 + \tau_2^2 \dot{\gamma}^2} \right] \tag{7.31}$$

$$\frac{N_1}{G_p \phi} = \frac{(50/9)\tau_1^2 \dot{\gamma}^2}{1 + \tau_2^2 \dot{\gamma}^2} \tag{7.32}$$

$$\frac{N_1 + 2N_2}{G_p \phi} = \frac{(50/21)\tau_1^2 \dot{\gamma}^2}{1 + \tau_2^2 \dot{\gamma}^2} \tag{7.33}$$

where $\tau_1 = 3\eta_c/2G_p$ and $\tau_2 = (3\eta_c + 2\eta_p)/2G_p$. Thus, a suspension of viscoelastic solid-like particles exhibits shear thinning and normal stress effects. If η_p is taken to be zero, Equation 7.31 to Equation 7.33 reduce to the corresponding expressions obtained earlier using Goddard and Miller's analysis [27].

In uniaxial elongational flow of a suspension, the spherical particles suffer a static deformation (no rotation of particle material) into spheroids provided that the rate of elongation $\dot{\gamma}_E$ is sufficiently small. For small values of $\dot{\gamma}_E$, Roscoe's analysis [28] gives the following expression for elongational viscosity:

$$\eta_E = 3\eta_c \left[1 + \frac{5}{2}\phi + \frac{25}{14}\phi\tau_1\dot{\gamma}_E \right] \tag{7.34}$$

This expression (the same as Equation 7.30) indicates that the suspension exhibits strain-thickening behavior.

NOTATION

$\overline{\overline{B}}$	arbitrary tensor of order 2
D/Dt	Jaumann derivative
$\overline{\overline{E}}$	rate of strain tensor
$\overline{\overline{E}}^*$	rate of strain tensor inside the particle
$\overline{\overline{e}}^*$	finite strain tensor inside the particle, Equation 7.10
G_p	elastic modulus of particle material
N_1	first normal stress difference

N_2	second normal stress difference
N_{1r}	reduced first normal stress difference
N_{2r}	reduced second normal stress difference
N_{se}	dimensionless group defined by Equation 7.29
$\hat{n}$	outward unit normal vector
P	pressure
P^*	mean pressure inside the particle
r	distance from the particle center of mass to the surface of the particle
$\vec{r}$	position vector
Sd	symmetric and traceless part of a tensor
t	time
tr	trace of a tensor
$\vec{u}$	velocity vector
$\vec{u}^*$	velocity vector inside the particle
$\vec{u}_\infty$	velocity vector far away from the particle
$\vec{x}$	current position of a material point in a deformed configuration
$\vec{X}$	initial position of a material point in a reference (unstressed) configuration
x_1, x_2, x_3	position coordinates of the surface of a particle

GREEK

$\dot{\gamma}$	shear rate
$\dot{\gamma}_E$	rate of elongation
∇	del operator
$\bar{\bar{\delta}}$	unit tensor
$\bar{\bar{\Gamma}}_\infty$	velocity gradient tensor far away from the particle
η_c	continuous-phase shear viscosity
η_E	elongation viscosity
η_p	viscosity of particle material
η_r	relative viscosity
$[\eta]$	intrinsic viscosity
$\bar{\bar{\sigma}}$	total stress tensor
$\bar{\bar{\sigma}}^*$	total stress tensor inside the particle
τ_1, τ_2	time constants
$\bar{\bar{\tau}}$	deviatoric stress tensor

$\overline{\overline{\tau}}^*$ deviatoric stress tensor inside the particle

ϕ volume fraction of particles

$\overline{\overline{\Omega}}$ vorticity tensor

REFERENCES

1. S. Lagarrique, G. Alvarez, *J. Food Eng.*, 50: 189–202, 2001.
2. M.A. Rao, T. Tattiyakol, *Carbohydrate Polym.*, 38: 123–132, 1999.
3. J.B. Gluck-Hirsch, J.L. Kokini, *J. Rheol.*, 41: 129–139, 1997.
4. W.J. Firth, A. Lips, *Adv. Colloid Interface Sci.*, 61: 161–189, 1995.
5. I.D. Evans, A. Lips, *J. Texture Stud.*, 23: 69–86, 1992.
6. L.M. Hansen, R.C. Hoseney, J.M. Faubion, *Cereal Chem.*, 68: 347–351, 1991.
7. I.D. Evans, D.R. Haisman, E.L. Elson, C. Pasternak, W.B. McConnaughey, *J. Sci. Food Agric.*, 37: 573–590, 1986.
8. A.C. Eliasoon, *J. Texture Stud.*, 17: 253–265, 1986.
9. L.L. Nawickis, E.B. Bagley, *J. Rheol.*, 27: 519–536, 1983.
10. D.D. Christianson, E.B. Bagley, *Cereal Chem.*, 60: 116–121, 1983.
11. E.B. Bagley, D.D. Christianson, A.C. Beckwith, *J. Rheol.*, 27: 503–507, 1983.
12. E.B. Bagley, D.D. Christianson, *J. Texture Stud.*, 13: 115–126, 1982.
13. I.D. Evans, D.R. Haisman, *J. Texture Stud.*, 10: 347–370, 1979.
14. K. Mequanint, R.D. Sanderson, *Macromol. Symp.*, 178: 117–130, 2002.
15. H. Senff, W. Richtering, *Colloid Polym. Sci.*, 278: 830–840, 2000.
16. B.R. Saunders, B. Vincent, *Adv. Colloid Interface Sci.*, 80: 1–25, 1999.
17. H. Senff, W. Richtering, *J. Chem. Phys.*, 111: 1705–1711, 1999.
18. A. Bischoff, M. Kluppel, R.M. Schuster, *Polym. Bull.*, 40: 283–290. 1998.
19. S. Fridrikh, C. Raquois, J.F. Tassin, S. Rezaiguia, *J. Chim. Phys.*, 93: 941–959. 1996.
20. C. Raquois, J.F. Tassin, S. Rezaiguia, A.V. Gindre, *Prog. Org. Coatings*, 26: 239–250, 1995.
21. I.D. Evans, A. Lips, *J. Chem. Soc., Faraday Trans.*, 86: 3413–3417, 1990.
22. M.S. Wolfe, C. Scopazzi, *J. Colloid Interface Sci.*, 133: 265–277, 1989.
23. R.J. Ketz, R.K. Prud'homme, W.W. Graessley, *Rheol. Acta*, 27: 531–539, 1988.
24. R. Pal, in *Encyclopedia of Emulsion Technology*, Vol. 4, P. Becher, Ed., Marcel Dekker, New York, 1996.
25. P.R. Hornsby, *Adv. Polym. Sci.*, 139: 155–217, 1999.
26. J. Jancar, *Adv. Polym. Sci.*, 139: 1–65, 1999.
27. J.D. Goddard, C. Miller, *J. Fluid Mech.*, 28: 657–673, 1967.
28. R. Roscoe, *J. Fluid Mech.*, 28: 273–293, 1967.
29. H. Brenner, *Prog. Heat Mass Transfer*, 5: 89–129, 1972.
30. D. Barthes-Biesel, in *Rheological Measurement*, A.A. Collyer, D.W. Clegg, Eds., Elsevier, London, 1988.
31. D. Barthes-Biesel, A. Acrivos, *Int. J. Multiphase Flow*, 1: 1–24, 1973.

8 Dispersions of Liquid Droplets

8.1 INTRODUCTION

Dispersions consisting of droplets of one liquid distributed throughout another immiscible liquid (continuous phase) are termed *emulsions*. Emulsions can be classified into two broad groups: (1) water-in-oil (designated as W/O) and (2) oil-in-water (designated as O/W). W/O emulsions consist of water droplets dispersed in a continuum of oil phase, whereas the reverse is true of O/W emulsions; that is, oil droplets are dispersed in a continuum of water phase. In the petroleum industry, W/O emulsions are more commonly encountered and, therefore, O/W emulsions are usually called *reverse emulsions*, being the reverse of the better-known W/O types.

A significant portion of the world's crude oil is produced in the form of emulsions [1–7]. Large quantities of emulsion-based fluids are also used for oil-well drilling [8], as well as for fracturing and acidizing purposes [9]. In some situations where the crude oil or W/O emulsion being produced from a given oil well is highly viscous, an aqueous solution of surface-active chemicals is often injected into the production well bore to convert the high-viscosity oil (or W/O emulsion) to a low-viscosity O/W emulsion [10–15].

Another important application of emulsions in the petroleum industry involves the transportation of highly viscous crude oils in the form of O/W emulsions [14,16–21]. To transport the highly viscous crude oils (such as bitumen and heavy oils) via pipelines, it is necessary to either heat the oil to reduce its viscosity or dilute the oil with a low-viscosity hydrocarbon diluent. In the former case, heating and pipeline insulation costs are usually very high. In the latter case, the cost of diluent and its availability pose difficult problems. These problems, however, can be avoided if the crude oil is transported in the form of an O/W emulsion. The viscosity of O/W emulsions is, in general, very low compared to crude oil [21].

Interest has also been shown in the use of high-viscosity emulsions to displace oil in the secondary oil recovery processes [22,23]. Emulsions are also being considered as blocking agents to control the permeability of the formation [24].

The applications of emulsions in industries other than petroleum are numerous. Many products of commercial importance are sold in emulsion form. Several books have been written describing the commercial applications of emulsion systems [25–28]. The nonpetroleum industries in which emulsions are of considerable importance include food, medical and pharmaceutical, cosmetics, agriculture, explosives, polish, leather, textile, bitumen, paints, lubricants, polymer, and transport [25–46].

Examples of food emulsions are milk [30,31,33,35], mayonnaise [29,35], salad dressings [29,35], cake batters [34], meat emulsions [32], butter [35], and margarine [29,35]. An example of a medicinal emulsion is a mineral-O/W emulsion used for the treatment of constipation [36]; the saturated hydrocarbons in the oil are known to modify the consistency of the stool. Emulsions are also used as a liquid food replacement for patients who cannot ingest solid food [26,29]. Another important medicinal application of emulsions involves injection of a therapeutic agent in emulsion form [27,29]. Many of the antibacterial and antiallergic creams are also emulsions [36]. For example, antihistamines are widely used in O/W emulsion form for local application to the skin where allergic reactions are involved. The antihistamines are generally dissolved in the aqueous phase. A large number of cosmetics (foundation creams, vanishing creams, sunscreen lotions and creams, hair creams, shaving creams, hand lotions and creams, cold creams, etc.) are sold in emulsion form [38,39]. Both O/W and W/O emulsions are common, although a majority of the cosmetic emulsions marketed today are O/W emulsions.

Emulsions are often used as a vehicle for the application of insecticides to plants and animals [29,40,41]. In the explosives industry [42], certain explosives are used in emulsion form; emulsion explosives are basically highly concentrated (>90%) W/O emulsions in which the dispersed phase is a concentrated solution of nitrate salts (mixtures of ammonium, sodium, and calcium nitrates) and the continuous phase is a mixture of wax and oil [42]. Emulsions also play an important role in the leather industry. Fat liquors employed in the treatment of leather (to lubricate the fibers) are O/W emulsions [29]. In the textile industry, antistatic and waterproofing agents are often used in emulsion form [29]. Also, the lubricants that reduce friction between the fibers and solid surfaces are used in emulsion form [29].

Emulsions find applications in the paints industry [29,44,45] as well. Emulsion paints have several important advantages over solvent-based paints, such as faster drying rates and little tendency to become brittle with aging. Lubricants for drilling and cutting tools are often used in emulsion form [46]. For example, cutting oil emulsions are O/W type. Emulsions of molten bitumen and water are widely used for the construction of roads and highways [29,43]. Emulsions also play a major role in emulsion polymerization [29]. The major advantages of emulsion polymerization over simple polymerization are as follows: (1) the viscosity of the reaction system is easily controllable, (2) heat transfer (cooling) is more efficient, and (3) mixing is more efficient. Emulsions find important applications in the transport industry [29] as well. For example, hazardous materials could be transported in emulsion form. Liquid explosives such as nitroglycerine can be transported from one location to another in the form of an O/W emulsion. Similarly, alkali metals at high temperatures can be handled in the form of metal-in-hydrocarbon emulsions [29].

In the handling, mixing, storage, and pipeline transportation of emulsions, knowledge of the rheological properties is required for the design, selection, and operation of the equipment involved.

8.2 DEFORMATION AND BREAKUP OF DROPLETS

The rheological behavior of emulsions is strongly influenced by the deformation and orientation of droplets caused by the flow field. When a droplet of one fluid is suspended in a continuum of a second fluid that is made to shear, the droplet will deform [47–65]. The equilibrium shape of the droplet, as well as its orientation with respect to the flow direction, is governed by two dimensionless parameters: (1) viscosity ratio λ, defined as the ratio of droplet fluid viscosity (η_d) to continuous-phase viscosity (η_c), and (2) capillary number N_{Ca}, defined as:

$$N_{Ca} = \frac{\eta_c \dot{\gamma} R}{\gamma} \qquad (8.1)$$

where $\dot{\gamma}$ is the shear rate, R is the radius of the undeformed droplet, and γ is the interfacial tension between the two fluids.

The hydrodynamic stress ($\sim \eta_c \dot{\gamma}$) tends to stretch the droplet out into a filamentary shape, whereas the interfacial stress ($\sim \gamma/R$) tends to contract the droplet into a sphere of radius R. The capillary number represents the ratio of these two competing stresses (that is, the ratio of hydrodynamic stress to interfacial stress). For small deformations, the droplet deforms into a triaxial ellipsoid [50,62–65] possessing semiaxes of lengths r_{max}, r_z, and r_{min} (Figure 8.1). The deformation of the droplet is often characterized by a dimensionless deformation parameter D, defined as:

$$D = \frac{r_{max} - r_{min}}{r_{max} + r_{min}} \qquad (8.2)$$

As the capillary number increases, the droplet becomes more elongated (D increases) and the angle θ between the major axis of the droplet and the flow direction decreases toward zero. The elongation and orientation of droplets results in shear-thinning behavior of emulsions. Emulsions also exhibit other non-Newtonian characteristics [66–72], such as (1) normal stresses in shear flow and (2) strain thickening in extensional flows. The "memory" or elastic effects (such as normal stress differences) in emulsions is generally attributable to interfacial tension; this tension tends to restore the original spherical shape that is altered by the hydrodynamic stress. Thus, the original spherical shape of the droplets serves as a "memory" for the emulsion.

When the capillary number exceeds a certain critical value ($N_{Ca,crit}$), the droplet becomes unstable and no equilibrium shape is possible. After some

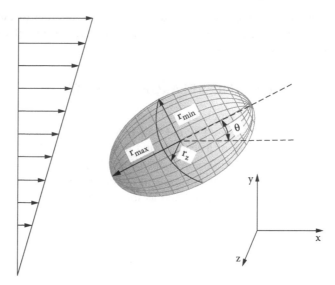

FIGURE 8.1 Deformation of a droplet suspended in simple shear flow. (Adapted from S. Guido, M. Villone, Three-dimensional shape of a drop under simple shear flow, *J. Rheol.*, 42: 395–415, 1998.)

transient elongation, the droplet undergoes breakup. The droplets are known to breakup usually by two different modes, shown in Figure 8.2. If $N_{Ca} > N_{Ca,crit}$, a simple breakup of an extended droplet into two droplets occurs. If $N_{Ca} \gg N_{Ca,crit}$, Rayleigh breakup of a highly extended filament into a large number of droplets occurs [73,74]. For a given type of flow field, the critical capillary number depends on the viscosity ratio λ [56]. Grace [56] conducted a thorough experimental investigation of droplet breakup over a wide range of λ. Figure 8.3 shows the variation of $N_{Ca,crit}$ with λ for shear and elongation flows.

In shear flow, breakup of droplets is easy if λ is in the range of 0.1 to 1.0. The critical capillary number tends to diverge above a λ value of about 4. Therefore, breakup of droplets with $\lambda > 4$ is difficult in shear flow. If λ is very small ($\lambda < 0.01$), extreme deformation of the droplet is possible without breakup in shear flow as $N_{Ca,crit}$ is large. For $\lambda < 0.01$, the experimental data generally fit the breakup criterion for low-viscosity droplets developed by Hinch and Acrivos [55]:

$$N_{Ca,crit} = 0.054\lambda^{-2/3} \tag{8.3}$$

For $10^{-3} < \lambda < 10$, the experimental data of $N_{Ca,crit}$ in shear flow can be fitted by the following empirical equation [75]:

$$\log N_{Ca,crit} = a + b \log \lambda + c(\log \lambda)^2 + \frac{d}{\log \lambda - \log \lambda_c} \tag{8.4}$$

where $a = -0.506$, $b = -0.0994$, $c = 0.124$, $d = -0.115$, and $\lambda_c = 4.08$.

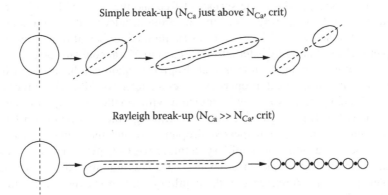

FIGURE 8.2 Deformation and breakup of droplets in simple shear. (Adapted from E.H. Lucassen-Reynders, Dynamic interfacial properties in emulsification, in *Encyclopedia of Emulsion Technology*, Vol. 4, P. Becher, Ed., Marcel Dekker, New York, 1996, chap. 2.)

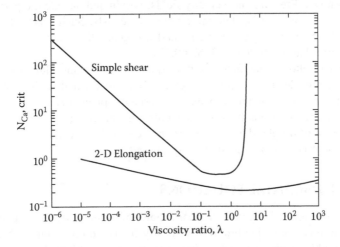

FIGURE 8.3 Variation of critical capillary number for droplet breakup with the viscosity ratio λ in simple shear and elongational flow fields. (Adapted from H.P. Grace, Dispersion phenomena in high viscosity immiscible fluid systems and application of static mixers as dispersion devices in such systems, *Chem. Eng. Commun.*, 14: 225–277, 1982.)

If the applied flow field is elongational, a different picture is observed. In elongational flows, the critical capillary number is generally small (Figure 8.3) and, therefore, droplet breakup is easier.

The preceding discussion on droplet deformation and breakup ignores the effects of surfactants. In the discussion, it is assumed that (1) the interfacial tension is constant and (2) the interfacial elasticity and viscosity are negligible so that the interface transmits tangential stresses from external fluid to internal fluid undiminished.

Droplets in the presence of surfactant molecules at their surface behave differently from pure droplets without surfactant [60,74,76]. Two droplets with the same λ and N_{Ca} values, one surfactant-free and the other with surfactant, do not deform in the same way. Thus, the lowering of interfacial tension is not the only effect of surfactant. The uneven distribution of surfactant molecules on the droplet surface, caused by external flow, leads to Marangoni stresses.

An important point to note is that droplet breakup can have a significant effect on the rheological properties of emulsions [68,70]. The process of droplet breakup creates a "new" emulsion with smaller average droplet size as compared to the "original" emulsion. Although the new emulsion has the same volume fraction of dispersed droplets (ϕ) and the same λ as that of the original emulsion, the capillary number $N_{Ca}\,(=\eta_c\dot{\gamma}R/\gamma)$ is now smaller because the average droplet size has decreased (assuming that coalescence of droplets is absent). A decrease in the capillary number is expected to alter the rheological properties.

It is difficult to model the behavior of deformable particles. One major source of difficulty lies in the satisfaction of the boundary conditions (continuity of velocities and force balance) on the surface of the deforming particle, which is itself unknown. Pressure and velocity fields, within and outside the particle, and particle shape must all be determined simultaneously. Such types of problems are called *free-surface-type* problems and are generally treated by means of a domain perturbation method [63–65,69,77].

In what follows, the fundamental equations governing fluid motion both inside and outside the droplets are given. It is assumed that the droplets are surfactant-free and that breakup and coalescence phenomena are absent. The rheological constitutive equations for emulsions (surfactant-free) in the absence and presence of droplet deformation are presented. Finally, the effect of surfactants on the rheology of emulsions is discussed.

8.3 FUNDAMENTAL EQUATIONS

Consider a neutrally buoyant spherical droplet of radius R freely suspended in a Newtonian incompressible liquid of viscosity η_c. The droplet fluid is Newtonian incompressible of viscosity $\lambda\eta_c$. When the system is subjected to a shear flow, the droplet deforms under the influence of the viscous stresses. In a set of axes fixed at the center of the droplet, the equation of the droplet surface at any instant of time t can be written as:

$$r=\left(x_1^2+x_2^2+x_3^2\right)^{1/2}=f\left(x_1,x_2,x_3\right) \tag{8.5}$$

where r is the distance from the droplet center of mass to the interface. The function f is *a priori* unknown and must be determined as part of the problem solution. Assuming that the droplet's Reynolds number in each phase is small, the fluid motion both inside and outside the droplet satisfies the creeping motion and continuity equations. Thus, with respect to a reference frame centered on the droplet's center of mass and translating with the droplet, the system can be described by:

$$\eta_c \nabla^2 \vec{u} - \nabla P = 0 \qquad \nabla \bullet \vec{u} = 0 \quad \text{for} \quad r > f \qquad (8.6)$$

$$\lambda \eta_c \nabla^2 \vec{u}^* - \nabla P^* = 0 \qquad \nabla \bullet \vec{u}^* = 0 \quad \text{for} \quad r < f \qquad (8.7)$$

with boundary conditions assigned at infinity and at the interface.

The boundary condition at infinity (far away from the droplet) is:

$$\vec{u} \to \vec{u}_\infty \quad \text{as} \quad r \to \infty \qquad (8.8)$$

The boundary conditions at the droplet surface $(r = f)$ are:

(1) Continuity of tangential velocities at the interface

$$(\vec{u} - \vec{u} \bullet \hat{n}\hat{n}) = (\vec{u}^* - \vec{u}^* \bullet \hat{n}\hat{n}) \qquad (8.9)$$

(2) Continuity of normal velocities at the interface

$$\vec{u} \bullet \hat{n} = \vec{u}^* \bullet \hat{n} = u_{\text{interface}} \qquad (8.10)$$

where $u_{\text{interface}}$ is the velocity of the interface.

(3) Continuity of tangential stresses at the interface

$$(\bar{\bar{\sigma}} \bullet \hat{n} - \hat{n} \bullet \bar{\bar{\sigma}} \bullet \hat{n}\hat{n}) = (\bar{\bar{\sigma}}^* \bullet \hat{n} - \hat{n} \bullet \bar{\bar{\sigma}}^* \bullet \hat{n}\hat{n}) \qquad (8.11)$$

(4) Normal stress discontinuity across the interface

$$\hat{n} \bullet (\bar{\bar{\sigma}} - \bar{\bar{\sigma}}^*) \bullet \hat{n} = -(\nabla_s \bullet \bar{\bar{P}}_s) \bullet \hat{n} \qquad (8.12)$$

where ∇_s is the surface gradient operator defined as $(\bar{\bar{\delta}} - \hat{n}\hat{n}) \bullet \nabla$ and $\bar{\bar{P}}_s$ is the surface-stress tensor given as:

$$\bar{\bar{P}}_s = \gamma(\bar{\bar{\delta}} - \hat{n}\hat{n}) \qquad (8.13)$$

The solution of Equation 8.6 and Equation 8.7 subject to the boundary conditions Equation 8.8 to Equation 8.12 gives the instantaneous velocity and pressure distributions both inside and outside the droplet. Although the problem is simple in principle, it is very difficult to solve because the boundary conditions need to be satisfied on the surface of the deforming drop. Analytical results have been obtained only under the condition of small drop deformation. The "domain perturbation" technique is utilized to solve the problem. This is an iterative technique. The first iteration deals with the calculation of the velocity and pressure distribution everywhere using the continuity of velocity and continuity of tangential stress boundary conditions at the undeformed, spherical interface. This gives zero-order deformation solution of the flow fields. The normal stress boundary condition is then used to determine the deviation of the interface from sphericity. This gives the first-order deformation of the interface. In the next iteration, the flow fields are determined by satisfying the boundary conditions at the deformed interface calculated from the first iteration.

8.4 RHEOLOGY OF EMULSIONS

The development of the rheological constitutive equation for emulsions requires knowledge of the flow fields (velocity, pressure, and stress) and shape of the droplet.

The constitutive equation for a dilute emulsion of identical droplets is given by:

$$\langle \bar{\bar{\sigma}} \rangle = -\langle P \rangle \bar{\bar{\delta}} + 2\eta_c \langle \bar{\bar{E}} \rangle + \frac{3\phi}{4\pi R^3} \bar{\bar{S}}^o \tag{8.14}$$

where $\langle \bar{\bar{\sigma}} \rangle$ is the bulk (average) stress tensor; $\langle P \rangle$ is the average pressure; $\bar{\bar{\delta}}$ is the unit tensor; η_c is the matrix viscosity; $\langle \bar{\bar{E}} \rangle$ is the bulk rate of strain tensor; ϕ is the volume fraction of droplets; R is the droplet radius; and $\bar{\bar{S}}^o$ is the dipole strength of a single droplet located in an infinite matrix fluid. $\bar{\bar{S}}^o$ is given by:

$$\bar{\bar{S}}^o = \int_{A_0} \left[\bar{\bar{\sigma}} \bullet \hat{n}\, \vec{x} - \frac{1}{3} (\vec{x} \bullet \bar{\bar{\sigma}} \bullet \hat{n}) \bar{\bar{\delta}} - \eta_c (\hat{n}\, \vec{u} + \vec{u}\, \hat{n}) \right] dA \tag{8.15}$$

where A_0 is the surface area of the droplet; $\hat{n}$ is a unit outward normal to the droplet surface; $\vec{u}$ is fluid velocity at the droplet surface; and $\bar{\bar{\sigma}}$ is the fluid stress tensor at the droplet surface.

To evaluate $\bar{\bar{S}}^o$, the velocity and stress distribution around the droplet is needed. The velocity and stress distribution around the droplet can be determined by solving Equation 8.6 and Equation 8.7 subject to the boundary conditions in Equation 8.8 to Equation 8.12.

8.4.1 ZERO-ORDER DEFORMATION SOLUTION

The zero-order deformation solution of Equation 8.6 and Equation 8.7 is obtained by assuming that the droplet remains spherical in shape. Except for the normal-stress-discontinuity boundary condition (Equation 8.12), all of the velocity and stress boundary conditions (Equation 8.8 to Equation 8.11) are satisfied. The interface velocity $u_{interface}$ is taken to be zero.

The velocity in the suspending fluid (matrix) surrounding the droplet turns out to be [78–80]:

$$\vec{u}(\vec{r}) = \overline{\overline{E}}_\infty \bullet \vec{r} + \vec{\omega}_o \times \vec{r} - \left(\frac{\lambda}{1+\lambda}\right)\left(\frac{R}{r}\right)^5 (\overline{\overline{E}}_\infty \bullet \vec{r})$$

$$+ \frac{\vec{r}(\vec{r} \bullet \overline{\overline{E}}_\infty \bullet \vec{r})}{r^2}\left[-\left(\frac{2+5\lambda}{2+2\lambda}\right)\left(\frac{R}{r}\right)^3 + \left(\frac{5\lambda}{2+2\lambda}\right)\left(\frac{R}{r}\right)^5\right]$$

(8.16)

where $\vec{r}$ is a vector to the point under consideration from the center of mass of the droplet, $\overline{\overline{E}}_\infty$ is the rate of strain tensor of the undisturbed flow, and $\vec{\omega}_o$ is half the vorticity of the undisturbed flow. Note that the first two terms on the right-hand side of Equation 8.16 represent undisturbed external flow with respect to a reference frame centered on the droplet's center of mass, and the remaining terms represent the disturbance motion caused by the droplet. As $r \to \infty$, the terms representing the disturbance motion decay to zero. The pressure distribution in the matrix surrounding the droplet is:

$$P = P_\infty - \left(\frac{2+5\lambda}{1+\lambda}\right)\eta_c\left(\frac{R}{r}\right)^3\left(\frac{\vec{r} \bullet \overline{\overline{E}}_\infty \bullet \vec{r}}{r^2}\right)$$

(8.17)

The local rate of strain tensor $\overline{\overline{E}}$ in the matrix fluid is as follows [78]:

$$\overline{\overline{E}}(\vec{r}) = \overline{\overline{E}}_\infty\left[1 - \left(\frac{\lambda}{1+\lambda}\right)\left(\frac{R}{r}\right)^5\right] + \left[\frac{\vec{r}(\overline{\overline{E}}_\infty \bullet \vec{r}) + (\vec{r} \bullet \overline{\overline{E}}_\infty)\vec{r}}{r^2} - \frac{2}{3}\left(\frac{\vec{r} \bullet \overline{\overline{E}}_\infty \bullet \vec{r}}{r^2}\right)\overline{\overline{\delta}}\right]$$

$$\times \left[\frac{-5}{2}\left(\frac{2+5\lambda}{5+5\lambda}\right)\left(\frac{R}{r}\right)^3 + 5\left(\frac{\lambda}{1+\lambda}\right)\left(\frac{R}{r}\right)^5\right] + \left[\left(\frac{\vec{r} \bullet \overline{\overline{E}}_\infty \bullet \vec{r}}{r^2}\right)\left(\frac{\vec{r}\vec{r}}{r^2} - \frac{1}{3}\overline{\overline{\delta}}\right)\right]$$

(8.18)

$$\times \left[\frac{25}{2}\left(\frac{2+5\lambda}{5+5\lambda}\right)\left(\frac{R}{r}\right)^3 - \frac{35}{2}\left(\frac{\lambda}{1+\lambda}\right)\left(\frac{R}{r}\right)^5\right]$$

The stress field in the matrix fluid can be determined from Equation 8.17 and Equation 8.18 using

$$\overline{\overline{\sigma}} = -P\overline{\overline{\delta}} + 2\eta_c\,\overline{\overline{E}} \tag{8.19}$$

Thus,

$$
\overline{\overline{\sigma}} = -P_\infty\overline{\overline{\delta}} + 2\eta_c\,\overline{\overline{E}}_\infty\left[1-\left(\frac{\lambda}{1+\lambda}\right)\left(\frac{R}{r}\right)^5\right] + 2\eta_c\left[\frac{\vec{r}(\overline{\overline{E}}_\infty\bullet\vec{r})+(\vec{r}\bullet\overline{\overline{E}}_\infty)\vec{r}}{r^2}\right]
$$
$$
\times\left[\frac{-5}{2}\left(\frac{2+5\lambda}{5+5\lambda}\right)\left(\frac{R}{r}\right)^3+5\left(\frac{\lambda}{1+\lambda}\right)\left(\frac{R}{r}\right)^5\right]+5\eta_c\left(\frac{\vec{r}\bullet\overline{\overline{E}}_\infty\bullet\vec{r}}{r^2}\right)\left(\frac{\lambda}{1+\lambda}\right)\left(\frac{R}{r}\right)^5\overline{\overline{\delta}} \tag{8.20}
$$
$$
-35\eta_c\left(\frac{\lambda}{1+\lambda}\right)\left(\frac{R}{r}\right)^5\left(\frac{\vec{r}\vec{r}}{r^2}\right)\left(\frac{\vec{r}\bullet\overline{\overline{E}}_\infty\bullet\vec{r}}{r^2}\right)+25\eta_c\left(\frac{2+5\lambda}{5+5\lambda}\right)\left(\frac{R}{r}\right)^3\left(\frac{\vec{r}\vec{r}}{r^2}\right)\left(\frac{\vec{r}\bullet\overline{\overline{E}}_\infty\bullet\vec{r}}{r^2}\right)
$$

Using the expression for local stress in the ambient fluid (Equation 8.20), the dipole strength $\overline{\overline{S}}^o$ can be evaluated from Equation 8.15. Equation 8.15 gives the following expression for the force dipole strength of a droplet:

$$\overline{\overline{S}}^o = \frac{4}{3}\pi R^3\eta_c\left[\frac{2+5\lambda}{1+\lambda}\right]\overline{\overline{E}}_\infty \tag{8.21}$$

For a dilute dispersion, the rate of strain tensor far away from the particles ($\overline{\overline{E}}_\infty$) can be equated to the bulk or imposed rate of strain tensor $\langle\overline{\overline{E}}\rangle$. Hence, for a dilute emulsion, the rheological constitutive equation (Equation 8.14) can be expressed as:

$$\langle\overline{\overline{\sigma}}\rangle = -\langle P\rangle\overline{\overline{\delta}} + 2\eta_c\left\langle\overline{\overline{E}}\right\rangle + \phi\eta_c\left[\frac{2+5\lambda}{1+\lambda}\right]\langle\overline{\overline{E}}\rangle \tag{8.22}$$

To simplify the notation, the angular brackets $\langle\ \rangle$ are dropped from Equation 8.22 so that:

$$\overline{\overline{\sigma}} = -P\overline{\overline{\delta}} + 2\eta_c\left[1+\frac{2+5\lambda}{2(1+\lambda)}\phi\right]\overline{\overline{E}} \tag{8.23}$$

The constitutive equation for an incompressible Newtonian fluid having a shear viscosity of η is given as:

$$\overline{\overline{\sigma}} = -P\overline{\overline{\delta}} + 2\eta\overline{\overline{E}} \tag{8.24}$$

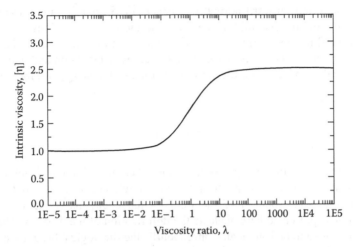

FIGURE 8.4 Variation of intrinsic viscosity of emulsion with viscosity ratio λ, predicted from the Taylor equation.

Clearly, Equation 8.23 is of the form of Equation 8.24, indicating that a dilute dispersion of spherical droplets is a Newtonian fluid of viscosity η given by:

$$\eta = \eta_c \left[1 + \frac{2 + 5\lambda}{2(1 + \lambda)} \phi \right] \tag{8.25}$$

Equation 8.25 is the celebrated Taylor equation [81] for the viscosity of a dilute emulsion of spherical droplets. Figure 8.4 shows the plot of intrinsic viscosity [η], defined as $(\eta - \eta_c)/\phi\, \eta_c$, vs. viscosity ratio λ, generated from the Taylor equation. The intrinsic viscosity has a constant value of unity at low values of λ; it increases with increase in λ for intermediate values of λ, and finally levels off to a value of 2.5.

Interestingly, Equation 8.25 can be derived using the energy dissipation approach [82]. The energy dissipation approach [83] involves comparison of the rates of mechanical energy dissipation in emulsion and in matrix fluid alone without droplets, when they are subjected to identical boundary conditions. If $\dot{Q}$ is the rate of energy dissipation in emulsion and $\dot{Q}_c$ is the rate of energy dissipation in matrix fluid alone (without droplets), under identical boundary conditions, then:

$$\frac{\eta}{\eta_c} = \frac{\dot{Q}}{\dot{Q}_c} = \frac{\displaystyle\int_{A'} \left(\hat{n} \bullet \bar{\bar{\sigma}} \bullet \vec{u} \right) dA}{\displaystyle\int_{A'} \left(\hat{n} \bullet \bar{\bar{\sigma}}_c \bullet \vec{u}_c \right) dA} \tag{8.26}$$

where $\hat{n}$ is a unit outward normal to surface A', $\bar{\bar{\sigma}}$ and $\bar{\bar{\sigma}}_c$ are stress tensors of emulsion and matrix fluid, respectively, $\vec{u}$ and $\vec{u}_c$ are velocities of emulsion and matrix fluid, respectively, and A' is the bounding surface of the region where fluid motion is taking place. The velocities of emulsion and matrix fluid at the boundary A' must be the same to compare the energy dissipation rates, that is,

$$\vec{u} = \vec{u}_c \quad \text{on} \quad A' \tag{8.27}$$

The energy dissipation approach has the inherent limitation of yielding only the viscosity of the emulsion and not the complete constitutive equation.

The zero-order solution of the fundamental equations (Equation 8.6 and Equation 8.7) presented here neglects the deformation of the droplets. The droplets are assumed to be spherical. Consequently, the rheological behavior of emulsion is predicted to be Newtonian (Equation 8.23).

8.4.2 FIRST-ORDER DEFORMATION SOLUTION

The first-order deformation solution is obtained by solving the fundamental equations (Equation 8.6 and Equation 8.7) with the boundary conditions now applied at the deformed interface. As only a small deformation of spherical droplets is considered, the droplet shape can be expressed in terms of surface spherical harmonics, defined as:

$$S_i = r^{i+1} \overbrace{\nabla .....\nabla}^{i \text{ times}} \left(\frac{1}{r}\right) \tag{8.28}$$

where S_i is the ith surface spherical harmonic. For first-order deformation, only the second-order harmonics are sufficient. Thus, the shape of a deformed droplet can be expressed as:

$$\frac{r}{R} = 1 + \varepsilon g \tag{8.29}$$

with

$$g = \bar{\bar{F}}'(t) : S_2 \tag{8.30}$$

where $\bar{\bar{F}}'(t)$ is a dimensionless, symmetric, and traceless tensor describing the instantaneous shape of a droplet, and ε is the deformation parameter, which is smaller than unity ($\varepsilon \ll 1$). As the physical factor that prevents a large deviation of droplet shape from the spherical is either high viscosity ratio λ ($\lambda \gg 1$) or weak flow strength ($N_{Ca} \ll 1$), ε can be set equal to either λ^{-1} when $\lambda \gg 1$ or N_{Ca} when the flow is weak.

The weak shear flow case, where interfacial stress dominates the hydrodynamic stress ($N_{Ca} \ll 1$), has received greater attention in the literature. If ε is taken to be N_{Ca}, Equation 8.29 for the instantaneous shape of the droplet can be written as:

$$\frac{r}{R} = 1 + N_{Ca}\left(\overline{\overline{F}}'(t):S_2\right) \tag{8.31}$$

Schowalter et al. [66] have pointed out that to satisfy the previously neglected normal stress boundary condition (Equation 8.12), $\overline{\overline{F}}':S_2$ at steady state should be set equal to:

$$\overline{\overline{F}}' : S_2 = \left[\frac{19\lambda+16}{8(1+\lambda)}\right]\left(\frac{\overline{\overline{E}}'_\infty : \vec{r}\,\vec{r}}{r^2}\right) \tag{8.32}$$

where $\overline{\overline{E}}'_\infty$ is dimensionless steady rate-of-strain tensor far away from the droplet. From Equation 8.31 and Equation 8.32:

$$\frac{r}{R} = 1 + N_{Ca}\left[\frac{19\lambda+16}{8(1+\lambda)}\right]\left(\frac{\overline{\overline{E}}'_\infty : \vec{r}\,\vec{r}}{r^2}\right) \tag{8.33}$$

Equation 8.33 represents first-order deformation of a droplet, originally obtained by Taylor [47]. Equation 8.33 leads to the following expressions for deformation D and orientation angle θ (Figure 8.1):

$$D = N_{Ca}\left(\frac{19\lambda+16}{16\lambda+16}\right) \tag{8.34a}$$

$$\theta = \frac{\pi}{4} \tag{8.34b}$$

Based on the first-order deformation of droplets, Schowalter et al. [66] have developed a rheological constitutive equation for dilute emulsions. Their analysis leads to the following expression for the force dipole strength $\bar{\bar{S}}^o$ for steady or weakly time-dependent flows:

$$\bar{\bar{S}}^o = \frac{4}{3}\pi R^3 \eta_c$$

$$\times \left\{ \left(\frac{5\lambda+2}{1+\lambda}\right)\bar{\bar{E}}_\infty + \left(\frac{\eta_c R}{\gamma}\right) \left[\begin{array}{c} \frac{3(19\lambda+16)(25\lambda^2+41\lambda+4)}{140(1+\lambda)^3}\left(\bar{\bar{E}}_\infty \bullet \bar{\bar{E}}_\infty - \frac{\bar{\bar{\delta}}}{3}\left(\bar{\bar{E}}_\infty : \bar{\bar{E}}_\infty\right)\right) \\ -\frac{1}{40}\left(\frac{19\lambda+16}{\lambda+1}\right)^2 \frac{D\bar{\bar{E}}_\infty}{Dt} \end{array} \right] \right\}$$

$$(8.35)$$

where $D\bar{\bar{E}}_\infty/Dt$ is the Jaumann time derivative of $\bar{\bar{E}}_\infty$. Hence, the rheological constitutive equation for dilute emulsions incorporating the first-order deformation of droplets can be expressed as:

$$\bar{\bar{\sigma}} = -P\bar{\bar{\delta}} + 2\eta_c \bar{\bar{E}} + \phi\eta_c$$

$$\times \left\{ \left(\frac{5\lambda+2}{1+\lambda}\right)\bar{\bar{E}} + \left(\frac{\eta_c R}{\gamma}\right) \left[\begin{array}{c} \frac{3(19\lambda+16)(25\lambda^2+41\lambda+4)}{140(1+\lambda)^3}\left(\bar{\bar{E}} \bullet \bar{\bar{E}} - \frac{\bar{\bar{\delta}}}{3}\left(\bar{\bar{E}} : \bar{\bar{E}}\right)\right) \\ -\frac{1}{40}\left(\frac{19\lambda+16}{\lambda+1}\right)^2 \frac{D\bar{\bar{E}}}{Dt} \end{array} \right] \right\}$$

$$(8.36)$$

where $\bar{\bar{\sigma}}$, P, and $\bar{\bar{E}}$ are average (bulk) quantities. This equation is valid for steady or weakly time-dependent flows for arbitrary λ (viscosity ratio) but small capillary numbers. Unlike the constitutive equation (Equation 8.23) based on zero-order flow fields (no deformation of droplets), Equation 8.36 predicts emulsion behavior to be non-Newtonian. In a simple steady shear flow, the emulsion exhibits nonzero normal stresses, although the viscosity is predicted to be independent of shear rate [66,84]:

$$\eta = \eta_c\left[1 + \frac{5\lambda+2}{2\lambda+2}\phi\right] \qquad (8.37a)$$

$$N_1 = \tau_{11} - \tau_{22} = \frac{\eta_c^2 \dot{\gamma}^2 R}{40\gamma}\left(\frac{19\lambda+16}{1+\lambda}\right)^2 \phi \qquad (8.37b)$$

$$N_2 = \tau_{22} - \tau_{33} = -\frac{\eta_c^2 \dot{\gamma}^2 R}{280\gamma}\left(\frac{(19\lambda+16)(29\lambda^2+61\lambda+50)}{(1+\lambda)^3}\right)^2 \phi \qquad (8.37c)$$

where N_1 is the first normal stress difference and N_2 is the second normal stress difference, and τ_{11}, τ_{22}, and τ_{33} are normal stresses (here, the subscript 1 denotes the direction of flow, the subscript 2 denotes the direction perpendicular to the flow — the direction pointing to the velocity gradient — and the subscript 3 denotes the neutral direction).

Frankel and Acrivos [67] have presented a more general constitutive equation, with no restriction on time dependence (such as steady or weakly), based on first-order deformation of droplets. Their equation is as follows:

$$
\bar{\bar{\sigma}} = -P\bar{\bar{\delta}} + 2\eta_c\bar{\bar{E}} + \eta_c\phi\left[\frac{10(\lambda-1)}{2\lambda+3}\bar{\bar{E}} + \frac{24}{2\lambda+3}\bar{\bar{F}} \right.
$$
$$
\left. + \frac{360(\lambda-1)^2}{7(2\lambda+3)^2}\left(\frac{R\eta_c}{\gamma}\right)Sd(\bar{\bar{F}}\bullet\bar{\bar{E}}) + \frac{288(\lambda-6)}{7(2\lambda+3)^2}\left(\frac{R\eta_c}{\gamma}\right)Sd(\bar{\bar{F}}\bullet\bar{\bar{F}})\right] \tag{8.38}
$$

where $\bar{\bar{F}}$ is a dimensional form of symmetric and traceless deformation tensor $\bar{\bar{F}}'$ describing the shape of the droplet. The time evolution of the deformation tensor $\bar{\bar{F}}$ is given as:

$$
\bar{\bar{F}} + \left(\frac{R\eta_c}{\gamma}\right)\frac{(2\lambda+3)(19\lambda+16)}{40(\lambda+1)}\frac{D\bar{\bar{F}}}{Dt} = \frac{19\lambda+16}{24(\lambda+1)}\bar{\bar{E}}
$$
$$
+ \frac{(4\lambda-9)(19\lambda+16)}{28(2\lambda+3)(\lambda+1)}\left(\frac{R\eta_c}{\gamma}\right)Sd(\bar{\bar{F}}\bullet\bar{\bar{E}}) \tag{8.39}
$$
$$
+ \frac{36(137\lambda^3+624\lambda^2+741\lambda+248)}{35(2\lambda+3)(19\lambda+16)(\lambda+1)}\left(\frac{R\eta_c}{\gamma}\right)Sd(\bar{\bar{F}}\bullet\bar{\bar{F}})
$$

In Equation 8.38 and Equation 8.39, Sd refers to the symmetric and traceless part of the indicated tensor. For any tensor $\bar{\bar{B}}$, Sd is given as:

$$
Sd\left(\bar{\bar{B}}\right) = \frac{1}{2}\left[\bar{\bar{B}} + \bar{\bar{B}}^T - \frac{2}{3}tr\left(\bar{\bar{B}}\right)\bar{\bar{\delta}}\right] \tag{8.40}
$$

For weakly time-dependent flows, Equation 8.39 can be solved by a method of successive substitutions and the following result is obtained:

$$
\bar{\bar{F}} = \frac{19\lambda+16}{24(1+\lambda)}\left[\bar{\bar{E}} - \frac{(2\lambda+3)(19\lambda+16)}{40(1+\lambda)}\left(\frac{\eta_c R}{\gamma}\right)\frac{D\bar{\bar{E}}}{Dt}\right.
$$
$$
\left. + \frac{1202\lambda^3+3589\lambda^2+3191\lambda+768}{140(2\lambda+3)(\lambda+1)^2}\times\left(\frac{\eta_c R}{\gamma}\right)Sd(\bar{\bar{E}}\bullet\bar{\bar{E}})\right] \tag{8.41}
$$

Using Equation 8.41, the constitutive equation (Equation 8.38) becomes:

$$\bar{\bar{\sigma}} = -P\bar{\bar{\delta}} + 2\eta_c\left[1 + \frac{5\lambda+2}{2(1+\lambda)}\phi\right]\bar{\bar{E}} + \phi\eta_c\left(\frac{\eta_c R}{\gamma}\right)\left[\frac{-(19\lambda+16)^2}{40(1+\lambda)^2}\frac{D\bar{\bar{E}}}{Dt}\right.$$

$$\left. + \frac{3(19\lambda+16)(25\lambda^2+41\lambda+4)}{140(1+\lambda)^3}Sd(\bar{\bar{E}}\bullet\bar{\bar{E}})\right] \tag{8.42}$$

This constitutive equation for weakly time-dependent or steady flows is identical to Equation 8.36 developed by Schowalter et al. [66]. It can also be recast into a different form as:

$$\left[1 + \tau_o\frac{D}{Dt}\right]\bar{\bar{\tau}} = 2\eta_c\left[1 + \frac{5\lambda+2}{2(1+\lambda)}\phi\right]\left[\bar{\bar{E}} + \tau_o\frac{D\bar{\bar{E}}}{Dt}\right]$$

$$+ \phi\eta_c\left(\frac{\eta_c R}{\gamma}\right)\left[-\frac{1}{40}\left(\frac{19\lambda+16}{1+\lambda}\right)^2\frac{D\bar{\bar{E}}}{Dt} + \frac{3(19\lambda+16)(25\lambda^2+41\lambda+4)}{140(1+\lambda)^3}\times Sd(\bar{\bar{E}}\bullet\bar{\bar{E}})\right] \tag{8.43a}$$

where $\bar{\bar{\tau}}$ is the deviatoric stress tensor defined as $(\bar{\bar{\sigma}} + P\bar{\bar{\delta}})$, and τ_o is the characteristic time constant for the emulsion given as:

$$\tau_o = \frac{(19\lambda+16)(2\lambda+3)}{40(1+\lambda)}\left(\frac{\eta_c R}{\gamma}\right) \tag{8.43b}$$

Equation 8.43a is a particular form of the general constitutive equation proposed by Oldroyd [85]. In a simple steady shear flow, Equation 8.43a predicts shear-thinning and nonzero normal stresses [84]:

$$\eta = \frac{\eta_c}{1+\tau_o^2\dot{\gamma}^2}\left\{1 + \left(\frac{5\lambda+2}{2\lambda+2}\right)\phi\right.$$

$$\left. + \tau_o^2\dot{\gamma}^2\left[1 + \left(\frac{5\lambda+2}{2\lambda+2}\right)\phi - \frac{(19\lambda+16)}{(2\lambda+2)(2\lambda+3)}\phi\right]\right\} \tag{8.44a}$$

$$N_1 = \tau_{11} - \tau_{22} = \frac{\eta_c^2\dot{\gamma}^2 R}{40\gamma(1+\tau_o^2\dot{\gamma}^2)}\left(\frac{19\lambda+16}{1+\lambda}\right)^2\phi \tag{8.44b}$$

$$N_2 = \tau_{22} - \tau_{33} = -\frac{\eta_c^2 \dot{\gamma}^2 R}{40\gamma(1+\tau_0^2\dot{\gamma}^2)}$$

$$\times \phi \left\{ \frac{1}{2}\left(\frac{19\lambda+16}{\lambda+1}\right)^2 - \left[\frac{3(19\lambda+16)(25\lambda^2+41\lambda+4)}{14(\lambda+1)^3}\right](1+\tau_0^2\dot{\gamma}^2)\right\}$$

(8.44c)

Equation 8.44a to Equation 8.44c can be rewritten in terms of intrinsic or reduced quantities as:

$$[\eta] = \underset{\phi\to 0}{Lim}\left(\frac{\eta-\eta_c}{\phi\eta_c}\right) = \frac{1}{1+(hN_{Ca})^2}\left\{\left(\frac{5\lambda+2}{2\lambda+2}\right)\right.$$

$$\left. \times \left[1+(hN_{Ca})^2 - \frac{(19\lambda+16)}{(5\lambda+2)(2\lambda+3)}(hN_{Ca})^2\right]\right\}$$

(8.44d)

where $h = (19\lambda+16)(2\lambda+3)/[40(1+\lambda)]$ and N_{Ca} is $\eta_c\dot{\gamma}R/\gamma$.

$$N_{1r} = \frac{N_1}{\eta_c\dot{\gamma}} = \frac{N_{Ca}}{40\left[1+(hN_{Ca})^2\right]}\left(\frac{19\lambda+16}{1+\lambda}\right)^2\phi$$

(8.44e)

$$N_{2r} = \frac{N_2}{\eta_c\dot{\gamma}} = -\frac{N_{Ca}}{40(1+(hN_{Ca})^2)}\phi\left\{\frac{1}{2}\left(\frac{19\lambda+16}{\lambda+1}\right)^2\right.$$

$$\left. -\left[\frac{3(19\lambda+16)(25\lambda^2+41\lambda+4)}{14(\lambda+1)^3}\right](1+h^2N_{Ca}^2)\right\}$$

(8.44f)

Figure 8.5 shows the plots of $[\eta]$ vs. N_{Ca} for different values of viscosity ratio λ. The plots are generated from Equation 8.44d. With the increase in the capillary number N_{Ca}, the intrinsic viscosity $[\eta]$ decreases, indicating shear-thinning behavior. At any given N_{Ca}, $[\eta]$ decreases with the decrease in λ. Interestingly, $[\eta]$ becomes zero at high N_{Ca} when $\lambda = 1$; when $\lambda < 1$, $[\eta]$ becomes negative at high N_{Ca}. At low N_{Ca}, $[\eta]$ is always greater than unity and increases with increase in λ.

Figure 8.6 shows the plots of reduced normal stress differences (N_{1r} and $-N_{2r}$) vs. N_{Ca} for different values of λ. The magnitude of the normal stress difference (N_{1r} or N_{2r}) increases initially with the increase in N_{Ca}, reaches a maximum value, and then decreases with further increase in N_{Ca}. The decrease in $-N_{2r}$ after reaching a maximum value is quite abrupt. The maximum in N_{1r} or $-N_{2r}$ shifts to lower values of N_{Ca} with increase in λ.

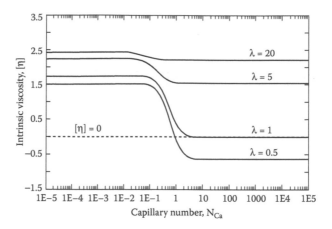

FIGURE 8.5 Variation of intrinsic viscosity of emulsion with capillary number for different values of viscosity ratio λ.

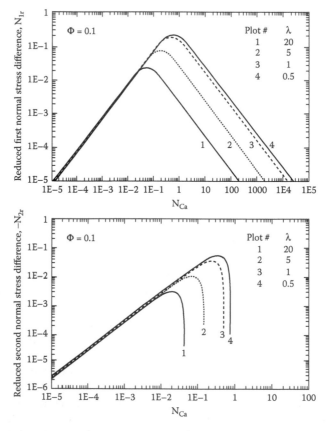

FIGURE 8.6 Variations of reduced normal stress differences (N_{1r} and $-N_{2r}$) with capillary number for different values of viscosity ratio λ.

In uniaxial elongational flow of emulsion, the constitutive equation (Equation 8.43a) predicts strain-thickening behavior with elongation viscosity η_E as follows:

$$\eta_E = 3\eta_c\left\{1+\left(\frac{5\lambda+2}{2\lambda+2}\right)\phi+\frac{\eta_c R\dot{\gamma}_E}{4\gamma}\left[\frac{3(19\lambda+16)(25\lambda^2+41\lambda+4)}{140(\lambda+1)^3}\right]\phi\right\} \tag{8.44g}$$

where $\dot{\gamma}_E$ is the rate of elongation.

8.4.3 SECOND-ORDER DEFORMATION SOLUTION

The second-order deformation of droplets has been studied by several authors [52,54,57,59]. For second-order deformation of droplets, surface spherical harmonics of higher order need to be included in the equation for the surface of a deformed droplet. Thus, for second-order deformation of droplets, the shape of a deformed droplet can be expressed as:

$$\frac{r}{R} = 1+\varepsilon\,\overline{\overline{F}}'(t):S_2 + \varepsilon^2\left[\left(\frac{-6}{5}\right)\overline{\overline{F}}'(t):\overline{\overline{F}}'(t)+H_4'(t)\vdots S_4\right] \tag{8.45}$$

where S_2 and S_4 are second-order and fourth-order surface spherical harmonics, respectively (see Equation 8.28), and $\overline{\overline{F}}'(t)$, H_4' are second-order and fourth-order dimensionless, symmetric, and traceless deformation tensors describing the instantaneous shape of the droplet. The equations for the time evolution of the deformation tensors $\overline{\overline{F}}'$ and H_4' are given as [54,59]:

$$\varepsilon\frac{D\overline{\overline{F}}'}{Dt'} = a_0\,\overline{\overline{E}}'+\varepsilon\left[\frac{a_1}{N_{Ca}}\,\overline{\overline{F}}'+a_2\,Sd(\overline{\overline{E}}'\bullet\overline{\overline{F}}')\right]$$

$$+\varepsilon^2\left[\frac{a_3}{N_{Ca}}\,Sd(\overline{\overline{F}}'\bullet\overline{\overline{F}}')+a_4\,\overline{\overline{E}}'\,\overline{\overline{F}}':\overline{\overline{F}}'+a_5\,\overline{\overline{F}}'\,\overline{\overline{E}}':\overline{\overline{F}}'\right.$$

$$\left.+a_7\,Sd\left(\overline{\overline{E}}'\bullet\overline{\overline{F}}'\bullet\overline{\overline{F}}'\right)+a_8\,H_4':\overline{\overline{E}}'\right] \tag{8.46}$$

$$+\varepsilon^3\left[\frac{a_6}{N_{Ca}}\,\overline{\overline{F}}'\overline{\overline{F}}':\overline{\overline{F}}'+\frac{a_9}{N_{Ca}}\,H_4':\overline{\overline{F}}'\right]$$

$$\varepsilon\frac{DH_4'}{Dt'} = b_1\,Sd_4(\overline{\overline{E}}'\overline{\overline{F}}')+\varepsilon\left[\frac{b_0}{N_{Ca}}\,H_4'+\frac{b_2}{N_{Ca}}\,Sd_4(\overline{\overline{F}}'\overline{\overline{F}}')\right.$$

$$\left.+b_3\,Sd_4(\overline{\overline{E}}'\bullet H_4')+b_4\,Sd_4\left(\overline{\overline{E}}'\bullet\overline{\overline{F}}'\overline{\overline{F}}'\right)\right] \tag{8.47}$$

where Sd_4 is the "symmetric deviator" of the fourth-order tensor (see Appendix D for definition). For the weak shear flow case where $N_{Ca} \ll 1$ and $\varepsilon = N_{Ca}$, Equation 8.46 and Equation 8.47 reduce to:

$$\varepsilon \frac{D\overline{\overline{F}}'}{Dt'} = a_0 \overline{\overline{E}}' + a_1 \overline{\overline{F}}' + \varepsilon \left[a_2\, Sd(\overline{\overline{E}}' \bullet \overline{\overline{F}}') + a_3\, Sd(\overline{\overline{F}}' \bullet \overline{\overline{F}}') \right]$$

$$+ \varepsilon^2 \left[a_4\, \overline{\overline{E}}'\overline{\overline{F}}' : \overline{\overline{F}}' + a_5\, \overline{\overline{F}}'\overline{\overline{E}}' : \overline{\overline{F}}' + a_6\, \overline{\overline{F}}'\overline{\overline{F}}' : \overline{\overline{F}}' \right. \qquad (8.48)$$

$$\left. + a_7\, Sd(\overline{\overline{E}}' \bullet \overline{\overline{F}}' \bullet \overline{\overline{F}}') + a_8\, H_4' : \overline{\overline{E}}' + a_9\, H_4' : \overline{\overline{F}}' \right]$$

$$\varepsilon \frac{DH_4'}{Dt'} = b_0 H_4' + b_1\, Sd_4(\overline{\overline{E}}'\overline{\overline{F}}') + b_2\, Sd_4(\overline{\overline{F}}'\overline{\overline{F}}') \qquad (8.49)$$

The coefficients a_i in Equation 8.48 are given as [52]:

$$a_0 = 5/3(2\lambda + 3) \qquad (8.50)$$

$$a_1 = -40(\lambda + 1)/(2\lambda + 3)(19\lambda + 16) \qquad (8.51)$$

$$a_2 = 10(4\lambda - 9)/7(2\lambda + 3)^2 \qquad (8.52)$$

$$a_3 = 288(137\lambda^3 + 624\lambda^2 + 741\lambda + 248)/7(2\lambda + 3)^2(19\lambda + 16)^2 \qquad (8.53)$$

$$a_4 = -\frac{2\left(11172\lambda^4 + 18336\lambda^3 + 17440\lambda^2 + 3499\lambda - 7572\right)}{49(2\lambda + 3)^3(19\lambda + 16)} \qquad (8.54)$$

$$a_5 = -\frac{2(\lambda - 1)\left(22344\lambda^3 + 52768\lambda^2 + 45532\lambda + 19356\right)}{49(2\lambda + 3)^3(19\lambda + 16)} \qquad (8.55)$$

$$a_6 = -48P(\lambda)/49(2\lambda+3)^3(19\lambda+16)^3(10\lambda+11)(17\lambda+16) \quad (8.56)$$

$$a_7 = 48(\lambda-1)(2793\lambda^3+7961\lambda^2+8474\lambda+3522)/49(2\lambda+3)^3(19\lambda+16) \quad (8.57)$$

$$a_8 = -400(43\lambda^2+79\lambda+53)/3(2\lambda+3)^2(19\lambda+16) \quad (8.58)$$

$$a_9 = 80Q(\lambda)/(2\lambda+3)^2(19\lambda+16)^2(10\lambda+11)(17\lambda+16) \quad (8.59)$$

where

$$P(\lambda)=2127976\lambda^7-16341920\lambda^6-38494964\lambda^5+1229425551\lambda^4$$
$$+47406831l\lambda^3+591515680\lambda^2+332123136\lambda+71700480 \quad (8.60)$$

$$Q(\lambda) = 405260\lambda^5+2366960\lambda^4+9142173\lambda^3+8595967\lambda^2$$
$$+3334160\lambda+693760 \quad (8.61)$$

The coefficients b_i in Equation 8.49 are given as [52]:

$$b_0 = -360(\lambda+1)/(17\lambda+16)(10\lambda+11) \quad (8.62)$$

$$b_1 = 1/7(2\lambda+3) \quad (8.63)$$

$$b_2 = \frac{16\left(-14\lambda^3+207\lambda^2+431\lambda+192\right)}{21\left(2\lambda+3\right)\left(19\lambda+16\right)\left(17\lambda+16\right)\left(10\lambda+11\right)} \quad (8.64)$$

The rheological constitutive equation for dilute emulsions incorporating the second-order deformation of droplets has been developed by Barthes-Biesel and Acrivos [69] for the weak-flow case ($\varepsilon = N_{Ca} \ll 1$). The equation for general time-dependent flows is given as:

$$\bar{\bar{\sigma}} = -P\bar{\bar{\delta}} + 2\eta_c\bar{\bar{E}} + \eta_c\phi\left\{\left[\left[\frac{10(\lambda-1)}{2\lambda+3} + \frac{24(\lambda-1)^2(611\lambda+579)}{49(2\lambda+3)^3}\left(\frac{\eta_c R}{\gamma}\right)^2\bar{\bar{F}}:\bar{\bar{F}}\right]\bar{\bar{E}}\right.\right.$$

$$+\left[\frac{24}{2\lambda+3} + \frac{240(\lambda-1)^2(121\lambda+159)}{49(2\lambda+3)^3}\left(\frac{\eta_c R}{\gamma}\right)^2(\bar{\bar{F}}:\bar{\bar{E}})\right]$$

$$-\frac{288(5912\lambda^3 + 48779\lambda^2 + 74931\lambda + 29628)}{245(2\lambda+3)^3(19\lambda+16)}$$

$$\times\left(\frac{\eta_c R}{\gamma}\right)^2\bar{\bar{F}}:\bar{\bar{F}}\bar{\bar{F}} + \left[\frac{360(\lambda-1)^2}{7(2\lambda+3)^2}\left(\frac{\eta_c R}{\gamma}\right)Sd(\bar{\bar{F}}\cdot\bar{\bar{E}})\right] \qquad (8.65)$$

$$+\left[\frac{288(\lambda-6)}{7(2\lambda+3)^2}\left(\frac{\eta_c R}{\gamma}\right)Sd(\bar{\bar{F}}\cdot\bar{\bar{F}})\right] - \left[\frac{720(\lambda-1)^2(79\lambda+96)}{49(2\lambda+3)^3}\left(\frac{\eta_c R}{\gamma}\right)^2Sd(\bar{\bar{E}}\cdot\bar{\bar{F}}\cdot\bar{\bar{F}})\right]$$

$$-\left[\frac{800(\lambda-1)^2}{(2\lambda+3)^2}\left(\frac{\eta_c R}{\gamma}\right)^2 H_4:\bar{\bar{E}}\right]$$

$$+\left[\frac{240(103020\lambda^4 + 481092\lambda^3 + 433959\lambda^2 + 549640\lambda + 136576)}{(2\lambda+3)^2(19\lambda+16)(17\lambda+16)(10\lambda+11)}\right]\left(\frac{\eta_c R}{\gamma}\right)^2 H_4:\bar{\bar{F}}\right\}$$

If we neglect $(\eta_c R/\gamma)^2$ terms, the Frankel and Acrivos constitutive equation (Equation 8.38) is recovered.

For weakly time-dependent flows, Equation 8.48 and Equation 8.49 can be solved by a method of successive approximations, which yield:

$$\bar{\bar{F}} = \frac{19\lambda+16}{24(\lambda+1)}\bar{\bar{E}} + \frac{19\lambda+16}{24(\lambda+1)}\left(\frac{\eta_c R}{\gamma}\right)\left[\frac{601\lambda^2 + 893\lambda + 256}{140(\lambda+1)^2}Sd(\bar{\bar{E}}\cdot\bar{\bar{E}}) - \frac{(2\lambda+3)(19\lambda+16)}{40(\lambda+1)}\left(\frac{D\bar{\bar{E}}}{Dt}\right)\right]$$

$$+\left(\frac{\eta_c R}{\gamma}\right)^2\frac{19\lambda+16}{24(\lambda+1)}\left[\frac{(19\lambda+16)^2(2\lambda+3)^2}{1600(\lambda+1)^2}\frac{D^2\bar{\bar{E}}}{Dt^2} - \frac{3(19\lambda+16)(1476\lambda^3 + 4837\lambda^2 + 4673\lambda + 1264)}{5600(\lambda+1)^3}\times\right. \qquad (8.66)$$

$$\left.Sd\left(\bar{\bar{E}}\cdot\frac{D\bar{\bar{E}}}{Dt}\right) + \frac{I(\lambda)\bar{\bar{E}}(\bar{\bar{E}}:\bar{\bar{E}})}{(2\lambda+3)(17\lambda+16)(10\lambda+11)(\lambda+1)^4}\right]$$

where

$$I(\lambda) = 3103.908\lambda^7 + 20684.86\lambda^6 + 72725.24\lambda^5 + 123993.98\lambda^4 \qquad (8.67)$$

$$+ 103839.12\lambda^3 + 41745.30\lambda^2 + 7148.42\lambda + 428.53$$

The expression for the fourth-order tensor H_4 is:

$$H_4 = \frac{751\lambda + 656}{544320(\lambda + 1)^2} Sd_4\left(\bar{\bar{E}}\,\bar{\bar{E}}\right) \tag{8.68}$$

Thus, the constitutive equation (Equation 8.65) becomes:

$$
\begin{aligned}
\bar{\bar{\sigma}} = -P\bar{\bar{\delta}} + 2\eta_c\bar{\bar{E}}&\left\{1 + \phi\left[\frac{5\lambda + 2}{2(\lambda + 1)} + \frac{(19\lambda + 16)Z(\lambda)\left(\dfrac{\eta_c R}{\gamma}\right)^2}{(2\lambda + 3)(17\lambda + 16)(10\lambda + 11)(\lambda + 1)^5}(\bar{\bar{E}} : \bar{\bar{E}})\right]\right\} \\[2mm]
&+ \eta_c\phi\left(\frac{\eta_c R}{\gamma}\right)\frac{(19\lambda + 16)}{20(\lambda + 1)^2}\left\{-\left[\frac{(19\lambda + 16)}{2}\frac{D\bar{\bar{E}}}{Dt}\right] + \left[\frac{3(25\lambda^2 + 41\lambda + 4)}{7(\lambda + 1)}Sd(\bar{\bar{E}} : \bar{\bar{E}})\right]\right. \\[2mm]
&\left. - \left[\left(\frac{\eta_c R}{\gamma}\right)\frac{(19\lambda + 16)(150\lambda^3 + 2179\lambda^2 + 2897\lambda + 724)}{280(\lambda + 1)^2}Sd\left(\bar{\bar{E}} \bullet \frac{D\bar{\bar{E}}}{Dt}\right)\right] \right. \\[2mm]
&\left. + \left[\left(\frac{\eta_c R}{\gamma}\right)\frac{(19\lambda + 16)^2(2\lambda + 3)}{80(\lambda + 1)}\left(\frac{D^2\bar{\bar{E}}}{Dt^2}\right)\right]\right\}
\end{aligned}
\tag{8.69}
$$

where

$$
\begin{aligned}
Z(\lambda) = &\ 422.28\lambda^7 + 3208.5\lambda^6 + 9960.91\lambda^5 + 11052.97\lambda^4 \\
&+ 5115.57\lambda^3 + 2848.07\lambda^2 + 1711.66\lambda + 100.86
\end{aligned}
\tag{8.70}
$$

8.5   EFFECT OF SURFACTANTS ON EMULSION RHEOLOGY

The presence of a surfactant at the droplet surface complicates the rheological behavior of emulsions. For instance, dilute emulsions can exhibit non-Newtonian rheology in the presence of surfactant at the droplet surface even when the capillary number (N_{Ca}) is very small ($N_{Ca} \to 0$) and the droplets are spherical [86,87].

Flow of continuous-phase fluid around the droplet results in nonuniform concentration of surfactant at the droplet surface. The resulting inhomogeneity in surfactant concentration at the droplet surface generates interfacial tension gradients (Marangoni stresses). Marangoni stresses produce a jump in tangential viscous tractions across the interface. Thus, the presence of surfactant at the droplet surface modifies the tangential stress boundary condition at the droplet surface. The normal stress boundary condition is also modified because of a reduction in the interfacial tension.

The Marangoni stresses can be neglected if the surface Peclet number (N_{Pe}^s) is very small $(N_{Pe}^s \to 0)$. The surface Peclet number is a measure of the relative importance of surface convection over surface diffusion of surfactant, and is defined as:

$$\left(N_{Pe}^s = \frac{R^2 \dot{\gamma}}{D_s} \right) \tag{8.71}$$

where D_s is the surface diffusivity of the surfactant. If $N_{Pe}^s \to 0$, the surfactant concentration over the droplet surface is basically uniform because surface diffusion of surfactant is able to counteract the convective transport of surfactant [88].

Marangoni phenomenon is not the only complication introduced by the presence of surfactant at the droplet surface. In general, the interfacial film of surfactant also possesses shear and dilational viscosities. These interfacial viscosities have a significant influence on the rheology of emulsions.

The rheology of emulsions of surfactant-covered droplets (in the absence of Marangoni stresses) is governed by four dimensionless groups: (1) the viscosity ratio (λ) defined as dispersed phase viscosity to continuous phase viscosity, (2) the capillary number (N_{Ca}) defined as $\eta_c \dot{\gamma} R / \gamma$, (3) the dilational Boussinesq number (N_{Bo}^k) defined as $\eta_s^k / \eta_c R$ (η_s^k is surface dilational viscosity), and (4) the shear Boussinesq number (N_{Bo}^η) defined as $\eta_s / \eta_c R$ (η_s is the surface shear viscosity).

In the limit $N_{Ca} \to 0$, the shear viscosity of a dilute emulsion of surfactant-covered droplets is given by the following Oldroyd expression [89]:

$$\eta = \eta_c \left[1 + \alpha \phi \right] \tag{8.72}$$

where α is defined as follows:

$$\alpha = \frac{1 + \frac{5}{2}\lambda + 2N_{Bo}^\eta + 3N_{Bo}^k}{1 + \lambda + \frac{2}{5}\left(2N_{Bo}^\eta + 3N_{Bo}^k\right)} \tag{8.73}$$

When $N_{Bo}^\eta \to 0$ and $N_{Bo}^k \to 0$, that is, the interface is inviscid, Equation 8.72 reduces to the Taylor expression (Equation 8.25) for emulsion viscosity.

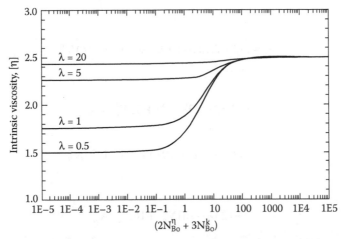

FIGURE 8.7 Variation of intrinsic viscosity with $(2\,N_{Bo}^{\eta}+3\,N_{Bo}^{k})$ for different values of λ.

When $(2N_{Bo}^{\eta}+3N_{Bo}^{k})\gg 1$, corresponding to a highly viscous interface, α becomes 5/2 and Equation 8.72 reduces to the Einstein expression for suspension (rigid particles) viscosity:

$$\eta=\eta_{c}\left(1+\frac{5}{2}\phi\right)\tag{8.74}$$

Figure 8.7 shows the plots of intrinsic viscosity $[\eta]$ vs. $(2\,N_{Bo}^{\eta}+3\,N_{Bo}^{k})$ for different values of viscosity ratio λ. For large values of λ, $[\eta]$ is 2.5 regardless of the value of $(2\,N_{Bo}^{\eta}+3\,N_{Bo}^{k})$, as expected. For low values of λ, $[\eta]$ increases with the increase in $(2\,N_{Bo}^{\eta}+3\,N_{Bo}^{k})$. For large values of $(2\,N_{Bo}^{\eta}+3\,N_{Bo}^{k})$, $[\eta]$ is always 2.5.

The complete constitutive equation for a dilute emulsion of surfactant-covered droplets in the absence of Marangoni effect is as follows:

$$\left(1+\tau_{1}\frac{D}{Dt}\right)\langle\bar{\bar{\tau}}\rangle=2\eta_{o}\left(1+\tau_{2}\frac{D}{Dt}\right)\langle\bar{\bar{E}}\rangle\tag{8.75}$$

This equation is based on Oldroyd's analysis [89,90]. The relaxation time τ_{1}, is defined as:

$$\tau_{1}=\left(\frac{\alpha_{1}-2\phi\alpha_{2}}{\beta_{1}-2\phi\beta_{2}}\right)\left(\frac{\eta_{c}R}{4\gamma}\right)\tag{8.76}$$

where

$$\alpha_1 = \left[(3+2\lambda)(16+19\lambda)+4(12+13\lambda)N_{Bo}^\eta \right.$$

(8.77)

$$\left. + 2(32+23\lambda)N_{Bo}^k + 32N_{Bo}^k \, N_{Bo}^\eta \right]$$

$$\alpha_2 = \left[(\lambda-1)(16+19\lambda)+2(8+13\lambda)N_{Bo}^\eta \right.$$

(8.78)

$$\left. + (23\lambda-16)N_{Bo}^k + 16N_{Bo}^k \, N_{Bo}^\eta \right]$$

$$\beta_1 = 10(1+\lambda)+8N_{Bo}^\eta + 12N_{Bo}^k \qquad (8.79)$$

$$\beta_2 = (2+5\lambda)+4N_{Bo}^\eta + 6N_{Bo}^k \qquad (8.80)$$

The retardation time τ_2 is defined as:

$$\tau_2 = \left[\frac{\alpha_1+3\phi\alpha_2}{\beta_1+3\phi\beta_2}\right]\left(\frac{\eta_c R}{4\gamma}\right) \qquad (8.81)$$

The zero-shear viscosity η_o is given by Equation 8.72 and D/Dt is the Jaumann derivative.

Oldroyd's analysis [89] assumes that the interfacial tension is uniform and that Marangoni stresses are negligible. In the limit $N_{Ca} \to 0$ (negligible deformation of droplets), Danov [91] has derived the following expression for the *zero shear viscosity* of dilute emulsions taking into account the Marangoni effect:

$$\eta = \eta_c \left[1+(1+\tfrac{3}{2}\varepsilon_m)\phi\right] \qquad (8.82)$$

where ε_m is the so-called *interfacial mobility parameter*, defined as:

$$\varepsilon_m = \left[\frac{\lambda+\dfrac{2}{5}\left(\dfrac{RE_G}{2\eta_c D}+2N_{Bo}^\eta+3N_{Bo}^k\right)}{1+\lambda+\dfrac{2}{5}\left(\dfrac{RE_G}{2\eta_c D}+2N_{Bo}^\eta+3N_{Bo}^k\right)}\right] \qquad (8.83)$$

where D is the effective diffusion coefficient of the surfactant and E_G is the Gibbs elasticity of the interface defined as:

$$E_G = -\left(\frac{\partial \gamma}{\partial \ell n \Gamma}\right) \qquad (8.84)$$

Γ in this equation is the surfactant concentration.

The value of the interfacial mobility parameter ranges from 0 to 1, that is, $0 \le \varepsilon_m \le 1$. For interfaces with high Gibbs elasticity (strong Marangoni effect), $E_G \to \infty$ and the parameter ε_m is unity. The expression for viscosity in this case reduces to the Einstein formula (Equation 8.74). For a surfactant-free interface: $E_G \to 0$, $N_{Bo}^\eta \to 0$, $N_{Bo}^k \to 0$; consequently, the interfacial parameter ε_m is $\lambda/(1+\lambda)$ and the expression for viscosity reduces to the Taylor formula (Equation 8.25).

Figure 8.8 shows the plots of intrinsic viscosity $[\eta]$ vs. $RE_G/2\eta_c D$ for different values of viscosity ratio λ. The surface shear and dilational viscosities of the interface are assumed to be zero. For a large value of λ, $[\eta]$ is 2.5 regardless of the value of $RE_G/2\eta_c D$, as expected. For low values of λ, $[\eta]$ increases with the increase in dimensionless film elasticity given by $RE_G/2\eta_c D$. For highly elastic film ($RE_G/2\eta_c D$ large), $[\eta]$ is always 2.5 regardless of the value of viscosity ratio λ.

Blawzdziewicz et al. [87] and Vlahovska et al. [86] have studied the effects of the Marangoni phenomenon on the rheology of dilute emulsion of surfactant-covered droplets under conditions of low capillary number ($N_{Ca} \to 0$, droplets are spherical) and arbitrary flow strength (any shear rate). The emulsion exhibits

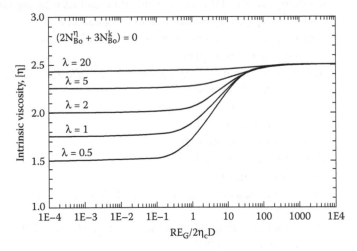

FIGURE 8.8 Variation of intrinsic viscosity with dimensionless group $RE_G/2\eta_c D$ for different values of viscosity ratio λ ($\left(2N_{Bo}^\eta + 3N_{Bo}^k\right) = 0$).

non-Newtonian rheology (shear thinning and normal stresses). As the capillary number is small $(N_{Ca} \to 0)$ and the droplets are spherical, the non-Newtonian behavior exhibited by the emulsion is entirely due to the Marangoni effect. The normalized shear stress and normalized normal stresses in shear flow of emulsion are found to depend on two dimensionless parameters: (1) the viscosity ratio λ and (2) the Marangoni number N_{Ma} defined as:

$$N_{Ma} = \frac{\Delta\gamma}{\eta_c \dot\gamma R} \tag{8.85}$$

N_{Ma} is the ratio of characteristic Marangoni stress ($\Delta\gamma/R$) to viscous stress $\left(\eta_c \dot\gamma\right)$. The quantity $\Delta\gamma$ in the definition of N_{Ma} is $\gamma_0 - \gamma_e$ where γ_0 is the interfacial tension in the absence of surfactant and γ_e is the equilibrium interfacial tension in the presence of surfactant. The authors neglected the effects of interfacial rheological properties such as η_s and η_s^k, that is, the Boussinesq numbers N_{Bo}^η and N_{Bo}^k were assumed to be zero; also, the surfactant was assumed to be nondiffusing.

With the increase in N_{Ma}^{-1} (inverse of the Marangoni number), the emulsion viscosity decreased, indicating shear thinning. The first and second normal stress differences (N_1 and $-N_2$) increased with the increase in N_{Ma}^{-1}.

NOTATION

A	surface area
A_o	surface area of a droplet
A'	bounding surface of the region where fluid motion is taking place
a_i	coefficients given by Equation 8.50 to Equation 8.59
b_i	coefficients given by Equation 8.62 to Equation 8.64
$\overline{\overline{B}}$	arbitrary second-order tensor
D	deformation parameter (Equation 8.2) or effective diffusion coefficient of surfactant
D_s	surface diffusivity of surfactant
D/Dt	Jaumann derivative
D^2/Dt^2	second-order Jaumann derivative
E_G	Gibbs elasticity of interface
$\overline{\overline{E}}$	rate of strain tensor
$\overline{\overline{E}}_\infty$	rate of strain tensor of the undisturbed flow far away from the droplet
$\overline{\overline{E}}'_\infty$	dimensionless rate of strain tensor of the undisturbed flow far away from the droplet
$\overline{\overline{F}}$	symmetric and traceless deformation tensor describing the shape of a droplet

$\bar{\bar{F}}'(t)$	dimensionless form of symmetric and traceless deformation tensor $\bar{\bar{F}}$ describing the shape of a droplet
g	droplet shape variable, defined by Equation 8.30
h	function of viscosity ratio λ; see Equation 8.44d
H_4	symmetric and traceless fourth-order deformation tensor describing the instantaneous shape of a droplet
H_4'	dimensionless form of symmetric and traceless fourth-order deformation tensor H_4 describing the shape of a droplet
$I(\lambda)$	function of viscosity ratio λ; see Equation 8.67
$\hat{n}$	outward unit normal vector
N_1	first normal stress difference
N_2	second normal stress difference
N_{1r}	reduced first normal stress difference
N_{2r}	reduced second normal stress difference
N_{Bo}^{η}	shear Boussinesq number
N_{Bo}^{k}	dilational Boussinesq number
N_{Ca}	capillary number
$N_{Ca,crit}$	critical capillary number
N_{Ma}	Marangoni number defined by Equation 8.85
N_{Pe}^{s}	surface Peclet number defined by Equation 8.71
P	pressure
P^*	pressure inside the droplet
P_{∞}	pressure far away from the droplet
$\bar{\bar{P}}_s$	surface stress tensor
$P(\lambda)$	function of viscosity ratio λ given by Equation 8.60
$\dot{Q}$	rate of energy dissipation in emulsion
$\dot{Q}_c$	rate of energy dissipation in matrix alone without droplets
$Q(\lambda)$	function of viscosity ratio λ given by Equation 8.61
R	droplet radius
r	distance from the droplet center of mass to the droplet surface
$\vec{r}$	position vector
r_{max}	semiaxis length of triaxial ellipsoid; see Figure 8.1
r_{min}	semiaxis length of triaxial ellipsoid; see Figure 8.1
r_z	semiaxis length of triaxial ellipsoid; see Figure 8.1
S_i	ith surface spherical harmonic
S_2	second-order surface spherical harmonic
S_4	fourth-order surface spherical harmonic
$\underline{\underline{Sd_4}}$	symmetric deviator of fourth-order tensor; see Appendix D
$\bar{\bar{S}}^o$	dipole strength of a single droplet located in an infinite matrix
t	time
tr	trace of a tensor

$\vec{u}$	fluid velocity vector
$\vec{u}_c$	fluid velocity vector in continuous phase
$\vec{u}_\infty$	fluid velocity vector far away from the droplet
$\vec{u}^*$	fluid velocity vector inside the droplet
$u_{interface}$	velocity of droplet interface
$\vec{x}$	position vector
x_1, x_2, x_3	position coordinates of droplet surface
$Z(\lambda)$	function of viscosity ratio λ given by Equation 8.70

GREEK

α	parameter given by Equation 8.73
α_1	parameter given by Equation 8.77
α_2	parameter given by Equation 8.78
β_1	parameter given by Equation 8.79
β_2	parameter given by Equation 8.80
γ	interfacial tension
γ_o	interfacial tension in the absence of surfactant
γ_e	equilibrium interfacial tension in the presence of surfactant
$\dot{\gamma}$	shear rate
$\dot{\gamma}_E$	rate of elongation
Γ	surfactant concentration
∇	del operator
∇_s	surface gradient operator
∇^2	Laplacian
$\bar{\bar{\delta}}$	unit tensor
ε	deformation parameter
ε_m	interfacial mobility parameter defined by Equation 8.83
η	shear viscosity
η_c	continuous phase viscosity
η_d	dispersed phase viscosity
η_E	elongation viscosity
η_o	zero shear viscosity
η_s	surface shear viscosity
η_s^k	surface dilational viscosity
$[\eta]$	intrinsic viscosity
θ	angle between the major axis of the droplet and flow direction
λ	viscosity ratio
$\bar{\bar{\sigma}}$	total stress tensor
$\bar{\bar{\sigma}}_c$	total stress tensor in matrix
$\bar{\bar{\sigma}}^*$	total stress tensor inside the droplet

τ_o	time constant given by Equation 8.43b
τ_1	relaxation time given by Equation 8.76
τ_2	retardation time given by Equation 8.81
$\bar{\bar{\tau}}$	deviatoric stress tensor
$\tau_{11}, \tau_{22}, \tau_{33}$	normal stress components
ϕ	volume fraction of droplets
$\vec{\omega}_o$	one half the vorticity of undisturbed flow
$\langle X \rangle$	volume average of quantity X

REFERENCES

1. B.W. Davis, in *Encyclopedia of Emulsion Technology*, Vol. 3, P. Becher, Ed., Marcel Dekker, New York, 1988, chap. 8.
2. R. Pal, *Ind. Eng. Chem. Res.*, 33: 1413–1435, 1994.
3. R. Pal, *Colloids Surf. A*, 84: 141–193, 1994.
4. F. Steinhauff, *Petroleum*, 25: 294–296, 1962.
5. K.J. Lissant, in *AIChE Symposium on Improved Oil Recovery by Surfactant and Polymer Flooding*, D.O. Shah, P.S. Schechter, Eds., AIChE, New York, 1978, pp. 93–100.
6. R. Raghavan, S.S. Marsden, *SPE J.*, 11: 153–161, 1971.
7. F.W. Jenkins, Oil in water emulsions: causes and treatment, in *Water Problems in Oil Production — An Operator's Manual*, 2nd ed., L.C. Case, Ed., Petroleum Publishing Co., Tulsa, OK, 1977, chap. 13.
8. J.L. Lummus, Water-in-Oil Emulsion Drilling Fluid, U.S. Patent 2,661,334, 1953.
9. B.B. Williams, J.L. Gidley, R.S. Schechter, *Acidizing Fundamentals*, Vol. 6, SPE Henry L. Doherty Series, SPE-AIME, Dallas, TX,1979, pp. 29–37, 58–59.
10. R. Simon, W.G. Poynter, Down-hole emulsification for improving viscous crude production. *J. Pet. Technol.*, 20(12), 1349–1353, 1968.
11. C.D. McAuliffe, R. Simon, C.E. Johnson, Method of Pumping Viscous Crude Oil, U.S. Patent 3,380,531, 1968.
12. C.D. McAuliffe, Fullerton, R. Simon, Whittier, C.E. Johnson, Pumping Viscous Crude, U.S. Patent 3,467,195, 1969.
13. A.H. Beyer, D.E. Osborn, Down-hole emulsification for improving paraffinic crude production, presented at the 44th Annual Fall Meeting of the SPE and AIME, Denver, CO, 1968, paper SPE 2676.
14. S.S. Marsden, R. Raghavan, A system for producing and transporting crude oil as oil-in-water emulsion, *J. Inst. Pet.*, 59(570): 273–278, 1973.
15. F.M. Mourits, L. Kurucz, R.J. Scouler, Heavy oil emulsion research at Saskatchewan Research Council, presented at the annual technical meeting of the south Saskatchewan section of Petroleum Society of CIM, Regina, Saskatchewan, 1987, paper no. 14.
16. J.L. Zakin, R. Pinaire, M.E. Borgmeyer, Transportation of oils as oil-in-water emulsions, *J. Fluids Eng.*, 101: 100–104, 1979.
17. E.V. Seymour, Process for Facilitating Pipeline Flow of Highly Viscous Fluids, U.S. Patent 3,520,313, 1970.

18. R. Simon, W.G. Poynter, Pipelining Oil/Water Mixtures, U.S. Patent 3,519,006, 1970.
19. R. Simon, C.D. McAuliffe, W. Poynter, H. Jennings, Pipelining Crude Oils, U.S. Patent 3,487,844, 1970.
20. W.C. Simpson, H.J. Sommer, Transportation of waxy oils. U.S. Patent 2,981,633, 1961.
21. D.P. Rimmer, A.A. Gregoli, J.A. Hamshar, E. Yildirim, Pipeline emulsion transportation for heavy oils, in *Emulsions: Fundamentals and Applications in the Petroleum Industry,* L.L. Schramm, Ed., American Chemical Society, Washington, D.C., 1992, pp. 295–312.
22. D.A. Alvarado, S.S. Marsden, Flow of oil-in-water emulsions through tubes and porous media. *SPE J.,* 19: 269–377, 1979.
23. G.G. Binder, N.J. Clark, C.D. Russel, Method of Secondary Recovery. U.S. Patent 3,203,517, 1965.
24. R.M. Decker, D.L. Flock, Thermal stability and application of emulsion composed blocking agents for steam flooding, *J. Can. Pet. Technol.,* 27: 69–78, 1988.
25. S. Torrey, *Emulsions and Emulsifier Applications — Recent Developments,* Noyes Data Corporation, Park Ridge, NJ, 1984.
26. S. Friberg, *Food Emulsions,* Marcel Dekker, New York, 1976.
27. K. J. Lissant, *Emulsions and Emulsion Technology,* Marcel Dekker, New York 1974.
28. H. Bennet, J. L. Bishop, M. F. Wulfinghoff, *Practical Emulsions — Applications,* Chemical Publishing Co., New York, 1968.
29. P. Becher, *Emulsions: Theory and Practice,* Krieger Publishing Co., Malabar, FL, 1977, chap. 8.
30. M.J. Lynch, W.C. Griffin, Food emulsions, in *Emulsions and Emulsion Technology,* Part I, K.J. Lissant, Ed., Marcel Dekker, New York, 1974, chap. 5.
31. S.E. Friberg, I. Kayoli, Surfactant association structure, microemulsions, and emulsions in foods — an overview, in *Microemulsions and Emulsions in Foods,* M. El-Nokaly, D. Cornell, Eds., ACS Symposium Series 448, American Chemical Society, Washington, D.C., 1990, chap. 2.
32. J. Schut, Meat emulsions, in *Food Emulsions,* S. Friberg, Ed., Marcel Dekker, New York, 1976, chap. 8.
33. E. Graf, H. Baver, Milk and milk products, in *Food Emulsions,* S. Friberg, Ed., Marcel Dekker, New York, 1976, chap. 5.
34. I.S. Shepperd, R.W. Yoell, Cake emulsions, in *Food Emulsions,* S. Friberg, Ed., Marcel Dekker, New York, 1976, chap. 5.
35. N.J. Krog, T.H. Riisom, K. Larsson, Applications in the food industry: I, in *Encyclopedia of Emulsion Technology,* Vol. 2, P. Becher, Ed., Marcel Dekker, New York, 1985, chap. 6.
36. B.A. Mully, Medicinal emulsions, in *Emulsions and Emulsion Technology,* Part I, K.J. Lissant, Ed., Marcel Dekker, New York, 1974, chap. 6.
37. S.S. Davis, J. Hadgraft, K.J. Palin, Medical and pharmaceutical applications of emulsions, in *Encyclopedia of Emulsion Technology,* Vol. 2, P. Becher, Ed., Marcel Dekker, New York, 1985, chap. 6.
38. H. Bennet, J.L. Bishop, M.F. Wulfinghoff, *Practical Emulsions — Applications,* Vol. 2, Chemical Publishing Co., New York, 1968, chap. 4.
39. C. Fox, Cosmetic emulsions, in *Emulsions and Emulsion Technology,* Part II, K.J. Lissant, Ed., Marcel Dekker, New York, 1974, chap. 13.

40. R.E. Ford, The use of emulsions in the application of pesticides, in *Theory and Practice of Emulsion Technology*, A.L. Smith, Ed., Academic Press, London, 1976, chap. 13.

41. H. Bennett, J.L. Bishop, M.F. Wulfinghoff, *Practical Emulsions — Applications*, Vol. 2, Chemical Publishing Co., New York, 1968, chap. 1.

42. H.A. Bampfield, J. Cooper, Emulsion explosives, in *Encyclopedia of Emulsion Technology*, Vol. 3, P. Becher, Ed., Marcel Dekker, New York, 1988, chap. 7.

43. H. Bennet, J.L. Bishop, M. F. Wulfinghoff, *Practical Emulsions — Applications*, Vol. 2, Chemical Publishing Co., New York, 1968, chap. 2.

44. G. Allyn, Emulsion paints, in *Emulsions and Emulsion Technology*, Part I, K.J. Lissant, Ed., Marcel Dekker, New York, 1974, chap. 7.

45. H. Bennet, J.L. Bishop, M.F. Wulfinghoff, *Practical Emulsions — Applications*, Vol. 2, Chemical Publishing Co., New York, 1968, chap. 12.

46. L. Benali, Rheological and granulometrical studies of a cutting oil emulsion, *J. Colloid Interface Sci.*, 156: 454–461, 1993.

47. G.I. Taylor, *Proc. Roy. Soc. Lond. A*, 146: 501–523, 1934.

48. R.G. Cox, *J. Fluid Mech.*, 37: 601–623, 1969.

49. C.E. Chaffey, H. Brenner, *J. Colloid Int. Sci.*, 24: 258–269, 1967.

50. H. Brenner, *Prog. Heat Mass Transfer*, 6: 509–574, 1972.

51. S. Torza, R.G. Cox, S.G. Mason, *J. Colloid Int. Sci.*, 38: 395–411, 1972.

52. D. Barthes-Biesel, A. Acrivos, *J. Fluid Mech.*, 61: 1–21, 1973.

53. F.S. Hakami, W.R. Schowalter, *J. Fluid Mech.*, 98: 635–645, 1980.

54. J.M. Rallison, *J. Fluid Mech.*, 98: 625–633, 1980.

55. E.J. Hinch, A. Acrivos, *J. Fluid Mech.*, 98: 305–328, 1980.

56. H.P. Grace, *Chem. Eng. Commun.*, 14: 225–277, 1982.

57. A. Acrivos, *Ann. N.Y. Acad. Sci.*, 404: 1–11, 1983.

58. J. M. Rallison, *Annu. Rev. Fluid Mech.*, 16: 45–66, 1984.

59. B.J. Bentley, L.G. Leal, *J. Fluid Mech.*, 167: 241–283, 1986.

60. H.A. Stone, *Annu. Rev. Fluid Mech.*, 26: 65–102, 1994.

61. V.T. Tsakolos, P. Navard, E. Peuvrel-Disdier, *J. Rheol.*, 42: 1403–1417, 1998.

62. S. Guido, M. Villone, *J. Rheol.*, 42: 395–415, 1998.

63. S. Guido, F. Greco, M. Villone, *J. Colloid Int. Sci.*, 219: 298–309, 1999.

64. S. Guido, F. Greco, *Rheol. Acta*, 40: 176–184, 2001.

65. F. Greco, *Phys. Fluids*, 14: 946–954, 2002.

66. W.R. Schowalter, C.E. Chaffey, H. Brenner, *J. Colloid Interface Sci.*, 26: 152–160, 1968.

67. N.A. Frankel, A. Acrivos, *J. Fluid Mech.*, 44: 65–78, 1970.

68. H. Brenner, *Prog. Heat Mass Transfer*, 5: 89–129, 1972.

69. D. Barthes-Biesel, A. Acrivos, *Int. J. Multiphase Flow*, 1: 1–24, 1973.

70. E.J. Hinch, in *Theoretical Rheology*, J.F. Hutton, J.R.A. Pearson, K. Walters, Eds., Applied Science Publishers Ltd., London, 1974.

71. S.J. Choi, W.R. Schowalter, *Phys. Fluids*, 18: 420–427, 1975.

72. M.R. Kennedy, C. Pozrikidis, R. Skolak, *Comput. Fluids*, 23: 251–278, 1994.

73. P.V. Puyvelde, S. Velankar, P. Moldenaers, *Curr. Opin. Colloids Interface Sci.*, 6: 457–463, 2001.

74. E.H. Lucassen-Reynders, in *Encyclopedia of Emulsion Technology*, Vol. 4, P. Becher, Ed., Marcel Dekker, New York, 1996, chap. 2.

75. R.G. Larson, *The Structure and Rheology of Complex Fluids*, Oxford University Press, New York, 1999, chap. 9.

76. T.G.M. Van de ven, *Colloidal Hydrodynamics*, Academic Press, San Diego, CA, 1989, chap. 3.

77. H.A. Stone, L.G. Leal, *J. Fluid Mech.*, 211: 123–156, 1990.

78. G.K. Batchelor, J.T. Green, *J. Fluid Mech.*, 56: 401–427, 1972.

79. G.K. Batchelor, *An Introduction to Fluid Dynamics*, Cambridge University Press, Cambridge, 1967.

80. A. Nadim, *Chem. Eng. Commun.*, 148–150: 391–407, 1996.

81. G.I. Taylor, *Proc. R. Soc. Lond. A*, 138: 41–48, 1932.

82. E. Wacholder, G. Hetsroni, *Isr. J. Technol.*, 8: 271–279, 1970.

83. J. Happel, H. Brenner, *Low Reynolds Number Hydrodynamics*, Noordhoff International Publishing, Leydon, the Netherlands, 1973.

84. C.D. Han, *Multiphase Flow in Polymer Processing*, Academic Press, New York, 1981.

85. J.G. Oldroyd, *Proc. R. Soc. Lond. A*, 245: 278–297, 1958.

86. P. Vlahovska, J. Blawzdziewicz, M. Loewenberg, *J. Fluid Mech.*, 463: 1–24, 2002.

87. J. Blawzdziewicz, P. Vlahovska, M. Loewenberg, *Physica A*, 276: 50–85, 2000.

88. D. Edwards, H. Brenner, D. Wasan, *Interfacial Transport Processes and Rheology*, Butterworth-Heinemann, Boston, MA, 1991.

89. J.G. Oldroyd, *Proc. R. Soc. Lond. A,* 232: 567–577, 1955.

90. J.G. Oldroyd, in *Rheology of Disperse Systems*, C.C. Mill, Ed., Pergamon Press, London, 1959.

91. K.D. Danov, *J. Colloid Interface Sci.* 235: 144–149, 2001.

9 Dispersions of Bubbles

9.1 INTRODUCTION

Two-phase bubbly suspensions are encountered in flotation columns, oil wells, and bioreactors. They are also encountered in food processing and processing of polymer melts. Bubbly suspensions are known to play a key role in volcanic eruptions [1–6]. During the ascent of magma from a magna chamber to the earth's surface, nucleation and growth of gas bubbles occur by exsolution of volatiles (mainly water and carbon dioxide) that are initially dissolved in magma at high pressures. The vesiculation of the rising magma drastically changes the physico-chemical properties and rheology of magma within the volcanic conduit and, therefore, the ascent rate of magma.

Understanding the effect of bubbles on the rheological properties of bubbly suspensions is important for the analysis and modeling of processes where such systems (bubbly suspensions) are encountered. As bubbly suspensions are a special case of emulsions (the viscosity ratio λ is now zero), the bulk stress tensor expressions developed in the previous chapter are expected to apply. However, there is one important difference between bubbly suspensions and emulsions. Unlike the liquid/liquid emulsions, the bubbly suspensions cannot be treated as incompressible. As bubbles can undergo changes in volume, bubbly suspensions behave as a compressible fluid with nonzero dilational viscosity.

9.2 DILATIONAL VISCOSITY OF BUBBLY SUSPENSIONS

The dilational properties of a bubbly suspension can be determined using the cell model approach [7,8]. Consider a volume V of a dilute bubbly suspension consisting of a large number N of small, same-size, bubbles dispersed uniformly in an incompressible Newtonian liquid. Let the total volume V of the suspension be subdivided into N identical spherical cells of radius R each consisting of a single bubble at the center. The cell radius R is chosen such that:

$$R = a/\phi^{1/3} \tag{9.1}$$

where a is the bubble radius and ϕ is the volume fraction of bubbles in the suspension. The suspension is subjected to spherically symmetric expansion with a uniform rate of expansion given by Δ:

$$\Delta = \frac{1}{V}\frac{dV}{dt} \tag{9.2}$$

As the expansion occurs wholly in the bubbles, Δ can be expressed as:

$$\Delta = 3\left(\frac{a^2}{R^3}\right)\frac{da}{dt} = \frac{3}{R}\frac{dR}{dt} \tag{9.3}$$

Thus

$$\frac{dR}{dt} = \left(\frac{a}{R}\right)^2\left(\frac{da}{dt}\right) \tag{9.4}$$

The dilational properties of a suspension can be determined by equating the normal force (rr component of the total stress tensor $\bar{\bar{\sigma}}$) at the boundary of the cell just described to the normal force at the boundary of the same-size (radius R) spherical region filled with a "homogeneous" compressible fluid having the same properties as that of the suspension under consideration and subjected to the same dilational motion.

9.2.1 NORMAL FORCE AT THE CELL BOUNDARY

To determine σ_{rr} at the cell boundary ($r = R$), we need velocity and pressure distributions in the continuous-phase liquid surrounding the bubble.

The continuity equation in spherical coordinates is (see Appendix A):

$$\frac{\partial \rho}{\partial t} + \frac{1}{r^2}\frac{\partial}{\partial t}\left(\rho\, r^2\, u_r\right) = 0 \tag{9.5}$$

Because the continuous-phase liquid is incompressible, $\partial \rho/\partial t = 0$ and the continuity equation reduces to:

$$\frac{\partial}{\partial r}\left(r^2\, u_r\right) = 0 \tag{9.6}$$

Upon integration of this equation with the boundary condition $u_r = dR/dt$ at $r = R$, one obtains:

$$u_r = \left(\frac{R}{r}\right)^2\frac{dR}{dt} = \left(\frac{a}{r}\right)^2\frac{da}{dt} \tag{9.7}$$

The r component of the Navier–Stokes equation neglecting inertial terms is (see Appendix A):

$$-\frac{\partial P}{\partial r} + \eta \left[\frac{1}{r^2} \frac{\partial}{\partial r} \left(r^2 \frac{\partial u_r}{\partial r} \right) - \frac{2u_r}{r^2} \right] = 0 \tag{9.8}$$

Substituting the velocity distribution into this equation gives:

$$\frac{\partial P}{\partial r} = 0 \tag{9.9}$$

Thus, pressure in the continuous phase liquid is uniform. From normal stress balance at the bubble/liquid interface:

$$\sigma_{rr}^c \big|_{r=a} = \sigma_{rr}^d \big|_{r=a} + \frac{2\gamma}{a} \tag{9.10}$$

where superscripts c and d refer to continuous and dispersed (bubble) phases, respectively. Equation 9.10 can be written as:

$$\left(-P + \tau_{rr} \right)^c \big|_{r=a} = \left(-P + \tau_{rr} \right)^d \big|_{r=a} + \frac{2\gamma}{a} \tag{9.11}$$

From Equation 9.11, pressure in the continuous-phase liquid at the interface (and elsewhere) is:

$$P^c \big|_{r=a} = -\left(-P + \tau_{rr} \right)^d \big|_{r=a} + \tau_{rr}^c \big|_{r=a} - \frac{2\gamma}{a} \tag{9.12}$$

The rr component of the viscous stress tensor $\overline{\overline{\tau}}$ is given as:

$$\tau_{rr}^c = 2\eta_c \left(\frac{\partial u_r}{\partial r} \right)^c$$
$$= -4\eta_c \left(\frac{dR}{dt} \right) \left(\frac{R^2}{r^3} \right) \tag{9.13}$$

Inside the bubble, the rr component of $\overline{\overline{\tau}}$ can be neglected, that is, $\tau_{rr}^d \simeq 0$. The pressure inside the bubble can be determined from the ideal gas law:

$$P^d = P_0^d (V_0^d / V^d) = P_0^d (a_0 / a)^3 \tag{9.14}$$

where P_0^d, V_0^d, and a_0 are the initial bubble pressure, volume, and radius, respectively. Thus, pressure in the continuous-phase liquid is:

$$P^c = P_0^d (a_0/a)^3 - 4\eta_c (R^2/a^3) \frac{dR}{dt} - \frac{2\gamma}{a} \tag{9.15}$$

and the rr component of $\bar{\bar{\sigma}}$ at the cell boundary ($r = R$) becomes:

$$\sigma_{rr}\big|_{r=R} = (-P + \tau_{rr})^c \big|_{r=R}$$

$$\tag{9.16}$$

$$= -P_0^d \left(\frac{a_0}{a} \right)^3 + \frac{2\gamma}{a} + \frac{4\eta_c R^2}{a^3} \left(\frac{dR}{dt} \right) \left(1 - \frac{a^3}{R^3} \right)$$

9.2.2 Normal Force at the Boundary of the Equivalent Homogeneous Fluid

Consider a spherical region of radius R filled with a homogeneous compressible fluid having the same properties as that of a bubbly suspension and subjected to the same spherically symmetric expansion.

Because $\rho = \rho_0 (R_0/R)^3$, the derivative $\partial\rho/\partial t$ in the equation of continuity is:

$$\frac{\partial\rho}{\partial t} = -3\rho_0 \frac{R_0^3}{R^4} \frac{dR}{dt} \tag{9.17}$$

Substituting this into the continuity equation and integrating with the boundary condition $u_r = 0$ at $r = 0$ gives:

$$u_r = \left(\frac{r}{R} \right) \left(\frac{dR}{dt} \right) \tag{9.18}$$

The viscous stress tensor for compressible fluid undergoing spherically symmetric flow is:

$$\bar{\bar{\tau}} = 2\eta \left[\left(\frac{\partial u_r}{\partial r} \right) \hat{\delta}_r \hat{\delta}_r + \left(\frac{u_r}{r} \right) \hat{\delta}_\theta \hat{\delta}_\theta + \left(\frac{u_r}{r} \right) \hat{\delta}_\phi \hat{\delta}_\phi \right] + \left(\eta^k - \frac{2}{3}\eta \right) (\nabla \bullet \vec{u}) \bar{\bar{\delta}} \tag{9.19}$$

and

$$\nabla \bullet \vec{u} = \frac{1}{r^2} \frac{\partial}{\partial r} \left(r^2 u_r \right) \tag{9.20}$$

Substituting the expression for u_r in the viscous stress tensor equation gives:

$$\bar{\bar{\tau}} = \frac{3\eta^k}{R}\left(\frac{dR}{dt}\right)\left(\hat{\delta}_r\hat{\delta}_r + \hat{\delta}_\theta\hat{\delta}_\theta + \hat{\delta}_\phi\hat{\delta}_\phi\right) \qquad (9.21)$$

Thus,

$$\tau_{rr} = \tau_{\theta\theta} = \tau_{\phi\phi} = \frac{3\eta^k}{R}\left(\frac{dR}{dt}\right) \qquad (9.22)$$

The r component of the equation of motion is (see Appendix A):

$$-\frac{\partial P}{\partial r} + \frac{1}{r^2}\frac{\partial}{\partial r}\left(r^2\tau_{rr}\right) - \left(\frac{\tau_{\theta\theta} + \tau_{\phi\phi}}{r}\right) = 0 \qquad (9.23)$$

Using the expressions for τ_{rr}, $\tau_{\theta\theta}$, and $\tau_{\phi\phi}$, this equation yields:

$$\frac{\partial P}{\partial r} = 0 \qquad (9.24)$$

That is, the pressure is uniform throughout the fluid.

The rr component of $\bar{\bar{\sigma}}$ at the boundary of the spherical region of radius R filled with the fluid is:

$$\sigma_{rr}\big|_{r=R} = \left(-P + \tau_{rr}\right)\big|_{r=R}$$
$$= -P + \frac{3\eta^k}{R}\left(\frac{dR}{dt}\right) \qquad (9.25)$$

9.2.3 EQUATION FOR DILATIONAL VISCOSITY

The equation for dilational viscosity of a bubbly suspension can be obtained by equating the rr component of $\bar{\bar{\sigma}}$ at the boundary ($r = R$) of the cell with the rr component of $\bar{\bar{\sigma}}$ at the boundary ($r = R$) of a spherical region filled with homogeneous compressible fluid having the same properties as that of a bubbly suspension. Thus,

$$-P_0^d\left(\frac{a_0}{a}\right)^3 + \frac{2\gamma}{a} + \frac{4\eta_c R^2}{a^3}\left(\frac{dR}{dt}\right)\left(1 - \frac{a^3}{R^3}\right) = -P + \frac{3\eta^k}{R}\frac{dR}{dt} \qquad (9.26)$$

Comparing terms of the same order of $\left(\dfrac{dR}{dt}\right)$ gives:

$$P = P_0^d \left(\frac{a_0}{a}\right)^3 - \frac{2\gamma}{a} \tag{9.27}$$

$$\eta^k = \frac{4}{3}\eta_c \left(\frac{R}{a}\right)^3 \left(1 - \frac{a^3}{R^3}\right) = \frac{4}{3}\eta_c \left(\frac{1-\phi}{\phi}\right) \tag{9.28}$$

In the limit $\phi \to 0$, the dilational viscosity of a bubbly suspension is:

$$\eta^k = \frac{4\eta_c}{3\phi} \tag{9.29}$$

This is the well-known result first derived by Taylor [9].

9.3 CONSTITUTIVE EQUATION FOR BUBBLY SUSPENSIONS

Assuming that the deformation of bubbles is negligible, the constitutive equation for a dilute bubbly suspension can be written as:

$$\bar{\bar{\sigma}} = -P\bar{\bar{\delta}} + \left(\eta^k - \frac{2}{3}\eta\right)\left(tr\bar{\bar{E}}\right)\bar{\bar{\delta}} + 2\eta\bar{\bar{E}} \tag{9.30}$$

where the dilational viscosity η^k is $4\eta_c/3\phi$ and the shear viscosity η is given by $\eta = \eta_c(1+\phi)$. Equation 9.30 predicts suspension behavior to be Newtonian.

Under the condition of small first-order deformation of bubbles, the total stress tensor for a dilute bubbly suspension can be written as:

$$\bar{\bar{\sigma}} = -P\bar{\bar{\delta}} + \left(\eta^k - \frac{2}{3}\eta\right)(tr\bar{\bar{E}})\,\bar{\bar{\delta}} + 2\eta_c\bar{\bar{E}}$$

$$\tag{9.31}$$

$$+\phi\eta_c\left[2\bar{\bar{E}} + \left(\frac{\eta_c a}{\gamma}\right)\left\{\frac{48}{35}\left(\bar{\bar{E}}\bullet\bar{\bar{E}} - \frac{\bar{\bar{\delta}}}{3}\bar{\bar{E}}:\bar{\bar{E}}\right) - \frac{32}{5}\frac{D\bar{\bar{E}}}{Dt}\right\}\right]$$

This equation can also be recast into the following form:

$$\bar{\bar{\tau}}+\beta\frac{D\bar{\bar{\tau}}}{Dt}=2\eta_c[1+\phi]\left[\bar{\bar{E}}+\beta\frac{D\bar{\bar{E}}}{Dt}\right]+\left(\eta^k-\frac{2}{3}\eta\right)(tr\bar{\bar{E}})\bar{\bar{\delta}}$$

$$+\phi\eta_c\left(\frac{\eta_c a}{\gamma}\right)\left[\frac{48}{35}\left(\bar{\bar{E}}\bullet\bar{\bar{E}}-\frac{\bar{\bar{\delta}}}{3}\bar{\bar{E}}:\bar{\bar{E}}\right)-\frac{32}{5}\frac{D\bar{\bar{E}}}{Dt}\right]$$

(9.32)

where $\bar{\bar{\tau}}$ is the viscous stress tensor and β is equal to $6\eta_c a/5\gamma$. Equation 9.32 predicts bubbly suspensions to be non-Newtonian.

In steady viscometric flow expressed as $\vec{u}=(\dot{\gamma}x_2,0,0)$, where $\dot{\gamma}$ is the shear rate, the constitutive equation of bubbly suspension (Equation 9.32) can be written as [1,6]:

$$\begin{pmatrix}\tau_{11}&\tau_{12}&0\\\tau_{12}&\tau_{22}&0\\0&0&\tau_{33}\end{pmatrix}+\frac{\beta\dot{\gamma}}{2}\begin{pmatrix}-2\tau_{12}&\tau_{11}-\tau_{22}&0\\\tau_{11}-\tau_{22}&2\tau_{12}&0\\0&0&0\end{pmatrix}$$

$$=2\eta_c(1+\phi)\left[\frac{\dot{\gamma}}{2}\begin{pmatrix}0&1&0\\1&0&0\\0&0&0\end{pmatrix}+\frac{\beta\dot{\gamma}^2}{2}\begin{pmatrix}-1&0&0\\0&1&0\\0&0&0\end{pmatrix}\right]$$

(9.33)

$$+\left(\frac{\eta_c^2 a}{\gamma}\phi\right)\left[-\frac{16}{5}\dot{\gamma}^2\begin{pmatrix}-1&0&0\\0&1&0\\0&0&0\end{pmatrix}+\frac{4}{35}\dot{\gamma}^2\begin{pmatrix}1&0&0\\0&1&0\\0&0&-2\end{pmatrix}\right]$$

Comparing terms yields the following set of equations:

$$\tau_{11}-\beta\dot{\gamma}\tau_{12}=\eta_c\dot{\gamma}^2\beta\left(-1+\frac{37}{21}\phi\right)$$

(9.34)

$$\tau_{12}+\frac{\beta\dot{\gamma}}{2}(\tau_{11}-\tau_{22})=\eta_c\dot{\gamma}[1+\phi]$$

(9.35)

$$\tau_{22}+\beta\dot{\gamma}\tau_{12}=\eta_c\dot{\gamma}^2\beta\left[1-\frac{11}{7}\phi\right]$$

(9.36)

$$\tau_{33}=\frac{-4\eta_c\dot{\gamma}^2\beta\phi}{21}$$

(9.37)

Equation 9.34 to Equation 9.37 give the following expression for shear stress:

$$\tau_{12} = \frac{\eta_c \dot{\gamma}}{1+\beta^2\dot{\gamma}^2}(1+\phi) + \frac{\eta_c\beta^2\dot{\gamma}^3}{1+\beta^2\dot{\gamma}^2}\left(1-\frac{5}{3}\phi\right) = \eta_c\dot{\gamma}\left[1+\phi\left(\frac{1-\frac{5}{3}\beta^2\dot{\gamma}^2}{1+\beta^2\dot{\gamma}^2}\right)\right] \qquad (9.38)$$

Using the definitions of viscosity (ratio of shear stress to shear rate), relative viscosity (ratio of suspension viscosity η to continuous-phase viscosity η_c), and β ($=6\eta_c$ $a/5\gamma$), Equation 9.38 yields:

$$\eta_r = \frac{\eta}{\eta_c} = 1+\phi\left[\frac{1-\left(\frac{12}{5}N_{Ca}^2\right)}{1+\left(\frac{6}{5}N_{Ca}\right)^2}\right] \qquad (9.39)$$

where N_{Ca} is the capillary number defined as $\eta_c\dot{\gamma}a/\gamma$. The first and second normal stress differences are as follows:

$$N_1 = \tau_{11} - \tau_{22} = \frac{\left(\frac{32}{5}\right)N_{Ca}\phi\eta_c\dot{\gamma}}{1+\left(\frac{6}{5}N_{Ca}\right)^2} \qquad (9.40a)$$

$$N_2 = \tau_{22} - \tau_{33} = \frac{-\left(\frac{16}{5}\right)\phi N_{Ca}\eta_c\dot{\gamma}}{1+\left(\frac{6}{5}N_{Ca}\right)^2}\left[1-\frac{3}{28}\left\{1+\left(\frac{6}{5}N_{Ca}\right)^2\right\}\right] \qquad (9.40b)$$

Equation 9.40a and Equation 9.40b can be rewritten in terms of the reduced normal stress differences (N_{1r} and N_{2r}) as:

$$N_{1r} = \frac{N_1}{\eta_c\dot{\gamma}} = \frac{\left(\frac{32}{5}\right)N_{Ca}\phi}{1+\left(\frac{6}{5}N_{Ca}\right)^2} \qquad (9.41a)$$

$$N_{2r} = \frac{N_2}{\eta_c \dot{\gamma}} = \frac{-\left(\dfrac{16}{5}\right)\phi N_{Ca}}{1+\left(\dfrac{6}{5}N_{Ca}\right)^2}\left[1-\frac{3}{28}\left\{1+\left(\frac{6}{5}N_{Ca}\right)^2\right\}\right] \qquad (9.41b)$$

Figure 9.1 shows the plots of η_r, N_{1r}, and $-N_{2r}$ generated from Equation 9.39, Equation 9.41a, and Equation 9.41b. The suspension exhibits shear thinning in

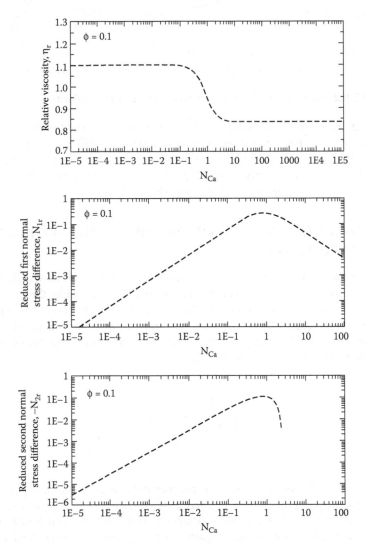

FIGURE 9.1 Variations of relative viscosity (η_r) and reduced normal stress differences (N_{1r} and N_{2r}) with capillary number for a bubbly suspension (bubble volume fraction $\phi = 0.10$).

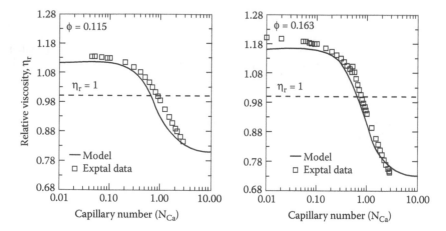

FIGURE 9.2 Comparison of model predictions with experimental data.

that η_r decreases with the increase in N_{Ca}. The suspension also exhibits viscoelastic properties in that the first and second normal stress differences are nonzero. Figure 9.2 compares the relative viscosity predictions of Equation 9.39 with experimental data on bubbly suspensions [4]. The model (Equation 9.39) predictions are in good agreement with the experimental data.

9.4 EFFECT OF SURFACTANTS ON THE RHEOLOGY OF BUBBLY SUSPENSIONS

The presence of surfactant film on the surface of bubbles can have a significant influence on the rheology of bubbly suspensions. First, consider the case where the Marangoni effect is absent and the interfacial film is purely viscous with nonzero interfacial shear viscosity η_s and nonzero interfacial dilational viscosity η_s^k. In the limit $N_{Ca} \rightarrow 0$, the shear viscosity of such a bubbly suspension is given by the Oldroyd formula [10]:

$$\eta = \eta_c \left[1 + \left(\frac{1 + 2N_{B0}^\eta + 3N_{B0}^k}{1 + \frac{2}{5}\left(2N_{B0}^\eta + 3N_{B0}^k\right)} \right) \phi \right] \qquad (9.42)$$

where N_{B0}^η is the shear Boussinesq number defined as $\eta_s/\eta_c a$ and N_{B0}^k is the dilational Boussinesq number defined as $\eta_s^k/\eta_c a$. For a highly viscous interface, $(2N_{B0}^\eta + 3N_{B0}^k) \gg 1$, the bubbles behave as rigid particles and Equation 9.42 reduces to the Einstein viscosity equation for suspensions of rigid particles. For an inviscid interface, $(2N_{B0}^\eta + 3N_{B0}^k) \ll 1$, Equation 9.42 gives an intrinsic viscosity $[\eta]$, defined as $(\eta - \eta_c)/\eta_c \phi$, of unity.

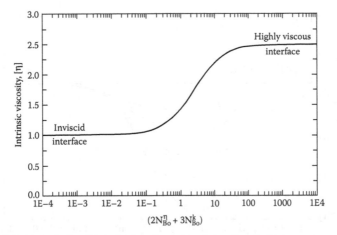

FIGURE 9.3 Variation of intrinsic viscosity [η] of a bubbly suspension with $(2\,N_{Bo}^{\eta} + 3\,N_{Bo}^{k})$.

Figure 9.3 shows the variation of intrinsic viscosity of a bubbly suspension with $(2N_{Bo}^{\eta} + 3N_{Bo}^{k})$. The intrinsic viscosity of the suspension increases with increase in the viscous properties of the interface.

Interestingly, the dilational viscosity of dilute bubbly suspension now depends on the dilational viscosity of the interface as well, and is given as [8]:

$$\eta^{k} = \frac{4\eta_{c}}{3\phi}\left[1 + \frac{\eta_{s}^{k}}{\eta_{c}a}\right] \tag{9.43}$$

In cases where the Marangoni effect is important and interfacial tension gradients are present along the bubble interface, one needs to consider the Gibbs elasticity of the interface. In the limit $N_{Ca} \to 0$ (negligible deformation of bubbles), the *zero-shear viscosity* of a dilute bubbly suspension in the presence of the Marangoni effect is given by [11]:

$$\eta = \eta_{c}\left[1 + \left(1 + \frac{3}{2}\varepsilon_{m}\right)\phi\right] \tag{9.44}$$

where ε_{m}, the interfacial mobility parameter, is given as:

$$\varepsilon_{m} = \frac{\dfrac{2}{5}\left(\dfrac{aE_{G}}{2\eta_{c}D} + 2N_{B0}^{\eta} + 3N_{B0}^{k}\right)}{1 + \dfrac{2}{5}\left(\dfrac{aE_{G}}{2\eta_{c}D} + 2N_{B0}^{\eta} + 3N_{B0}^{k}\right)} \tag{9.45}$$

where E_{G} is Gibbs elasticity of interface and D is effective diffusion coefficient of the surfactant.

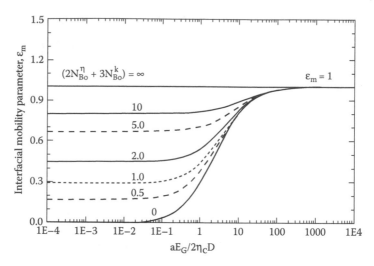

FIGURE 9.4 Variation of interfacial mobility parameter ε_m with the dimensionless elasticity $aE_G/2\eta_cD$.

For interfaces with high surface elasticity ($E_G \to \infty$), $\varepsilon_m = 1$ and the bubbly suspension behaves more as a suspension of rigid particles. When $E_G \to 0$, the preceding expression for bubbly suspension viscosity reduces to the Oldroyd equation (Equation 9.42).

Figure 9.4 shows the variation of the interfacial mobility parameter ε_m with the dimensionless elasticity of the interface given by $aE_G/2\eta_cD$. At high values of $aE_G/2\eta_cD$, $\varepsilon_m = 1$ regardless of the value of the viscous properties of the interface expressed in terms of $(2N_{Bo}^{\eta} + 3N_{Bo}^{k})$. At low values of $aE_G/2\eta_cD$, the interfacial mobility parameter ε_m varies from 0 to 1, depending on the value of $(2N_{Bo}^{\eta} + 3N_{Bo}^{k})$.

NOTATION

a	bubble radius
D	effective diffusion coefficient of the surfactant
D/Dt	Jaumann derivative
E_G	Gibbs elasticity of interface
$\overline{\overline{E}}$	rate of strain tensor
N	number of bubbles
N_1	first normal stress difference
N_2	second normal stress difference
N_{1r}	reduced first normal stress difference
N_{2r}	reduced second normal stress difference
N_{Bo}^{k}	dilational Boussinesq number

N_{Bo}^{η}	shear Boussinesq number
N_{Ca}	capillary number
P	pressure
R	cell radius
r	radial coordinate
t	time
tr	trace of a tensor
$\vec{u}$	velocity vector
u_r	radial component of velocity vector
V	volume
x_2	position coordinate in direction 2

GREEK

β	parameter defined as $6\eta_c a/5\gamma$; see Equation 9.32
γ	interfacial tension
$\dot{\gamma}$	shear rate
Δ	rate of expansion of suspension
∇	del operator
$\hat{\delta}_r, \hat{\delta}_\theta, \hat{\delta}_\phi$	unit vectors
$\bar{\bar{\delta}}$	unit tensor
ε_m	interfacial mobility parameter; see Equation 9.45
η	shear viscosity
η_c	continuous phase viscosity
η^k	dilational viscosity of suspension
η_r	relative viscosity
η_s^k	surface dilational viscosity
ρ	density
$\bar{\bar{\sigma}}$	total stress tensor
σ_{rr}	normal stress component of total stress tensor
$\bar{\bar{\tau}}$	deviatoric stress tensor
$\tau_{rr}, \tau_{\theta\theta}, \tau_{\phi\phi}$	normal stress components of deviatoric stress tensor
τ_{12}	shear stress
$\tau_{11}, \tau_{22}, \tau_{33}$	normal stress components of deviatoric stress tensor
ϕ	volume fraction of bubbles

REFERENCES

1. R. Pal, *Earth Planet. Sci. Lett.*, 207: 165–179, 2003.
2. A.M. Lejeune, Y. Bottinga, T.W. Trull, P. Richet, *Earth Planet. Sci. Lett.*, 166: 71–84, 1999.

3. D.J. Stein, F.J. Spera, *J. Volcanol. Geotherm. Res.*, 113: 243–258, 2002.

4. A.C. Rust, M. Manga, *J. Non-Newtonian Fluid Mech.*, 104: 53–63, 2002.

5. M. Manga, M. Loewenberg, *J. Volcanol. Geotherm. Res.*, 105: 19–24, 2001.

6. R. Pal, *Ind. Eng. Chem. Res.*, 43: 5372–5379, 2004.

7. R.K. Prud'homme, R.B. Bird, *J. Non-Newtonian Fluid Mech.*, 3: 261–279, 1978.

8. D.A. Edwards, H. Brenner, D.T. Wasan, *Interfacial Transport Processes and Rheology*, Butterworth-Heinemann, Boston, MA, 1991.

9. G. I. Taylor, *Proc. Roy. Soc. Lond. A*, 226: 34–39, 1954.

10. J.G. Oldroyd, *Proc. R. Soc. Lond. A*, 232: 567–577, 1955.

11. K.D. Danov, *J. Colloid Int. Sci.*, 235: 144–149, 2001.

10 Dispersions of Capsules

10.1 INTRODUCTION

The term *capsule* refers to a particle consisting of a drop of liquid surrounded by a thin deformable elastic-solid membrane [1]. Droplets coated with a polymerized interface formed by interfacial cross-linking polymerization [2] are a good example of capsules. Capsules are also thought to be models of red blood cells (RBCs) [3–5]. Capsules are frequently used in medical or pharmaceutical processes for the transport of drugs [6]; however, in these applications, the membrane is generally porous so that the transport of the internal capsule content (drug) to the external environment can take place.

The rheological behavior of suspensions of capsules (liquid drops enclosed by elastic membranes) differs from that of suspensions of solid particles or liquid droplets. At low shear rates the capsules behave as rigid particles and rotate as a rigid body independently of the viscosity of the enclosed liquid. At high shear rates, however, they deform and attain a steady-state orientation, and their membrane undergoes so-called tank-treading motion; the tank-treading motion of the membrane also induces an internal flow.

The tank-treading motion of capsules has been observed experimentally in the case of RBCs [7–9]. The RBCs consist of a thin elastic membrane filled with a hemoglobin solution. In the unstressed state, RBCs are biconcave discs about 8 μm in diameter and 1 μm in thickness at the center; the maximum thickness is about 2.5 μm [10]. Figure 10.1 shows a photomicrograph of RBCs in the unstressed state.

Fischer and co-workers [8] carried out microscopic observations of RBC suspensions under shear, using a rheoscope. The cells suspended in viscous solutions are deformed to flat ellipsoids when subjected to shear. Also, the deformed RBCs assume a stationary orientation in the shear field, as shown schematically in Figure 10.2. In the rheoscope, the particles are observed from the top by focusing on the stationary layer of the shear field, that is, at the midpoint of two plates moving in opposite directions. Thus, only the projection of the RBC in the x-y plane can be seen. Figure 10.3 shows a typical photomicrograph of stressed RBCs subjected to shear. Clearly, they are ellipsoidal in shape although the motion of the membrane cannot be seen directly. To observe the motion of the membrane, Fischer and co-workers [8] used latex particles as membrane markers (these particles adhere to the membrane). Figure 10.4 shows the motion of a latex membrane marker. The linear

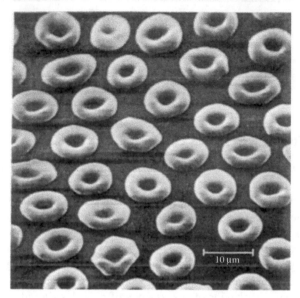

FIGURE 10.1 Photomicrograph of red blood cells (RBCs) in an unstressed state. (From S. Sutera, M. Mehrjardi, N. Mohandas, Deformation of erythrocytes under shear, *Blood Cells*, 1: 369–374, 1975. Reprinted with kind permission of Springer Science and Business Media.)

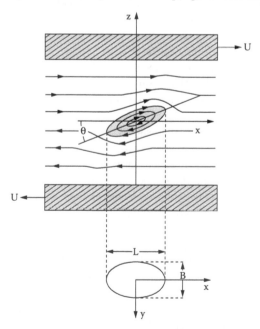

FIGURE 10.2 Orientation of a red blood cell (RBC) in a shear field. (Adapted from T.M. Fischer, M. Stohr-Liesen, H. Schmid-Schonbein, Red blood cell (RBC) microrheology: comparison of the behavior of single RBC and liquid droplets in shear flow, *AIChE Symp. Ser.* 182, 74: 38–45, 1978.)

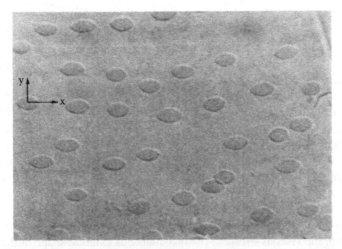

FIGURE 10.3 Typical photomicrograph of stressed RBCs subjected to shear. (From T.M. Fischer, M. Stohr-Liesen, H. Schmid-Schonbein, Red blood cell (RBC) microrheology: comparison of the behavior of single RBC and liquid droplets in shear flow, *AIChE Symp. Ser.* 182, 74: 38–45, 1978.)

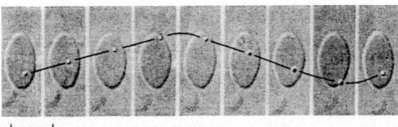

FIGURE 10.4 Motion of a latex membrane marker (the marker is adhered to the membrane of a red blood cell). (From T.M. Fischer, H. Schmid-Schonbein, Tank tread motion of red cell membrances in viscometric flow: Behavior of intracellular and extracellular markers (with film), *Blood Cells*, 3: 351–365, 1977. Reprinted with kind permission from Springer Science and Business Media.)

back and forth motion of the marker, observed experimentally in Figure 10.4, is the motion of the marker on closed paths in the x-z plane (Figure 10.2). The tank-treading frequency increases linearly with the shear rate.

10.2 FUNDAMENTAL EQUATIONS

The rheology of suspensions of deformable particles is intimately linked to the deformation and orientation of particles in the shear field. Knowledge of the motion and deformation of a single representative particle in a viscous shear flow is required to calculate the dipole strength term contributing to the bulk stress of the suspension.

As pointed out in Chapter 8, it is difficult to model the behavior of deformable particles. The main difficulty lies in satisfying the boundary conditions (continuity of velocities and force balance) on the surface of the deforming particle, which is itself unknown. Pressure and velocity fields, within and outside the particle, and particle shape must all be determined simultaneously. Such "free-surface type" problems are generally treated with a domain perturbation method [1,3,11].

Consider a spherical capsule of radius R freely suspended in a Newtonian incompressible liquid of viscosity η_c. The capsule is filled with a Newtonian incompressible liquid of viscosity $\lambda\eta_c$, where λ is the ratio of internal fluid viscosity to external fluid viscosity. An infinitely thin, impermeable, elastic membrane surrounds the internal fluid. The membrane material is assumed to be isotropic, and the membrane is treated as a two-dimensional surface with E_s as the characteristic surface elastic modulus [12].

When the system is subjected to a shear flow, the capsule deforms under the influence of the viscous stresses. The instantaneous external shape of the capsule can be expressed as:

$$r = (x_1^2 + x_2^2 + x_3^2)^{1/2} = f(x_1, x_2, x_3) \tag{10.1}$$

where r is the distance from the particle center of mass to the interface. The function f is *a priori* unknown and must be determined as part of the problem solution. Assuming that the particle Reynolds number is small, the fluid motion (inside and outside the capsule) can be described by the Stokes equations with respect to a reference frame centered on particle's center of mass and translating with the particle [12]. Thus:

$$\eta_c \nabla^2 \vec{u} - \nabla P = 0 \qquad \nabla \bullet \vec{u} = 0 \quad \text{for} \quad r \geq f \tag{10.2}$$

$$\lambda\eta_c \nabla^2 \vec{u}^* - \nabla P^* = 0 \qquad \nabla \bullet \vec{u}^* = 0 \quad \text{for} \quad r \leq f \tag{10.3}$$

The boundary condition at infinity (far away from the capsule) is:

$$\vec{u} \to \vec{u}_\infty \text{ as } r \to \infty \tag{10.4}$$

The boundary conditions at the interface of the capsule are:

$$\vec{u}^* = \vec{u} = \vec{u}_M \tag{10.5}$$

where $\vec{u}_M$ represents the velocity of the membrane, and

$$\hat{n} \bullet (\bar{\bar{\sigma}} - \bar{\bar{\sigma}}^*) = -(\bar{\bar{\delta}} - \hat{n}\hat{n}) \bullet \nabla \bullet \bar{\bar{P}}_s \tag{10.6}$$

where $\bar{\bar{\sigma}}$ is the stress tensor, $\hat{n}$ is the outward unit normal vector at the interface, and $\bar{\bar{P}}_s$ is the surface-stress (tension) tensor. The final boundary condition, given in Equation 10.6, corresponds to the dynamic equilibrium of the membrane.

To close the formulation of the problem, the stresses (tensions) in the membrane must be related to the displacement of its material points. Thus, knowledge of membrane mechanics is needed to solve the capsule problem.

10.3   MEMBRANE MECHANICS

The coupling between fluid mechanics and membrane mechanics arises through Equation 10.5 and Equation 10.6. Equation 10.5 represents kinematic coupling, indicating the continuity of velocities between the internal and external liquids and the membrane. Equation 10.6 represents the dynamic coupling between the fluid and the membrane.

The membrane is assumed to be purely elastic and of vanishingly small thickness so that it has no bending resistance. It is further assumed that the membrane reacts instantaneously to stress (the membrane is hyperelastic), and it is isotropic. Because the membrane material is purely elastic, its mechanical behavior is completely determined by a strain energy function W, defined as the internal strain energy per unit of initial membrane area. The surface-stress tensor $\bar{\bar{P}}_s$ is related to the strain energy function as [13]:

$$\bar{\bar{P}}_s = \frac{1}{J_s}\, \bar{\bar{F}}_s \bullet \frac{\partial W}{\partial \bar{\bar{\ell}}_s} \bullet \bar{\bar{F}}_s^{T} \tag{10.7}$$

where J_s is the ratio between the deformed and undeformed local surface areas, $\bar{\bar{F}}_s$ is the surface deformation gradient tensor, and $\bar{\bar{\ell}}_s$ is the surface-strain tensor. $\bar{\bar{F}}_s$ is defined as [13,14]:

$$\bar{\bar{F}}_s = (\bar{\bar{\delta}} - \hat{n}\hat{n}) \bullet \frac{\partial \vec{x}}{\partial \bar{\bar{X}}} \bullet (\bar{\bar{\delta}} - \hat{N}\hat{N}) \tag{10.8}$$

where $\hat{n}$ and $\hat{N}$ are unit vectors normal to the membrane surface in the deformed and reference configurations, respectively, and $\bar{\bar{\delta}}$ is a unit tensor. The material points in the membrane are labeled by $\bar{X}$ in the reference (unstressed) configuration. Upon application of a load, the material points are displaced to a new position $\bar{x}(\bar{X},t)$.

The surface strain tensor $\bar{\bar{\ell}}_s$ is defined as:

$$\bar{\bar{\ell}}_s = \frac{1}{2}\left[\bar{\bar{F}}_s^{T} \bullet \bar{\bar{F}}_s - (\bar{\bar{\delta}} - \hat{N}\hat{N})\right] \tag{10.9}$$

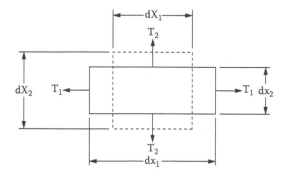

FIGURE 10.5 Diagram showing tensions and deformation in a membrane material.

In the reference configuration, $\overline{\overline{\ell}}_s = 0$.

The principal strains of surface strain tensor are:

$$\ell_{11} = \frac{1}{2}(\lambda_1^2 - 1) \tag{10.10}$$

$$\ell_{22} = \frac{1}{2}(\lambda_2^2 - 1) \tag{10.11}$$

where λ_1 and λ_2 are extension ratios along the principal directions, defined as:

$$\lambda_1 = \frac{dx_1}{dX_1}, \quad \lambda_2 = \frac{dx_2}{dX_2} \tag{10.12}$$

The dXs refer to line elements in the reference state and the dxs refer to line elements in the deformed state (Figure 10.5). Note that J_s, the ratio between the deformed and undeformed local surface areas, can be expressed as:

$$J_s = dA / dA_0 = \lambda_1 \lambda_2 \tag{10.13}$$

10.3.1 STRAIN ENERGY FUNCTION OF MEMBRANE MATERIAL

Depending on the nature of the membrane material, different strain energy functions (W) have been proposed in the literature. When deformations are small, the strain energy function can be expressed as [4]:

$$W = W_0 + \alpha_1' \, \ell n(\lambda_1 \lambda_2) + \frac{1}{2}(\alpha_1' + \alpha_2')(\ell n \lambda_1 \lambda_2)^2$$

$$+ \alpha_3' \left[\frac{1}{2}(\lambda_1^2 + \lambda_2^2) - 1 - \ell n \lambda_1 \lambda_2 \right] \tag{10.14}$$

where W_0 is a constant (which may be assigned an arbitrary value) and α_i' are material elastic coefficients.

10.3.1.1 Mooney–Rivlin (MR) Type Membrane

A membrane made of an infinitely thin sheet of a three-dimensional, isotropic, volume-incompressible elastomer is referred to as an MR membrane [4,15]. The elastic coefficients pertaining to strain energy function (Equation 10.14) of this material, in nondimensional form, are as follows:

$$\alpha_1 = \frac{\alpha_1'}{E_s} = 0, \alpha_2 = \frac{\alpha_2'}{E_s} = \frac{2}{3}, \alpha_3 = \frac{\alpha_3'}{E_s} = \frac{1}{3} \tag{10.15}$$

where E_s is the characteristic surface elastic modulus of the membrane.

10.3.1.2 RBC-Type Membrane

The membrane of RBCs consists of a lipid bilayer lined by a protein network that imparts elasticity to the membrane [15]. RBC membrane has the characteristic property that it is easily shearable but strongly resistant to any increase of local surface area [15,16]. The elastic coefficients of RBC membrane in dimensionless form (nondimensionalized by characteristic surface elastic modulus E_s of membrane) are:

$$\alpha_1 = \frac{\alpha_1'}{E_s} = 0, \alpha_2 = \frac{\alpha_2'}{E_s} = 1000, \alpha_3 = \frac{\alpha_3'}{E_s} = \frac{1}{2} \tag{10.16}$$

10.4 INSTANTANEOUS SHAPE OF INITIALLY SPHERICAL CAPSULE

Assuming that the deformation of the capsule from its initially spherical shape is small, the displacement of the material points of the membrane $\vec{X}$ in the reference configuration to new positions $\vec{x}(\vec{X}, t)$ can be written as [14]:

$$\vec{x} = \vec{X} + \in \overline{\overline{B}} \bullet \vec{X} + \in \vec{X} \bullet (\overline{\overline{A}} - \overline{\overline{B}}) \bullet \vec{X}\vec{X}/(\vec{X} \bullet \vec{X}) \tag{10.17}$$

where $\in$ is a measure of the deviation of shape from the initial sphere ($\in \ll 1$). As the material points undergo both radial and tangential (in-plane) displacements, two second-order symmetric and traceless tensors $\overline{\overline{A}}$ and $\overline{\overline{B}}$ are required to specify the displacement. $\overline{\overline{A}}$ specifies the radial displacement, and $\overline{\overline{B}}$ measures the tangential in-plane displacement of material points.

To a first-order deformation in $\in$, the instantaneous external shape of the capsule is given by [4]:

$$r = (\vec{x} \bullet \vec{x})^{1/2} = R(1 + \in \vec{X} \bullet \overline{\overline{A}} \bullet \vec{X}/R^2) = R\left[1 + \in \vec{x} \bullet \overline{\overline{A}} \bullet \vec{x}/r^2\right] \qquad (10.18)$$

with unit normal vector $\hat{n}$

$$\hat{n} = \frac{\vec{x}}{r} + 2\in \vec{x}\,\vec{x} \bullet \overline{\overline{A}} \bullet \vec{x}/r^3 - 2\in \overline{\overline{A}} \bullet \vec{x}/r \qquad (10.19)$$

Thus, to a first-order deformation in $\in$, the overall distortion of the profile of the capsule, given by Equation 10.18, depends only on $\overline{\overline{A}}$; the shape of the capsule is independent of $\overline{\overline{B}}$.

The kinematic coupling between the fluid and the membrane (Equation 10.5) leads to two differential equations, giving the time evolution of the capsule shape through the variation of $\overline{\overline{A}}$ and $\overline{\overline{B}}$ in time [4]:

$$\in \frac{D\overline{\overline{A}}}{Dt} = a_0 \overline{\overline{E}} + \in \left\{ \frac{3}{7} a_0 S_d\,(\overline{\overline{E}} \bullet \overline{\overline{B}}) + \left(a_1 + \frac{4}{7} a_0 \right) S_d\,(\overline{\overline{E}} \bullet \overline{\overline{A}}) \right\}$$

$$+ \frac{\in}{(\eta_c\,\dot{\gamma}\,R/E_s)}\ \left\{ b_0 \overline{\overline{L}} + (b_1 + b_2)\,\overline{\overline{M}} \right\} \qquad (10.20)$$

$$+ \in \overline{\overline{\delta}} \left\{ \frac{1}{2} a_0 \overline{\overline{E}} : \overline{\overline{B}} + \frac{1}{3} a_0 \overline{\overline{E}} : \overline{\overline{A}} \right\}$$

$$\in \frac{D}{Dt}(\overline{\overline{A}} - \overline{\overline{B}}) = \frac{2}{7}\in a_0 \left[S_d\,(\overline{\overline{E}} \bullet \overline{\overline{A}}) - S_d\,(\overline{\overline{E}} \bullet \overline{\overline{B}}) \right] + \frac{\in b_2}{\left(\dfrac{\eta_c\,\dot{\gamma}\,R}{E_s} \right)}\,\overline{\overline{M}} \qquad (10.21)$$

where

$$S_d\,(\overline{\overline{E}} \bullet \overline{\overline{A}}) = \frac{1}{2}\left(\overline{\overline{E}} \bullet \overline{\overline{A}} + \overline{\overline{A}} \bullet \overline{\overline{E}} - \frac{2}{3} \overline{\overline{E}} : \overline{\overline{A}}\,\overline{\overline{\delta}} \right) \qquad (10.22)$$

$$S_d\,(\overline{\overline{E}} \bullet \overline{\overline{B}}) = \frac{1}{2}\left(\overline{\overline{E}} \bullet \overline{\overline{B}} + \overline{\overline{B}} \bullet \overline{\overline{E}} - \frac{2}{3} \overline{\overline{E}} : \overline{\overline{B}}\,\overline{\overline{\delta}} \right) \qquad (10.23)$$

$$a_0 = 5/(2\lambda + 3) \qquad (10.24)$$

$$a_1 = 60\,(\lambda - 1)\Big/\left[7\,(2\lambda + 3)^2\right] \qquad (10.25)$$

$$b_0 = 1/(2\lambda + 3) \qquad (10.26)$$

$$b_1 = 2\,(3\lambda + 2)\Big/\left[(19\lambda + 16)\,(2\lambda + 3)\right] \qquad (10.27)$$

$$b_2 = 2/(19\lambda + 16) \qquad (10.28)$$

D/Dt is a Jaumann derivative, defined as a time derivative relative to axes rotating with the capsule:

$$\frac{D\bar{\bar{A}}}{Dt} = \frac{\partial \bar{\bar{A}}}{\partial t} + \bar{\bar{A}} \bullet \bar{\bar{\Omega}} - \bar{\bar{\Omega}} \bullet \bar{\bar{A}} \qquad (10.29)$$

$$\frac{D\bar{\bar{B}}}{Dt} = \frac{\partial \bar{\bar{B}}}{\partial t} + \bar{\bar{B}} \bullet \bar{\bar{\Omega}} - \bar{\bar{\Omega}} \bullet \bar{\bar{B}} \qquad (10.30)$$

The tensors $\bar{\bar{L}}$ and $\bar{\bar{M}}$ in Equation 10.20 and Equation 10.21 are symmetric and traceless and are linear functions of $\bar{\bar{A}}$ and $\bar{\bar{B}}$:

$$\bar{\bar{L}} = 4\left(\alpha_2 + \alpha_3\right)\bar{\bar{A}} - \left(6\,\alpha_2 + 10\,\alpha_3\right)\bar{\bar{B}} \qquad (10.31)$$

$$\bar{\bar{M}} = -4\left(\alpha_1 + 2\alpha_2 + 2\alpha_3\right)\bar{\bar{A}} + \left(12\,\alpha_2 + 16\,\alpha_3\right)\bar{\bar{B}} \qquad (10.32)$$

In Equation 10.31 and Equation 10.32, the coefficients α_i correspond to the coefficients present in the strain energy function of the membrane material (see Equation 10.14). The coefficients in Equation 10.31 and Equation 10.32 are in dimensionless form (see Equation 10.15 and Equation 10.16).

For weak flows where the elastic capillary number, defined as $\eta_c\,\dot{\gamma}\,R/E_s$, is small (less than one), certain terms in Equation 10.20 and Equation 10.21 can be neglected and ϵ can be set equal to $(\eta_c\,\dot{\gamma}\,R/E_s)$. Consequently,

$$\in \frac{D \bar{\bar{A}}}{D t} = a_0 \, \bar{\bar{E}} + \left\{ b_0 \, \bar{\bar{L}} + \left(b_1 + b_2 \right) \bar{\bar{M}} \right\} \qquad (10.33)$$

$$\in \frac{D \bar{\bar{B}}}{D t} = a_0 \, \bar{\bar{E}} + \left\{ b_0 \, \bar{\bar{L}} + b_1 \, \bar{\bar{M}} \right\} \qquad (10.34)$$

The analysis of the deformation of the initial spherical capsule presented here is first order in $\in$ (deformation parameter). Higher-order terms in $\in$ are neglected in the analysis. This analysis predicts a tank-treading motion of the membrane about the inner contents of the capsule in a shear field; the capsule deforms into an ellipsoid and orients itself at 45° with respect to the stream lines [14,15]. The deformation of the capsule increases with the increase in $(\eta_c \dot{\gamma} \, R/E_s)$, but the orientation remains constant at 45°. However, this is true only if the membrane is purely elastic in nature. For viscoelastic membranes (not considered here), the deformation increases with increase in shear rate, and the orientation angle decreases from 45° to 0° with increasing shear rate [14].

Barthes-Biesel [1] and Barthes-Biesel and Chhim [3] have extended the analysis of the deformation of a capsule with purely elastic membrane to second order in $\in$, that is, terms involving $\in^2$ have been included. However, they have considered only steady-state behavior, and their analysis is restricted to very weak steady flows.

10.4.1 DEFORMATION OF A CAPSULE IN STEADY IRROTATIONAL FLOW

If vorticity is neglected, then at steady state Equation 10.33 and Equation 10.34 give:

$$\bar{\bar{M}} = 0, \quad \bar{\bar{L}} = \left(\frac{-a_0}{b_0} \right) \bar{\bar{E}} = -5 \, \bar{\bar{E}} \qquad (10.35)$$

From Equation 10.31, Equation 10.32, and Equation 10.35, it follows that

$$\bar{\bar{A}} = \frac{5}{2} \left[\frac{3\alpha_2 + 4\alpha_3}{\alpha_1 \left(3\alpha_2 + 5\alpha_3 \right) + 2\alpha_3 \left(\alpha_2 + \alpha_3 \right)} \right] \bar{\bar{E}} \qquad (10.36)$$

$$\bar{\bar{B}} = \frac{5}{2} \left[\frac{\alpha_1 + 2\alpha_2 + 2\alpha_3}{\alpha_1 \left(3\alpha_2 + 5\alpha_3 \right) + 2\alpha_3 \left(\alpha_2 + \alpha_3 \right)} \right] \bar{\bar{E}} \qquad (10.37)$$

According to Equation 10.36 and Equation 10.37, $\bar{\bar{A}}$ and $\bar{\bar{B}}$ are independent of the viscosity ratio and depend on the elastic properties of the membrane material. Furthermore, they are proportional to and coaxial with $\bar{\bar{E}}$.

For MR membrane material, the coefficients α_i are (see Equation 10.15): $\alpha_1 = 0$, $\alpha_2 = 2/3$, $\alpha_3 = 1/3$. On substitution of these α_i values in Equation 10.36 and Equation 10.37, we obtain:

$$\bar{\bar{A}} = \frac{25}{2}\bar{\bar{E}}, \quad \bar{\bar{B}} = \frac{15}{2}\bar{\bar{E}} \tag{10.38}$$

For RBC membrane material, the coefficients α_i are (see Equation 10.16): $\alpha_1 = 0$, $\alpha_2 = 1000$, $\alpha_3 = 1/2$. On substitution of these α_i in Equation 10.36 and Equation 10.37, the following result is obtained:

$$\bar{\bar{A}} = \frac{15}{2}\bar{\bar{E}}, \quad \bar{\bar{B}} = 5\bar{\bar{E}} \tag{10.39}$$

10.4.2 Deformation of a Capsule in Steady Simple Shear Flow

In steady simple shear flow,

$$E_{12} = E_{21} = \Omega_{12} = -\Omega_{21} = \dot{\gamma}/2 \tag{10.40}$$

and

$$\partial/\partial t = 0 \tag{10.41}$$

10.4.2.1 Capsule with a Mooney–Rivlin-Type Membrane

In this case, Equation 10.33 and Equation 10.34 need to be solved for the components of $\bar{\bar{A}}$ and $\bar{\bar{B}}$ using Equation 10.31 and Equation 10.32 for $\bar{\bar{L}}$ and $\bar{\bar{M}}$ and the following values of α_i:

$$\alpha_1 = 0, \alpha_2 = 2/3, \alpha_3 = 1/3 \tag{10.42}$$

The solution for components B_{12} and A_{12} are as follows [17,18]:

$$B_{12} = \frac{a_0}{8\delta_2}\left\{1 - \frac{(\delta_1 + \epsilon^2/8)\left[\beta_1 - \beta_3 + (\delta_1 + \epsilon^2/8)(1/4\delta_2)\right]}{(\delta_2\,\epsilon^2/16) + (\delta_1 + \epsilon^2/8)^2(1/4\delta_2)}\right\} \tag{10.43}$$

$$A_{12} = \frac{a_0}{16\beta_1} \left\{ \frac{\beta_2}{3\delta_2} - 1 \right.$$

$$\left. + \frac{\left[-4\beta_2\delta_1 + (3\delta_2 - \beta_2)(\epsilon^2/2) \right]\left[\beta_1 - \beta_3 + (\delta_1 + \epsilon^2/8)(1/4\delta_2) \right]}{3(\delta_2^2 \epsilon^2/4) + 3(\delta_1 + \epsilon^2/8)^2} \right\}$$

(10.44)

where

$$2\beta_1 = b_0 - 2b_1; \quad 2\beta_2 = 11b_0 - 20b_1 \tag{10.45}$$

$$2\beta_3 = b_0 - 2b_1 - 2b_2; \quad 2\beta_4 = 11b_0 - 20b_1 - 20b_2 \tag{10.46}$$

$$\delta_1 = \frac{4}{3}(\beta_2\beta_3 - \beta_1\beta_4); \quad \delta_2 = \frac{\beta_2}{3} - 2\beta_3 \tag{10.47}$$

10.4.2.2 Capsule with an RBC-Type Membrane

The elastic properties of an RBC type membrane are such that its area is preserved to a close approximation. The constancy of area requires that [4]:

$$\overline{\overline{A}} = \frac{3}{2}\overline{\overline{B}} \tag{10.48}$$

Consequently, only one equation is necessary to describe the deformation of the capsule. Equation 10.31 to Equation 10.34 and Equation 10.48 reduce to a single differential equation for $\overline{\overline{A}}$:

$$\epsilon \frac{D\overline{\overline{A}}}{Dt} = \frac{60}{(23\lambda + 32)}\overline{\overline{E}} - \frac{8}{(23\lambda + 32)}\overline{\overline{A}} \tag{10.49}$$

This equation gives the following expression for the component A_{12} [17,18]:

$$A_{12} = \frac{3}{2}B_{12} = \frac{60}{\left[16 + (23\lambda + 32)^2(\epsilon^2/4) \right]} \tag{10.50}$$

10.5 RHEOLOGICAL CONSTITUTIVE EQUATION FOR DISPERSION OF CAPSULES

The constitutive equation for a dilute dispersion of identical capsules is given as:

$$\bar{\bar{\sigma}} = -P\bar{\bar{\delta}} + 2\eta_c \bar{\bar{E}} + \frac{\phi}{V_p} \bar{\bar{S}}^0 \tag{10.51}$$

where V_p is the volume of a single capsule, $\bar{\bar{\sigma}}$ is the average stress tensor, P is the average pressure, $\bar{\bar{E}}$ is average rate of strain tensor, ϕ is volume fraction of capsules, and $\bar{\bar{S}}^0$ is the dipole strength of a single capsule located in an infinite matrix fluid.

Barthes-Biesel and Rallison [4] obtained an expression for the dipole strength $\bar{\bar{S}}^0$. Only first-order deformation of capsule was considered. Their expression for $\bar{\bar{S}}^0$ is as follows:

$$\bar{\bar{S}}^0 = 2V_p \eta_c \left\{ \frac{5(\lambda - 1)}{2\lambda + 3} \bar{\bar{E}} - \frac{5}{2}\left(\frac{1}{2\lambda + 3}\right)\bar{\bar{L}} - \left(\frac{1}{2\lambda + 3}\right)\bar{\bar{M}} \right\} \tag{10.52}$$

Thus, the bulk stress for a dilute suspension of identical capsules is as follows:

$$\bar{\bar{\sigma}} = -P\bar{\bar{\delta}} + 2\eta_c \bar{\bar{E}} + 2\eta_c \phi \left\{ \frac{5(\lambda - 1)}{2\lambda + 3} \bar{\bar{E}} - \frac{5}{2}\left(\frac{1}{2\lambda + 3}\right)\bar{\bar{L}} - \left(\frac{1}{2\lambda + 3}\right)\bar{\bar{M}} \right\} \tag{10.53}$$

Equation 10.53 can be used to predict the rheological behavior of dilute suspensions of initially spherical capsules. Although the equation is derived using a first-order deformation of capsules, it seems to give reasonable predictions even for large values of $(\eta_c \dot{\gamma} R/E_s)$ [17,18].

10.5.1 STEADY IRROTATIONAL FLOW

In steady irrotational flow,

$$\bar{\bar{M}} = 0; \quad \bar{\bar{L}} = -\frac{a_0}{b_0}\bar{\bar{E}} = -5\bar{\bar{E}} \tag{10.54}$$

Consequently, Equation 10.53 reduces to:

$$\bar{\bar{\sigma}} = -P\bar{\bar{\delta}} + 2\eta_c \left(1 + \frac{5}{2}\phi\right)\bar{\bar{E}} \tag{10.55}$$

In uniaxial extensional flow, Equation 10.55 gives the following expression for the elongational viscosity η_E:

$$\eta_E = 3\eta_c \left(1 + \frac{5}{2}\phi\right) \tag{10.56}$$

Thus, the extensional (elongational) viscosity is independent of the rate of extension, indicating Newtonian behavior.

10.5.2 STEADY SIMPLE SHEAR FLOW

In steady simple shear flow, the shear stress τ_{12} can be determined from Equation 10.53:

$$\tau_{12} = \sigma_{12} + P = 2\eta_c E_{12}$$

$$+ 2\eta_c \phi \left[\frac{5(\lambda-1)}{2\lambda+3} E_{12} - \frac{5}{2}\left(\frac{1}{2\lambda+3}\right)L_{12} - \left(\frac{1}{2\lambda+3}\right)M_{12} \right] \tag{10.57}$$

where

$$L_{12} = 4\left(\alpha_2 + \alpha_3\right)A_{12} - \left(6\alpha_2 + 10\alpha_3\right)B_{12} \tag{10.58}$$

$$M_{12} = -4\left(\alpha_1 + 2\alpha_2 + 2\alpha_3\right)A_{12} + \left(12\alpha_2 + 16\alpha_3\right)B_{12} \tag{10.59}$$

From the shear stress, the intrinsic viscosity $[\eta]$ of the suspension can be calculated. The intrinsic viscosity $[\eta]$ is defined as:

$$[\eta] = \underset{\phi \to 0}{Lim} \frac{\left(\tau_{12}/2E_{12}\right) - \eta_c}{\eta_c \phi} \tag{10.60}$$

10.5.2.1 Capsule with a Mooney–Rivlin-Type Membrane

For Mooney–Rivlin-type membranes, α_i are given by Equation 10.15 and components B_{12} and A_{12} are given by Equation 10.43 and Equation 10.44. In this case, the intrinsic viscosity $[\eta]$ is [17,18] :

$$[\eta] = a_0(\lambda - 1) - \frac{a_0^2}{20\beta_1}\left(\frac{\beta_2}{3\delta_2} - 1\right) + \frac{a_0^2}{4\delta_2} - I \qquad (10.61)$$

where I is given as follows:

$$I = \frac{a_0^2\left[\beta_1 - \beta_3 + (1/4\delta_2)(\delta_1 + \epsilon^2/8)\right]\left[2\delta_1(15\beta_1 - \beta_2) + (3\delta_2 - \beta_2 + 15\beta_1)(\epsilon^2/4)\right]}{30\beta_1\left[(\delta_2^2 \epsilon^2/4) + (\delta_1 + \epsilon^2/8)^2\right]} \qquad (10.62)$$

The low and high shear limits of the intrinsic viscosity obtained from Equation 10.61 are:

$$[\eta]_0 = \frac{5}{2} \qquad (10.63)$$

$$[\eta]_\infty = \frac{5(\lambda - 1)}{2\lambda + 3} \qquad (10.64)$$

When $\epsilon \to 0$, the intrinsic viscosity $[\eta]_0$ is 5/2, corresponding to suspension of rigid spherical particles. When $\epsilon \to \infty$, the intrinsic viscosity $[\eta]_\infty$ depends on the viscosity ratio λ.

Figure 10.6 shows the plots of intrinsic viscosity $[\eta]$ vs. ϵ (defined as $\eta_c \dot{\gamma} R/E_s$) for different values of viscosity ratio λ. The dispersion exhibits shear thinning in that $[\eta]$ decreases with the increase in ϵ. At low values of ϵ,

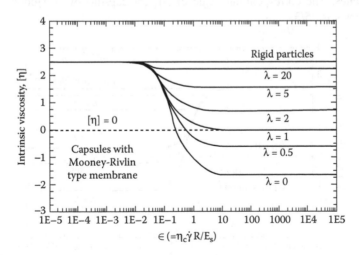

FIGURE 10.6 Plots of intrinsic viscosity $[\eta]$ vs. dimensionless group ϵ, defined as $\eta_c \dot{\gamma} R/E_s$, for different values of viscosity ratio λ (capsules with Mooney–Rivlin membrane).

capsules behave similar to rigid particles and $[\eta] = 2.5$. At high values of ϵ, $[\eta]$ varies from $-5/3$ to 2.5 depending on the viscosity ratio λ.

10.5.2.2 Capsule with an RBC-Type Membrane

For RBC-type membranes α_i are given by Equation 10.16 and components A_{12} and B_{12} are given by Equation 10.50. In this case, the intrinsic viscosity $[\eta]$ is [17,18]:

$$[\eta] = \frac{5(23\lambda - 16)}{2(23\lambda + 32)} + \frac{7680}{\left(23\lambda + 32\right)\left[64 + \left(23\lambda + 32\right)^2 \epsilon^2\right]} \tag{10.65}$$

The low and high shear limits of $[\eta]$ are as follows:

$$\left[\eta\right]_0 = \frac{5}{2} \tag{10.66}$$

$$\left[\eta\right]_\infty = \frac{5(23\lambda - 16)}{2(23\lambda + 32)} \tag{10.67}$$

Figure 10.7 shows $[\eta]$ vs. ϵ plots for dispersion of capsules with RBC-type membranes. The behavior is similar to that of dispersions of capsules with Mooney–Rivlin-type membrane, although the values of $[\eta]$ are not the same. For example, $[\eta]$ is $-5/4$ at high values of ϵ when $\lambda = 0$ for capsules with RBC-type membranes. The corresponding value of $[\eta]$ for capsules with Mooney–Rivlin membranes is $-5/3$.

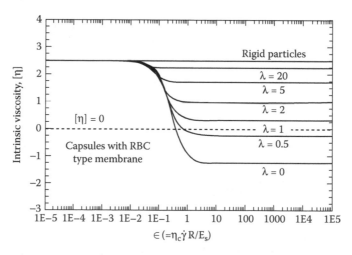

FIGURE 10.7 Plots of intrinsic viscosity $[\eta]$ vs. dimensionless group ϵ, defined as $\eta_c\dot{\gamma}R/E_s$, for different values of viscosity ratio λ (capsules with RBC-type membrane).

NOTATION

a_0, a_1	coefficients given by Equation 10.24 and Equation 10.25
$\overline{\overline{A}}$	second-order, symmetric, and traceless tensor that specifies the radial displacement of material
A_{12}	component of tensor $\overline{\overline{A}}$
dA	differential area in the deformed configuration
dA_0	differential area in the reference (unstressed) configuration
b_0, b_1, b_2	coefficients given by Equation 10.26 to Equation 10.28
$\overline{\overline{B}}$	second-order, symmetric, and traceless tensor that specifies the tangential in-plane displacement of material points
B_{12}	component of tensor $\overline{\overline{B}}$
D/Dt	Jaumann derivative
E_{12}, E_{21}	components of tensor $\overline{\overline{E}}$
$\overline{\overline{E}}$	rate of strain tensor
E_s	characteristic surface elastic modulus of membrane
$\overline{\overline{F}}_s$	surface deformation gradient tensor defined by Equation 10.8
$\overline{\overline{F}}_s^T$	transpose of tensor $\overline{\overline{F}}_s$
I	parameter given by Equation 10.62
J_s	ratio between the deformed and undeformed local surface areas
$\overline{\overline{L}}$	symmetric and traceless tensor; see Equation 10.31
L_{12}	component of tensor $\overline{\overline{L}}$
$\overline{\overline{\ell}}_s$	surface strain tensor defined by Equation 10.9
ℓ_{11}, ℓ_{22}	principal strains of surface strain tensor; see Equation 10.10 and Equation 10.11.
$\overline{\overline{M}}$	symmetric and traceless tensor; see Equation 10.32
M_{12}	component of tensor $\overline{\overline{M}}$
$\hat{n}$	outward unit normal vector at the interface
N	outward unit normal vector at the membrane surface in the reference configuration
P	pressure
P^*	pressure inside the capsule
$\overline{\overline{P}}_s$	surface stress tensor
R	radius of capsule
r	radial distance
S_d	symmetric and traceless part of a tensor defined by Equation 10.22 and Equation 10.23
$\overline{\overline{S}}^o$	dipole strength of a single capsule located in an infinite matrix fluid
t	time
$\vec{u}$	fluid velocity vector

$\vec{u}^*$	fluid velocity vector inside the capsule
$\vec{u}_M$	velocity vector of membrane
$\vec{u}_\infty$	fluid velocity vector far away from the capsule
V_p	volume of a single capsule
W	strain energy function of membrane material
$\vec{x}$	material point of the membrane in deformed configuration
$\vec{X}$	material point of the membrane in reference (unstressed) configuration
x_1, x_2, x_3	coordinates of the surface of the capsule

GREEK

$\alpha_1, \alpha_2, \alpha_3$	elastic coefficients (dimensionless form) pertaining to membrane strain energy function
$\alpha_1', \alpha_2', \alpha_3'$	elastic coefficients (dimensional form) pertaining to membrane strain energy function
$\beta_1, \beta_2, \beta_3, \beta_4$	coefficients given by Equation 10.45 and Equation 10.46
$\dot{\gamma}$	shear rate
δ_1, δ_2	coefficients given by Equation 10.47
$\bar{\bar{\delta}}$	unit tensor
∇	del operator
∇^2	Laplacian
$\in$	deformation parameter, measure of the deviation of shape from the initial sphere
η_c	continuous-phase viscosity
η_E	elongation viscosity
$[\eta]$	intrinsic viscosity
$[\eta]_o$	low shear intrinsic viscosity
$[\eta]_\infty$	high shear intrinsic viscosity
λ	viscosity ratio
λ_1, λ_2	extension ratios along the principal directions
$\bar{\bar{\sigma}}$	total stress tensor
$\bar{\bar{\sigma}}^*$	total stress tensor inside the capsule
σ_{12}	component of tensor $\bar{\bar{\sigma}}$
τ_{12}	shear stress
ϕ	volume fraction of capsules
$\bar{\bar{\Omega}}$	vorticity tensor
Ω_{12}, Ω_{21}	components of vorticity tensor $\bar{\bar{\Omega}}$

REFERENCES

1. D. Barthes-Biesel, *J. Fluid Mech.*, 100: 831–853, 1980.
2. M. Bredimas, M. Veyssie, D. Barthes-Biesel, V. Chhim, *J. Colloid Interface Sci.*, 93: 513–520, 1983.
3. D. Barthes-Biesel, V. Chhim, *Int. J. Multiphase Flow*, 7: 493–505, 1981.
4. D. Barthes-Biesel, J.M. Rallison, *J. Fluid Mech.*, 113: 251–267, 1981.
5. G. Breyiannis, C. Pozrikidis, *Theor. Comput. Fluid Mech.*, 13: 327–347, 2000.
6. G. Pieper, H. Rehage, D. Barthes-Biesel, *J. Colloid Interface Sci.*, 202: 293–300, 1998.
7. T.M. Fischer, H. Schmid-Schonbein, *Blood Cells*, 3: 351–365, 1977.
8. T.M. Fischer, M. Stohr-Liesen, H. Schmid-Schonbein, *AIChE Symp. Ser.* 182, 74: 38–45, 1978.
9. S. Sutera, M. Mehrjardi, N. Mohandas, *Blood Cells*, 1: 369–374, 1975.
10. R. Skalak, N. Ozkaya, T.C. Skalak, *Annu. Rev. Fluid Mech.*, 21: 167–204, 1989.
11. F. Greco, *Phys. Fluids*, 14: 946–954, 2002.
12. D. Barthes-Biesel, *Prog. Colloid Polym, Sci.*, 111: 58–64, 1998.
13. D. Barthes-Biesel, A. Diaz, E. Dhenin, *J. Fluid Mech.*, 460: 211–222, 2002
14. D. Barthes-Biesel, H. Sgaier, *J. Fluid Mech.*, 160: 119–135, 1985.
15. D. Barthes-Biesel, *Physica A*, 172: 103–124, 1991.
16. R. Skalak, A. Tozeren, R.P. Zarda, S. Chien, *Biophys. J.*, 13: 245–264, 1973.
17. A. Drochon, D. Barthes-Biesel, C. Lacombe, J.C. Lelievre, *J. Biomech. Eng., Trans. ASME*, 112: 241–249, 1990.
18. A. Drochon, *Eur. Phys. J. Appl. Phys.*, 22: 155–162, 2003.

11 Dispersions of Core–Shell Particles

11.1 INTRODUCTION

Dispersions of *core–shell* particles are encountered in a wide variety of applications. Three types of core–shell particles are particularly important: *solid core–hairy shell* particles, *solid core–liquid shell* particles, and *liquid core–liquid shell* particles.

Particles of the *solid core–hairy shell* type consist of a solid inner core and a porous outer layer of polymer or surfactant molecules (Figure 11.1). The layer of polymer or surfactant molecules is either physically adsorbed or chemically grafted on to the core surface. The presence of a hairy shell on the surface of the solid colloidal particles imparts stability to the dispersion against flocculation of particles. This technique of steric stabilization is widely used in different industries to control the stability of concentrated dispersions [1–9]. Another important application of solid core–hairy shell particles is related to filled polymeric materials [10–13]. Colloidal nanoparticles are often added as fillers to a polymer matrix to enhance the mechanical properties of the polymeric materials. One well-known application of this technique is the enhancement of the mechanical stability of tire rubber by adding carbon black to the rubber as a filler. For maximum effect, the colloidal fillers should be dispersed homogeneously in the polymer matrix. As most colloidal particles are entropically incompatible with linear polymer chains, they exhibit macroscopic phase separation because of depletion demixing [10–13] when they are added to a polymer matrix. To overcome this depletion demixing effect, colloid particles that have a hairy shell of polymer molecules are employed. The polymer molecules (either identical or compatible with the matrix polymer) are chemically grafted on to the core particle surface. The surface-modified hairy colloidal particles can be dispersed homogeneously in the polymer matrix without any phase separation. Hairy microgels also fall under the category of solid core–hairy shell particles. Hairy microgels are polymeric particles consisting of a core of cross-linked polymers and an external hairy layer of polymer chains (Figure 11.2). Suspensions of microgel particles are of importance to the coatings industry; microgel dispersions provide an interesting and industrially important means of thickening and texturing [14–19]. An interesting feature of suspensions of microgels and solid core–hairy shell particles, in general, is that the shell thickness can be increased or decreased by varying the temperature or nature of the dispersion medium [20–32].

FIGURE 11.1 Solid core–hairy shell type particle.

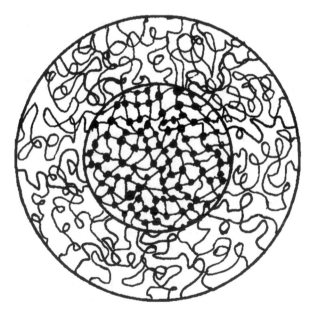

FIGURE 11.2 Hairy microgel (cross-linked polymeric core covered by a layer of polymer chains).

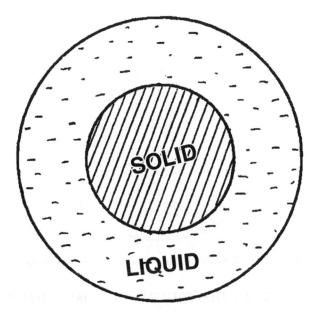

FIGURE 11.3 Solid core–liquid shell type particle.

Particles of the *solid core–liquid shell* type consist of a solid core encapsulated with a uniform layer of liquid that is immiscible in the dispersion medium of the suspension (Figure 11.3). The encapsulation of a solid particle with a liquid shell is desirable to modify the surface properties of particles, enhance the stability of the dispersions, improve the dispersibility of particles, and enhance the mechanical properties of the dispersion. For example, rigid particles are extensively used as fillers to enhance the stiffness, hardness, and abrasion resistance of polymeric materials. However, rigid fillers reduce the toughness and impact strength of the polymer composite. It is indeed a challenging task to create a composite both tougher and stiffer than the matrix material. Several researchers have suggested that rigid particles encapsulated with a uniform layer of a low-modulus elastomer, when dispersed in a polymer matrix, lead to a material that is stiffer and tougher than the matrix alone [33–37]. (Note that during processing of polymer composites, the polymeric materials are in liquid form.)

The third type of core–shell particle system of practical interest is *liquid core–liquid shell* particles, also referred to as *double-emulsion droplets* [38–44]. In double-emulsion droplets, the inner liquid core is completely engulfed by a second immiscible liquid (Figure 11.4), and this core–shell droplet is itself suspended in a third immiscible liquid (i.e., the continuous phase). Double emulsions have been shown to be efficient systems for mass transfer processes because of the high interfacial area that they provide. They are particularly suitable for controlled release of drugs.

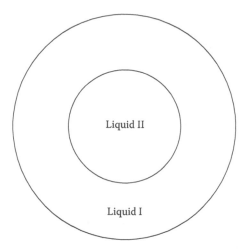

FIGURE 11.4 Liquid core–liquid shell particle (double-emulsion droplet).

11.2 RHEOLOGY OF DISPERSIONS OF CORE–SHELL PARTICLES

The rheology of dilute suspensions of core–shell particles has been investigated by several researchers [38,39,45–47]. Zackrisson and Bergenholtz [47] have recently derived a constitutive equation for dilute dispersion of *solid core–hairy shell* particles. As their analysis does not consider deformation of the outer porous shell of the particles, Newtonian behavior is predicted for the dispersions. Davis and Brenner [45] developed the expressions for velocity and pressure distribution in simple shear flow of *solid core–liquid shell* particles. The expression for the zero-shear-rate (Newtonian) viscosity of dilute dispersion of solid core–liquid shell particles was derived. A perturbation analysis of the shell deformation was also carried out. Stone and Leal [39] investigated deformation and rheology of dilute dispersions of double-emulsion droplets, that is, *liquid core–liquid shell* particles. An expression for the zero-shear-rate viscosity of a dilute system was developed. Ha and Yang [38] investigated the fluid dynamics of a double-emulsion droplet in an electric field. Their analysis shows that an electric field can have a significant influence on the rheology of dispersions of double-emulsion droplets.

11.2.1 Dispersions of Solid Core–Hairy Shell Particles

Consider the solid core–hairy shell particle shown in Figure 11.1. The solid core has a radius of a and the outer permeable (hairy) shell has a thickness of $(b - a)$ and permeability K. The core–shell particle is suspended in a continuous liquid phase having a viscosity of η_c. The coordinate system is fixed at the center of the inner solid core, and the system is subjected to linear shear flow. Thus, far away from the particle, the unperturbed velocity $\vec{u}_\infty$ is given by:

$$\vec{u}_\infty = \bar{\bar{\Gamma}}_\infty \bullet \vec{r} \tag{11.1}$$

where $\bar{\bar{\Gamma}}_\infty$ is the velocity gradient tensor of the unperturbed flow, given as:

$$\bar{\bar{\Gamma}}_\infty = \left(\nabla \vec{u}_\infty \right)^T \tag{11.2}$$

Assuming that the particle Reynolds number is small (negligible inertial effects), the fluid motion outside the porous shell of the particle can be described by the Stokes equations:

$$\nabla^2 \vec{u} = \frac{1}{\eta_c} \nabla P \qquad \text{for} \qquad r \geq b \tag{11.3}$$

$$\nabla \bullet \vec{u} = 0 \qquad \text{for} \qquad r \geq b \tag{11.4}$$

In the interior of the porous hairy shell, the fluid velocity $\vec{u}^*$ and pressure P^* can be described by the Brinkman equations [47–51]:

$$\nabla^2 \vec{u}^* - \frac{1}{K} \vec{u}^* = \frac{1}{\eta_c} \nabla P^* \qquad \text{for} \quad a \leq r \leq b \tag{11.5}$$

$$\nabla \bullet \vec{u}^* = 0 \qquad \text{for} \qquad a \leq r \leq b \tag{11.6}$$

To solve for the velocity and pressure distributions, the boundary conditions are:

1. Far away from the particle ($r \to \infty$):

$$\vec{u} \to \vec{u}_\infty \tag{11.7}$$

2. On the outer surface of the composite particle ($r = b$):

$$\vec{u} = \vec{u}^* \tag{11.8}$$

$$\left(\bar{\bar{\sigma}} - \bar{\bar{\sigma}}^* \right) \bullet \hat{n} = 0 \tag{11.9}$$

where $\hat{n}$ is the outward unit normal at the outer surface $r = b$, and $\overline{\overline{\sigma}}, \overline{\overline{\sigma}}^*$ are stress tensors given by:

$$\overline{\overline{\sigma}} = -P\overline{\overline{\delta}} + \eta_c \left[\nabla \vec{u} + \left(\nabla \vec{u} \right)^T \right] \tag{11.10}$$

$$\overline{\overline{\sigma}}^* = -P^*\overline{\overline{\delta}} + \eta_c \left[\nabla \vec{u}^* + \left(\nabla \vec{u}^* \right)^T \right] \tag{11.11}$$

$\overline{\overline{\sigma}}$ is the stress tensor outside the porous shell $(r > b)$, and $\overline{\overline{\sigma}}^*$ is the stress tensor inside the porous layer.

3. At the surface on the inner solid core $(r = a)$:

$$\vec{u}^* = 0 \tag{11.12}$$

Zackrisson and Bergenholtz [47] have solved Equation 11.3 to Equation 11.6 subject to the given boundary conditions: Equation 11.7 to Equation 11.9 and Equation 11.12. The velocity and pressure fields outside the porous layer $(r > b)$ are as follows:

$$P = P_\infty + \eta_c A(r)\vec{r} \bullet \overline{\overline{E}}_\infty \bullet \vec{r} \tag{11.13}$$

$$\vec{u} = \left[1 - B(r) \right] \left(\overline{\overline{E}}_\infty \bullet \vec{r} \right) + C(r)\vec{r} \left(\vec{r} \bullet \overline{\overline{E}}_\infty \bullet \vec{r} \right) \tag{11.14}$$

where $A(r)$, $B(r)$, and $C(r)$ are functions of r given by :

$$A(r) = C_1 r^{-5} \tag{11.15}$$

$$B(r) = \frac{2}{35} C_2 r^{-5} \tag{11.16}$$

$$C(r) = \frac{1}{2} C_1 r^{-5} - \frac{1}{7} C_2 r^{-7} \tag{11.17}$$

The constants C_1 and C_2 are given as follows:

$$C_1 = b^3 \left\{ -5 + \frac{15}{J(\alpha\beta)^3} \left(20\alpha^6\beta^4 + A' \sinh(\alpha - \beta) + B' \cosh(\alpha - \beta) \right) \right\} \tag{11.18}$$

$$C_2 = b^5 \left\{ -\frac{35}{2} + \frac{175}{J\beta^2} \left(5\alpha^3\beta^3 + C'\sinh(\alpha-\beta) + D'\cosh(\alpha-\beta) \right) \right\} \quad (11.19)$$

where

$$\alpha = a/\sqrt{K} \quad (11.20)$$

$$\beta = b/\sqrt{K} \quad (11.21)$$

$$A' = -30\alpha^6\beta - 12\alpha^8\beta + 30\alpha^3\beta^4 + 7\alpha^3\beta^6 + 30\alpha^7\beta^2$$
$$+ 2\alpha^9\beta^2 + 3\alpha^4\beta^7 - 30\alpha^4\beta^5 \quad (11.22)$$

$$B' = -30\alpha^6\beta^2 - 12\alpha^8\beta^2 - 3\alpha^3\beta^7 - 30\alpha^4\beta^4 - 7\alpha^4\beta^6$$
$$+ 30\alpha^7\beta + 2\alpha^9\beta + 30\alpha^3\beta^5 \quad (11.23)$$

$$C' = -6\alpha^5 - 15\alpha^3 - (3/2)\beta^5 + \alpha^6\beta + 15\alpha^4\beta + (3/2)\alpha\beta^6 \quad (11.24)$$

$$D' = -6\alpha^5\beta + 15\alpha^4 + (3/2)\alpha\beta^5 + \alpha^6 - 15\alpha^3 - (3/2)\beta^6 \quad (11.25)$$

$$J = 60\alpha^3\beta + 3(10\alpha^3 + 4\alpha^5 + 30\beta + \beta^5 + 10\beta^3 - 30\alpha\beta^2)\sinh(\alpha-\beta)$$
$$- (90\alpha\beta + 30\alpha\beta^3 + 30\alpha^4 + 2\alpha^6 + 3\alpha\beta^5 - 90\beta^2)\cosh(\alpha-\beta) \quad (11.26)$$

From the velocity and pressure fields given by Equation 11.13 and Equation 11.14, the stress tensor $\bar{\bar{\sigma}}$ in the ambient fluid is evaluated from:

$$\bar{\bar{\sigma}} = -P\bar{\bar{\delta}} + \eta_c \left[\nabla\vec{u} + \left(\nabla\vec{u}\right)^T \right] \quad (11.27)$$

Using the stress tensor of the ambient fluid at the surface of the particle ($r = b$), one can obtain the dipole strength $\bar{\bar{S}}^o$ defined in Chapter 3 as:

$$\bar{\bar{S}}^o = \int_{A_o} \left[\bar{\bar{\sigma}} \bullet \hat{n}\vec{x} - \frac{1}{3}(\vec{x} \bullet \bar{\bar{\sigma}} \bullet \hat{n})\bar{\bar{\delta}} - \eta_c(\hat{n}\vec{u} + \vec{u}\hat{n}) \right] dA \quad (11.28)$$

Zackrisson and Bergenholtz [47] have shown that $\bar{\bar{S}}^o$ is:

$$\bar{\bar{S}}^o = -\frac{4}{3}\pi\,\eta_c\,C_1\,\bar{\bar{E}}_\infty \tag{11.29}$$

where $\bar{\bar{E}}_\infty$ is the rate of strain tensor far away from the particle and C_1 is given by Equation 11.18. In the limit $K \rightarrow 0$ (i.e., the shell is completely impermeable), $C_1 = -5\,b^3$ and, hence, $\bar{\bar{S}}^o$ of the particle becomes equal to $\bar{\bar{S}}^o$ of a solid spherical particle of radius b. In the limit $K \rightarrow \infty$ (i.e., the shell is completely permeable), $C_1 = -5a^3$ and, hence, $\bar{\bar{S}}^o$ becomes equal to that of a solid spherical particle of radius a.

The full constitutive equation of a dilute dispersion of identical spheres of radius b is given as:

$$\langle\bar{\bar{\sigma}}\rangle = -\langle P\rangle\bar{\bar{\delta}} + 2\,\eta_c\left\langle\bar{\bar{E}}\right\rangle + \frac{3\phi}{4\pi b^3}\,\bar{\bar{S}}^o \tag{11.30}$$

From Equation 11.29 and Equation 11.30 and taking $\bar{\bar{E}}_\infty$ equal to the bulk or imposed rate of strain tensor $\langle\bar{\bar{E}}\rangle$, the following constitutive equation is obtained:

$$\bar{\bar{\sigma}} = -P\,\bar{\bar{\delta}} + 2\,\eta_c\left[1 + \left(\frac{-C_1\,b^{-3}}{2}\right)\phi\right]\bar{\bar{E}} \tag{11.31}$$

The angular brackets $\langle\,\rangle$ on quantities $\bar{\bar{\sigma}}, P,$ and $\bar{\bar{E}}$ have been dropped to simplify the notation. The form of Equation 11.31 indicates that the dispersion is a Newtonian fluid of viscosity given by:

$$\eta = \eta_c\left[1 + \left(\frac{-C_1 b^{-3}}{2}\right)\phi\right] \tag{11.32}$$

The intrinsic viscosity $[\,\eta\,]$ of the dispersion, defined as:

$$[\eta] = \lim_{\phi\to 0}\left[\frac{\eta/\eta_c - 1}{\phi}\right] \tag{11.33}$$

is equal to $(-C_1\,b^{-3}/2)$, that is,

$$[\eta] = -C_1\,b^{-3}/2$$

$$= \frac{5}{2}\left\{1 - 3J^{-1}(\alpha\beta)^{-3}(20\alpha^6\beta^4 + A'\sinh(\alpha-\beta) + B'\cosh(\alpha-\beta))\right\} \tag{11.34}$$

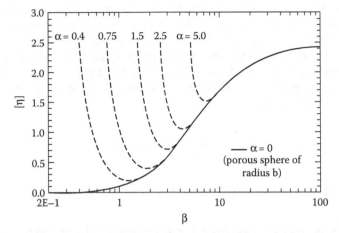

FIGURE 11.5 Intrinsic viscosity [η] of a dispersion of solid core–porous shell particles as a function of β for different values of α. Note that $\beta = b/\sqrt{K}$ and $\alpha = a/\sqrt{K}$, where a is the radius of the inner solid core, b is the radius of the outer porous shell, and K is shell permeability.

For a completely impermeable shell, $K \to 0$, $[\eta] = 2.5$, that is, the intrinsic viscosity is equal to the Einstein value. For a completely permeable shell, $K \to \infty$, $[\eta]$ is equal to $2.5(a/b)^3$.

Figure 11.5 shows the plot of intrinsic viscosity $[\eta]$ of a dispersion of solid core–porous shell particles as a function of β for different values of α. The following points should be noted from the figure: (1) $[\eta]$ never exceeds the Einstein value of 2.5, regardless of the porous layer thickness and its permeability; (2) the lowest value of $[\eta]$ is given by a porous sphere limit (where the solid core is absent and $\alpha = 0$); (3) $[\eta]$ is 2.5 when $\alpha = \beta$ (the porous layer is absent); and (4) when β is increased at a given α, a sharp drop in $[\eta]$ occurs from a value of 2.5 (at $\alpha = \beta$) to a porous sphere limit. An increase in β at constant α can be interpreted as an increase in the thickness of the porous layer.

11.2.2 DISPERSIONS OF SOLID CORE–LIQUID SHELL PARTICLES

Consider the solid core–liquid shell particle shown in Figure 11.3. The solid core has a radius of a, and the outer liquid shell has a thickness of $(b - a)$. The viscosities of the shell and matrix liquids are η_2 and η_1, respectively. The coordinate system is fixed at the center of the inner solid core, and the system is subjected to a linear shear flow.

When the system is subjected to a shear flow, the core–shell particle deforms because of the influence of viscous stresses on the outer liquid shell. The instantaneous external shape of the particle can be expressed as:

$$r = \left(x_1^2 + x_2^2 + x_3^2\right)^{\frac{1}{2}} = f\left(x_1, x_2, x_3\right) \tag{11.35}$$

where r is the distance from the particle center to the interface between the outer shell liquid and the matrix liquid. The function f is *a priori* unknown and must be determined as part of the problem solution. Assuming that the particle Reynolds number is small (negligible inertial effects), the fluid motion in the shell and external liquid can be described by the Stokes equations with respect to a coordinate system fixed at the center of the core–shell particle (and translating with the particle). Thus,

$$\eta_1 \, \nabla^2 \, \vec{u} \, - \, \nabla P = 0 \qquad \nabla \bullet \vec{u} \, = \, 0 \qquad \text{for} \qquad r > f \qquad (11.36)$$

$$\lambda \eta_1 \, \nabla^2 \, \vec{u}^* \, - \, \nabla P^* = 0 \qquad \nabla \bullet \vec{u}^* \, = \, 0 \qquad \text{for} \qquad a < r < f \qquad (11.37)$$

where λ is the viscosity ratio (η_2/η_1), $\vec{u}$ and P denote velocity and pressure in the matrix (continuous phase), and $\vec{u}^*$ and P^* denote velocity and pressure in the outer liquid shell.

The boundary conditions are:

1. Far away from the particle $(r \rightarrow \infty)$

$$\vec{u} \, \rightarrow \vec{u}_\infty \qquad (11.38)$$

2. On the outer surface of the liquid shell $(r = f)$

$$\vec{u} \, = \, \vec{u}^* \qquad (11.39)$$

$$\left(\overline{\overline{\sigma}} - \overline{\overline{\sigma}}^*\right) \, \bullet \, \hat{n} \, = \, \gamma \, \hat{n} \left(\nabla_s \bullet \hat{n}\right) \qquad (11.40)$$

where ∇_s is the surface gradient operator given as $\nabla = (\overline{\overline{\delta}} - \hat{n}\hat{n}) \bullet \nabla$, $\hat{n}$ is the outward unit normal at the interface $r = f$, and γ is the interfacial tension between the two liquids (continuous-phase liquid and shell liquid). Note that $\nabla_s \bullet \hat{n}$ is the local mean curvature of the interface.

3. On the surface of the inner solid core $(r = a)$

$$\vec{u}^* \, = \, 0 \qquad (11.41)$$

The mathematical treatment of simultaneously solving Equation 11.36 and Equation 11.37 subject to the boundary conditions, Equation 11.38 to Equation 11.41, is difficult. As pointed out in previous chapters, such "free surface" type problems are treated by means of the "domain perturbation procedure." This is an iterative procedure. First, the particle is treated as spherical (outer boundary of liquid shell at $r = b$ is assumed to be spherical), and the flow fields are determined using all the stated boundary conditions, except for the normal stress discontinuity across the interface at $r = b$. This solution is called a *zero-order deformation solution*. The normal stress boundary condition is then used to determine the deviation of the interface from sphericity. This gives the first-order deformation of the interface. In the next iteration, the flow fields are determined by satisfying the boundary conditions at the deformed interface.

Davis and Brenner [45] obtained a zero-order solution of the flow fields (velocity, pressure, and stress distributions) for a *solid core–liquid shell* particle. From the zero-order flow field results, the dipole strength $\bar{\bar{S}}^{o}$ (defined by Equation 11.28) was shown to be equal to:

$$\bar{\bar{S}}^{o} = \left(5\,\psi\,V_{p}\,\eta_{1}\right)\bar{\bar{E}}_{\infty} \tag{11.42}$$

where ψ (a/b, λ) is a dimensionless factor, function of radii ratio (a/b) and viscosity ratio (λ), V_{p} is the volume of a single core–shell particle (= $4\pi b^{3}/3$), and $\bar{\bar{E}}_{\infty}$ is the rate of deformation tensor far away from the particle. The expression for ψ is given as follows:

$$\psi = \frac{2}{5} + \frac{3}{5}\left[1 + \frac{4 - 25(a/b)^{3} + 42(a/b)^{5} - 25(a/b)^{7} + 4(a/b)^{10}}{2\lambda\{2 - 5(a/b)^{3} + 5(a/b)^{7} - 2(a/b)^{10}\}}\right]^{-1} \tag{11.43}$$

In the limit $a/b \to 0$,

$$\psi = \frac{\lambda + \dfrac{2}{5}}{\lambda + 1} + \frac{9\lambda\left(a/b\right)^{3}}{4\left(\lambda + 1\right)^{2}} \tag{11.44}$$

In the limit $a/b \to 1.0$,

$$\psi = 1 - \frac{3}{4\lambda}\left(1 - \frac{a}{b}\right) \tag{11.45}$$

From Equation 11.30 and Equation 11.42, the rheological constitutive equation for a dilute dispersion of identical spherical particles of *solid core–liquid shell* type is given by:

$$\bar{\bar{\sigma}} = -P\bar{\bar{\delta}} + 2\,\eta_{1}\,\bar{\bar{E}} + 5\,\eta_{1}\,\psi\,\phi\,\bar{\bar{E}} \tag{11.46}$$

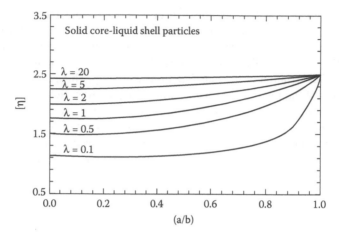

FIGURE 11.6 Intrinsic viscosity [η] of a dispersion of solid core–liquid shell particles as a function of a/b for different values of viscosity ratio λ (a is the radius of inner solid core and b is the radius of outer liquid shell).

where $\bar{\bar{\sigma}}, P$, and $\bar{\bar{E}}$ are average (bulk) quantities (angular brackets have been dropped to simplify the notation). The relative viscosity predicted from Equation 11.46 is:

$$\eta_r = \frac{\eta}{\eta_1} = 1 + \frac{5}{2}\,\psi\,\phi \qquad (11.47)$$

When $a/b = 1.0$, Equation 11.45 gives $\psi = 1$ and Equation 11.47 reduces to the well-known Einstein equation [52,53]. When $a/b = 0$, Equation 11.44 gives $\psi = (\lambda + (2/5))/(1+\lambda)$ and Equation 11.47 reduces to the Taylor viscosity equation for dilute emulsions [54]. The dispersion intrinsic viscosity increases from the Taylor limit at $a/b = 0$ to the Einstein viscosity limit of 2.5 at $a/b = 1.0$. This can clearly be seen in Figure 11.6. Note that [η] at $a/b = 0$ (Taylor's limit) is a function of the viscosity ratio.

11.2.3 Dispersions of Liquid Core–Liquid Shell (Double-Emulsion) Droplets

Consider the liquid core–liquid shell type double-emulsion droplet shown in Figure 11.7. In the unstressed state, the inner core drop has a radius of a and the outer liquid shell has a uniform thickness of $(b − a)$. The continuous-phase (fluid 1) has a viscosity of η_1, the outer shell liquid (fluid 2) has a viscosity of η_2, and the inner core liquid (fluid 3) has a viscosity of η_3. The interfacial tensions between different fluids are γ_{12} between fluid 1 and fluid 2, and γ_{23} between fluid 2 and fluid 3.

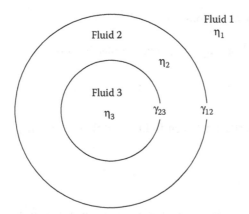

FIGURE 11.7 Liquid core–liquid shell particle (double-emulsion droplet).

When the system is subjected to a shear flow, the compound droplet deforms because of the influence of viscous stresses. The instantaneous shapes of the fluid-fluid interfaces of the double-emulsion droplet can be expressed as:

$$r = (x_1^2 + x_2^2 + x_3^2)^{1/2} = f_{12}(x_1, x_2, x_3) \quad (11.48)$$

where r is the distance from the center of the inner core drop to the interface between fluid 1 and fluid 2 (continuous-phase and outer shell fluid), and

$$r = (x_1^2 + x_2^2 + x_3^2)^{1/2} = f_{23}(x_1, x_2, x_3) \quad (11.49)$$

where r is the distance from the center of the inner core drop to the interface between fluid 2 and fluid 3 (shell and core fluids). The functions f_{12} and f_{23} are *a priori* unknown and must be determined as part of the problem solution.

Assuming that the inertial effects are negligible, the fluid motion in different fluids can be described by the Stokes equations with respect to a coordinate system fixed at the center of the double-emulsion droplet and translating with the droplet. Thus,

$$\eta_1 \nabla^2 \vec{u}_1 = \nabla P_1, \quad \nabla \bullet \vec{u}_1 = 0 \quad \text{for} \quad r > f_{12} \quad (11.50)$$

$$\eta_2 \nabla^2 \vec{u}_2 = \nabla P_2, \quad \nabla \bullet \vec{u}_2 = 0 \quad \text{for} \quad f_{23} < r < f_{12} \quad (11.51)$$

$$\eta_3 \nabla^2 \vec{u}_3 = \nabla P_3, \quad \nabla \bullet \vec{u}_3 = 0 \quad \text{for} \quad 0 \leq r < f_{23} \quad (11.52)$$

The boundary conditions are:

1. Far away from the compound droplet ($r \to \infty$)

$$\vec{u}_1 \to \vec{u}_\infty \tag{11.53}$$

2. At the interface between fluid 1 and fluid 2 ($r = f_{12}$)

$$\vec{u}_1 = \vec{u}_2 \tag{11.54}$$

$$(\bar{\bar{\sigma}}_1 - \bar{\bar{\sigma}}_2) \bullet \hat{n}_2 = \gamma_{12}\,\hat{n}_2\,(\nabla_s \bullet \hat{n}_2) \tag{11.55}$$

where $\hat{n}_2$ is the unit normal at interface $r = f_{12}$ directed outward from fluid 2 (Figure 11.7).
3. At the interface between fluid 2 and fluid 3 ($r = f_{23}$)

$$\vec{u}_2 = \vec{u}_3 \tag{11.56}$$

$$(\bar{\bar{\sigma}}_2 - \bar{\bar{\sigma}}_3) \bullet \hat{n}_3 = \gamma_{23}\,\hat{n}_3\,(\nabla_s \bullet \hat{n}_3) \tag{11.57}$$

where $\hat{n}_3$ is the unit normal at interface $r = f_{23}$ directed outward from fluid 3.

Similar to other "free surface" type problems, this problem is solved using the "domain perturbation" technique. Stone and Leal [39] determined the zero-order velocity and pressure distribution interior and exterior to the double-emulsion droplet. They also calculated the first-order deformation of the interfaces.

Figure 11.8 and Figure 11.9 exhibit streamline patterns in and around the droplet for $a/b = 0.2$ and 0.8, respectively, when $\eta_1 = \eta_2 = \eta_3$. Clearly, there exist two vortical flow patterns interior to the double-emulsion droplet. The external fluid motion creates a recirculating flow in the outer shell liquid that drives the motion in the inner core droplet in the opposite direction of rotation. An important point to note from the streamline patterns shown in Figure 11.8 and Figure 11.9 is that uniaxial external flow tends to deform the overall compound droplet into a prolate spheroidal shape, whereas the biaxial extensional flow generated in the outer shell liquid tends to deform the inner core droplet into an oblate spheroidal shape (Figure 11.10).

FIGURE 11.8 Streamline patterns in and around the double-emulsion droplet for $a/b = 0.20$. (From H.A. Stone, L.G. Leal, Breakup of concentric double-emulsion droplets in linear flows, *J. Fluid Mech.*, 211: 123–156, 1990. Reprinted with the permission of Cambridge University Press.)

FIGURE 11.9 Streamline patterns in and around the double-emulsion droplet for $a/b = 0.80$. (From H.A. Stone, L.G. Leal, Breakup of concentric double-emulsion droplets in linear flows, *J. Fluid Mech.*, 211: 123–156, 1990. Reprinted with the permission of Cambridge University Press.)

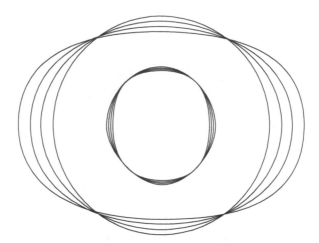

FIGURE 11.10 Deformation of a double-emulsion droplet. (From H.A. Stone, L.G. Leal, Breakup of concentric double-emulsion droplets in linear flows, *J. Fluid Mech.*, 211: 123–156, 1990. Reprinted with the permission of Cambridge University Press.)

The dipole strength $\bar{\bar{S}}^o$ of a single liquid core–liquid shell droplet located in an infinite matrix fluid (fluid 1) can be determined from the defining Equation 11.28 using the zero-order flow fields obtained by Stone and Leal [39]. It is given as:

$$\bar{\bar{S}}^o = 5\left(1 - \frac{3}{5}I\right)V_p \, \eta_1 \, \bar{\bar{E}}_\infty \tag{11.58}$$

where $I(a/b, \lambda_{21}, \lambda_{32})$ is a dimensionless factor, function of radii ratio a/b and viscosity ratios $\lambda_{21}(=\eta_2/\eta_1)$ and $\lambda_{32}(=\eta_3/\eta_2)$, V_p is the volume of a single double-emulsion droplet, and $\bar{\bar{E}}_\infty$ is the rate of deformation tensor far away from the droplet. The expression for I is as follows:

$$I = \frac{\delta_1 + \dfrac{(a/b)^3}{1+\lambda_{32}}\delta_2}{\delta_1 + \lambda_{21}\,\delta_3 + \dfrac{(a/b)^3}{1+\lambda_{32}}\,(\delta_2 + \lambda_{21}\,\delta_4)} \tag{11.59}$$

The coefficients δ_i depend on the a/b ratio only and are given as:

$$\delta_1 = 4 - 25\,(a/b)^3 + 42\,(a/b)^5 - 25\,(a/b)^7 + 4\,(a/b)^{10} \tag{11.60}$$

$$\delta_2 = 15 - 42\,(a/b)^2 + 35\,(a/b)^4 - 8\,(a/b)^7 \tag{11.61}$$

$$\delta_3 = 4 - 10\,(a/b)^3 + 10\,(a/b)^7 - 4\,(a/b)^{10} \qquad (11.62)$$

$$\delta_4 = 6 - 14\,(a/b)^4 + 8\,(a/b)^7 \qquad (11.63)$$

From Equation 11.30 and Equation 11.58, the rheological constitutive equation for a dilute dispersion of identical spherical particles of *liquid core–liquid shell* type is given by:

$$\overline{\overline{\sigma}} = -P\overline{\overline{\delta}} + 2\eta_1\,\overline{\overline{E}} + 5\left(1 - \frac{3}{5}I\right)\eta_1\,\phi\,\overline{\overline{E}} \qquad (11.64)$$

The relative viscosity predicted from Equation 11.64 is:

$$\eta_r = \frac{\eta}{\eta_1} = 1 + \frac{5}{2}\left(1 - \frac{3}{5}I\right)\phi \qquad (11.65)$$

In the case of a very viscous outer shell liquid ($\lambda_{21} \to \infty$), $I \to 0$ (see Equation 11.59) and Equation 11.65 reduces to the Einstein equation for the viscosity of a dilute suspension of rigid spheres [52,53]. Note that $I \to 0$ for any values of radii ratio (a/b) and λ_{32}, indicating that the inner core fluid has no influence on the viscosity of the dispersion if the outer shell liquid is much more viscous than the continuous phase (fluid 1).

In the limit $\lambda_{32} \to \infty$, Equation 11.65 reduces to Equation 11.47 for the viscosity of a dilute suspension of *solid core–liquid shell* particles. In the limit $a/b \to 0$, $I \to 1/(1 + \lambda_{21})$ and Equation 11.65 reduces to the Taylor equation for the viscosity of dilute emulsions [54].

It is interesting to note that in the case of a very thin outer shell layer ($a/b \to 1.0$), I approaches zero even when the viscosity ratio λ_{21} is not too large and is of the order of unity. This result is independent of the viscosity of the inner core liquid. Thus, suspensions of double-emulsion droplets with a thin outer shell (not necessarily of high viscosity) behave as suspensions of rigid particles regardless of the viscosity of the inner core liquid. This surprising result can be appreciated by examining the streamlines [55] (Figure 11.11). The motion of the external continuous-phase liquid sets up internal circulation. The direction of the streamlines is shown in Figure 11.11. As the outer shell thickness approaches a small value, all internal circulation (inside the shell and core liquid) is suppressed; otherwise, the shell fluid would have to change directions over a very short distance. Therefore, in the

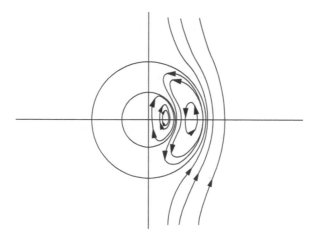

FIGURE 11.11 Streamline patterns in and around the double-emulsion droplet. (From P.O. Brunn, T. Roden, On the deformation and drag of a type – A multiple drop at low Reynolds number, *J. Fluid Mech.*, 160: 211–234, 1985. Reprinted with the permission of Cambridge University Press.)

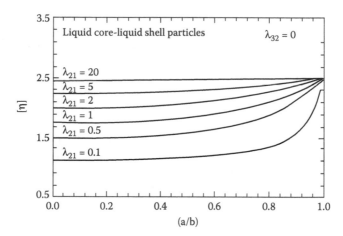

FIGURE 11.12 Plots of intrinsic viscosity [η] vs. *a/b* ratio for dispersions of liquid core–liquid shell droplets at different values of viscosity ratio λ_{21} (λ_{32} is taken to be zero).

limit $a/b \rightarrow 1.0$, the double-emulsion droplet consists of a stationary core liquid encapsulated by an immobile liquid shell and the double-emulsion droplet exhibits rigid sphere behavior [55].

Figure 11.12 and Figure 11.13 show the plots of intrinsic viscosity [η] of dispersions of liquid core–liquid shell droplets, generated from Equation 11.65 under different conditions. The intrinsic viscosity is sensitive to the value of λ_{21} (shell-to-matrix viscosity ratio). The effect of viscosity ratio λ_{32} (core-to-shell viscosity ratio) on the intrinsic viscosity of dispersion is marginal.

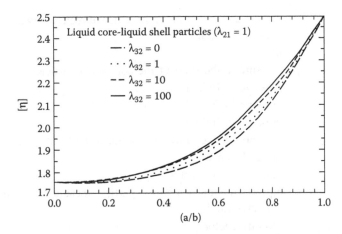

FIGURE 11.13 Plots of intrinsic viscosity [η] vs. *a/b* ratio for dispersions of liquid core–liquid shell droplets at different values of viscosity ratio λ_{32} (λ_{21} is taken to be unity).

NOTATION

A	surface area
A_o	surface area of a particle
A'	parameter given by Equation 11.22
$A(r)$	function of r given by Equation 11.15
a	radius of core of a core–shell particle
$B(r)$	function of r given by Equation 11.16
B'	parameter given by Equation 11.23
b	outer radius of shell of a core–shell particle
$C(r)$	function of r given by Equation 11.17
C'	parameter given by Equation 11.24
C_1	parameter given by Equation 11.18
C_2	parameter given by Equation 11.19
D'	parameter given by Equation 11.25
$\underline{\underline{E}}$	rate of strain tensor
$\underline{\underline{E}}_\infty$	rate of strain tensor of unperturbed flow far away from the particle
I	parameter given by Equation 11.59
J	parameter given by Equation 11.26
K	permeability
$\hat{n}$	outward unit normal vector
$\hat{n}_2$	unit normal vector at the interface, directed outward from fluid 2 (Figure 11.7)

$\hat{n}_3$	unit normal vector at the interface, directed outward from fluid 3 (Figure 11.7)
P	pressure
P^*	pressure in the shell of the core–shell particle
P_1	pressure in fluid 1
P_2	pressure in fluid 2
P_3	pressure in fluid 3
P_∞	pressure of unperturbed flow far away from the particle
r	radial distance
$\vec{r}$	radial position vector
$\overline{\overline{S}}^o$	dipole strength of a single particle located in an infinite matrix fluid
$\vec{u}$	velocity vector
$\vec{u}^*$	velocity vector in the interior of the shell of a core–shell
$\vec{u}_1$	velocity vector in fluid 1
$\vec{u}_2$	velocity vector in fluid 2
$\vec{u}_3$	velocity vector in fluid 3
$\vec{u}_\infty$	velocity vector of unperturbed flow far away from the particle
V_p	volume of a single particle
$\vec{x}$	position vector
x_1, x_2, x_3	position coordinates of the surface of a particle

GREEK

α	parameter defined by Equation 11.20
β	parameter defined by Equation 11.21
γ	interfacial tension
γ_{12}	interfacial tension between fluid 1 and fluid 2
γ_{23}	interfacial tension between fluid 2 and fluid 3
$\overline{\overline{\Gamma}}_\infty$	velocity gradient tensor of unperturbed flow
∇	del operator
∇_s	surface gradient operator
$\overline{\overline{\delta}}$	unit tensor
η	viscosity
η_1	viscosity of fluid 1
η_2	viscosity of fluid 2
η_3	viscosity of fluid 3
η_c	continuous-phase (matrix) viscosity
η_r	relative viscosity
$[\eta]$	intrinsic viscosity

λ	viscosity ratio
λ_{21}	viscosity ratio (η_2/η_1)
λ_{32}	viscosity ratio (η_3/η_2)
$\overline{\overline{\sigma}}$	total stress tensor
$\overline{\overline{\sigma}}^*$	total stress tensor inside the shell of a core–shell particle
$\overline{\overline{\sigma}}_1$	total stress tensor in fluid 1
$\overline{\overline{\sigma}}_2$	total stress tensor in fluid 2
$\overline{\overline{\sigma}}_3$	total stress tensor in fluid 3
ϕ	volume fraction of particles
ψ	dimensionless factor given by Equation 11.43
$\langle X \rangle$	volume average of quality X

REFERENCES

1. D.H. Napper, *Polymeric Stabilization of Colloidal Dispersions*, Academic Press, London, 1983.
2. R. Pal, in *Encyclopedia of Emulsion Technology*, Vol. 4, P. Becher, Ed., Marcel Dekker, New York, 1996, pp. 93–263.
3. T. Sato, R. Ruch, *Stabilization of Colloidal Dispersions by Polymer Adsorption*, Marcel Dekker, New York, 1980.
4. S. Ross, I.D. Morrison, *Colloidal Dispersions*, John Wiley & Sons, New York, 2002.
5. E. Kissa, *Dispersions:Characterization, Testing, and Measurement*, Marcel Dekker, New York, 1999.
6. Th.F. Tadros, Ed., *Solid/Liquid Dispersions*, Academic Press, London, 1987.
7. P. Becher, Ed., *Encyclopedia of Emulsion Technology*, Vol. 1, Marcel Dekker, New York, 1983, chap. 3.
8. D.F. Evans, H. Wennerstrom, *The Colloidal Domain*, Wiley-VCH, New York, 1999.
9. W.B. Russel, D.A. Saville, W.R. Schowalter, *Colloidal Dispersions*, Cambridge University Press, Cambridge, 1989.
10. G. Lindenblatt, W. Schartl, T. Pakula, M. Schmidt, *Macromolecules*, 33: 9340–9347, 2000.
11. G. Lindenblatt, W. Schartl, T. Pakula, M. Schmidt, *Macromolecules*, 34: 1730–1736, 2001.
12. Y. Chen, K. Ghor, W. Schartl, M. Schmidt, L. Yezek, *Prog. Colloid Polym. Sci.*, 121: 28–33, 2002.
13. J.I. Martin, Z.G. Wang, *J. Phys. Chem.*, 99: 2833–2844, 1995.
14. C.L.A Berli, D. Quemada, *Langmuir*, 16: 7968–7974, 2000.
15. B.R. Saunders, B. Vincent, *Adv. Colloid Interface Sci.*, 80: 1–25, 1999.
16. C. Raquois, J.F. Tassin, S. Rezaiguia, A. V. Gindre, *Prog. Org. Coat.*, 26: 239–250, 1995.
17. S. Fridrikh, C. Raquois, J.F. Tassin, S. Rezaiguia, *J. Chim. Phys.*, 93: 941–959, 1996.
18. A. Bischoff, M. Kluppel, R.H. Schuster, *Polym. Bull.*, 40: 283–290, 1998.
19. C.L.A. Berli, D. Quemada, *Langmuir*, 16: 10509–10514, 2000.

20.   H. Kawaguchi, Y. Isono, R. Sasabe, in *Polymer Colloids:Science and Technology of Latex Systems*, E.S. Daniels, E.D. Sudol, M.S. El-Aasser, Eds., ACS Symposium Series 801, ACS, Washington, D.C., 2002, chap. 21.

21.   I.D. Evans, A. Lips, *J. Chem. Soc. Faraday Trans.*, 86: 3413–3417, 1990.

22.   C.D. Jones, L.A. Lyon, *Macromolecules*, 33: 8301–8306, 2000.

23.   J.H. So, S.M. Yang, C. Kim, J.C. Hyun, *Colloids Surf.*, 190: 89–98, 2001.

24.   I. Deike, M. Ballauff, N. Willenbacher, A. Weiss, *J. Rheol.*, 45: 709–720, 2001.

25.   S. Sellenmeyer, I. Deike, S. Rosenfeldt, Ch. Norhausen, N. Dingenouts, M. Ballanoff, T. Narayanan, P. Linder, *J. Chem. Phys.*, 114: 10471–10478, 2001.

26.   D. Gan, L.A. Lyon, *J. Am. Chem. Soc.*, 123: 8203–8209, 2001.

27.   A. Fernandez-Barbero, A. Fernandez-Nieves, I. Grillo, E. Lopez-Cabarcos, *Phys. Rev. E.*, 66: 051803/1–051803/10, 2002.

28.   H. Senff, W. Richtering, *Langmuir*, 15: 102–106, 1999.

29.   H. Senff, W. Richtering, *Colloid Polym. Sci.*, 278, 830–840, 2000.

30.   H. Senff, W. Richtering, *J. Chem.. Phys.*, 111: 1705–1711, 1999.

31.   G. Fritz, V. Schadler, N. Willenbacher, N.J. Wagner, *Langmuir*, 18: 6381–6390, 2002.

32.   M. Ballauff, *Macromol. Chem. Phys.*, 204: 220–234, 2003.

33.   Y.C. Ou, T.T. Guo, X.P. Fang, Z.Z. Yu, *J. Appl. Polym. Sci.*, 74: 2397–2403, 1999.

34.   F. Stricker, C. Friedrich, R. Mulhaupt, *J. Appl. Polym. Sci.*, 69: 2499–2506, 1998.

35.   Y.C. Ou, J. Zhu, Y.P. Feng, *J. Appl. Polym. Sci.*, 59: 287–294, 1996.

36.   L.J. Broutman, B.D. Agarwal, *Polym. Eng. Sci.*, 14: 581–588, 1974.

37.   V.A. Matonis, *Polym. Eng. Sci.*, 9: 90–99, 1969.

38.   J.W. Ha, S.M. Yang, *Phys. Fluid*, 11: 1029–1041, 1999.

39.   H.A. Stone, L.G. Leal, *J. Fluid Mech.*, 211: 123–156, 1990.

40.   E. Rushton, G.A. Davis, *Int. J. Multiphase Flow*, 9: 337–342, 1983.

41.   J.J. Ulbrecht, P. Stroeve, P. Prabodh, *Rheol. Acta*, 21: 593–597, 1982.

42.   N. Garti, *Acta Polym.*, 49: 606–616, 1998.

43.   J.W. Ha, S.M. Yang, *J. Colloid Interface Sci.*, 213: 92–100, 1999.

44.   N. Garti, A. Aserin, *Adv. Colloid Interface Sci.*, 65: 37–69, 1996.

45.   A.M.J. Davis, H. Brenner, *J. Eng. Mech.*, 107: 609–621, 1981.

46.   S.B. Chen, X. Ye, *J. Colloid Interface Sci.*, 221: 50–57, 2000.

47.   M. Zackrisson, J. Bergenholtz, *Colloids Surf. A*, 225:119–127, 2003.

48.   H.C. Brinkman, *Appl. Sci. Res.*, A1: 27, 1947.

49.   G. Neale, W. Nader, *Can. J. Chem. Eng.*, 52: 475–478, 1974.

50.   G. Neale, N. Epstein, W. Nader, *Chem. Eng. Sci.*, 28: 1865–1974, 1973.

51.   J.J. L. Higdon, M. Kojima, *Int. J. Multiphase Flow*, 7: 719–727, 1981.

52.   A. Einstein, *Ann. Phys.*, 19: 289–306, 1906.

53.   A. Einstein, *Ann. Phys.*, 34: 591–592, 1911.

54.   G.I. Taylor, *Proc. R. Soc. Lond. A*, 138: 41–48, 1932.

55.   P.O. Brunn, T. Roden, *J. Fluid Mech.*, 160: 211–234, 1985.

Part IV

Rheology of Composites

Part IV is devoted to the rheology of solid two-phase composite materials. It consists of two chapters, namely, Chapter 12 and Chapter 13.

Chapter 12 describes the elastic properties of isotropic particulate composites. The steps involved in the derivation of the equation for dipole strength of a single linearly elastic solid particle located in an infinite linearly elastic solid matrix are discussed. The expression for the dipole strength of a spherical particle is given. The single-particle solution is used to describe the bulk stress tensor in dilute particulate composites of spherical particles. Using the bulk stress tensor equation, the expressions for the elastic moduli (shear, dilational, and Young's) of composites of spherical particles are derived. The elastic moduli of composites of disk-shaped particles are also discussed. The chapter concludes with a section on the general lower and upper bounds for the shear and bulk moduli of isotropic two-phase composites.

Chapter 13 deals with the elastic properties of fiber-reinforced composites. The rule-of-mixtures formulas and improved models for the elastic moduli and Poisson's ratio of *unidirectional* continuous fiber composites are discussed. The upper and lower bounds on the effective elastic properties of unidirectional continuous fiber composites (with fibers of irregular shapes) are presented. The expressions for Young's modulus and Poisson's ratio of isotropic *randomly oriented* continuous fiber composites are also given. Finally, the chapter describes the elastic properties of unidirectional *short-fiber* composites and randomly oriented short-fiber composites.

12 Particulate Composites

12.1 INTRODUCTION

Composites are solid heterogeneous materials consisting of two or more phases. However, many composites of practical interest are composed of just two phases, namely, dispersed phase (particles) and continuous phase (matrix). Interest in particulate composites has expanded in recent years as these two-phase mixtures often provide an advantageous blend of properties of the individual materials. For example, most polymers in homogeneous form are glassy and brittle; the addition of rubber particles to a polymer matrix can greatly improve the impact resistance of the material. Likewise, the addition of rigid fillers (carbon black, for example) to rubberlike elastomers can greatly improve the stiffness and strength of the material. A practical example of particle-reinforced composites is the automobile tire, which has carbon black particles in a matrix of polyisobutylene elastomeric polymer. Another practical example of particle-reinforced composite is concrete, where particles of various sizes ranging from sand grains of diameter 100 μm to large rock particles of diameter 10 to 20 mm are dispersed in the matrix of cement paste. Polymer-based particulate composites are also widely used as dental restorative materials [1–8]. Dental composites consist of blends of fillers and resins, and often have a coupling agent between the filler and matrix phases. The filler volume fraction in dental composites is generally high (>0.50). The solid propellants commonly used in aerospace propulsion are particulate composites consisting of particles of solid oxidizer, typically ammonium perchlorate, dispersed in a matrix of plastic fuel such as polybutadiene [9–11]. Although fillers are commonly added to a polymer matrix to enhance the mechanical properties of the material, this is not the only reason for filler addition; in some situations, fillers are added to a polymer to improve other properties such as electrical conductivity. Electrically conducting polymers find many application areas such as in electromagnetic shielding and in electronic packaging.

The most common advanced composites are based on a polymer matrix (and are called polymer matrix composites). However, there are other important particulate composites based on a nonpolymeric matrix such as metal matrix composites and ceramic matrix composites [12].

According to Cohen and Ishai [13], two-phase particulate composites can be divided into three major classes in terms of their filler-to-matrix modulus ratio: (1) composites with high filler-to-matrix modulus ratio, larger than 20; these include most filled or reinforced polymer composites; (2) composites with low filler-to-matrix modulus ratio, in the range of 1 to 5; these include most cement concretes [14] and metal matrix composites; and (3) composites with soft particles (modulus ratio less than one) and composites with pores/voids (modulus ratio zero).

12.2 CONSTITUTIVE EQUATION FOR PARTICULATE COMPOSITES

The constitutive equation for a solid particulate composite, valid for all concentrations of dispersed particles, can be written as (see Chapter 3):

$$\langle \overline{\overline{\sigma}} \rangle = \langle \overline{\overline{\sigma}}_o \rangle + \frac{1}{V} \sum_{i=1}^{N} \overline{\overline{S}}_i \qquad (12.1)$$

where $\langle \overline{\overline{\sigma}} \rangle$ is the bulk (volume-average) stress in the composite material, $\langle \overline{\overline{\sigma}}_o \rangle$ is the stress tensor in pure matrix material under conditions corresponding to the imposed bulk strain $\langle \overline{\overline{e}} \rangle$ on the composite, V is a sample volume whose dimensions are large relative to the average spacing between the particles but small relative to the characteristic macroscale of the system, N is the number of particles within V, and $\overline{\overline{S}}$ is the dipole strength of a particle defined in Chapter 3 as [15–17]:

$$\overline{\overline{S}} = \int_{A_0} \left[\overline{\overline{\sigma}} \bullet \hat{n}\, \vec{x} - \lambda_c \left(\vec{\xi} \bullet \hat{n} \right) \overline{\overline{\delta}} - G_c \left(\vec{\xi}\hat{n} + \hat{n}\, \vec{\xi} \right) \right] dA \qquad (12.2)$$

where A_0 is the surface area of a particle, $\hat{n}$ is a unit outward normal to the particle surface, $\vec{\xi}$ is the surface displacement vector, $\overline{\overline{\sigma}}$ is the stress tensor at the particle surface, $\vec{x}$ is the position vector with respect to a fixed origin, and $\overline{\overline{\delta}}$ is the unit tensor. The matrix material is assumed to be isotropic and linearly elastic with Lamé constants λ_c and G_c, where G_c is also called the *shear modulus*. Note that for a linearly elastic matrix, the stress tensor $\langle \overline{\overline{\sigma}}_o \rangle$ is given as (see Chapter 2) [15–18]:

$$\langle \overline{\overline{\sigma}}_o \rangle = \lambda_c \left(tr \langle \overline{\overline{e}} \rangle \right) \overline{\overline{\delta}} + 2G_c \langle \overline{\overline{e}} \rangle \qquad (12.3)$$

For a particulate composite of identical particles (same shape, size, and internal constitution), Equation 12.1 can be rewritten as:

$$\langle \overline{\overline{\sigma}} \rangle = \langle \overline{\overline{\sigma}}_o \rangle + \frac{N}{V} \langle \overline{\overline{S}} \rangle \qquad (12.4)$$

where N/V is the number density of particles in the composite and $\langle \bar{\bar{S}} \rangle$ is the average value of $\bar{\bar{S}}$ for a reference particle.

If the particulate composite is dilute so that the particles are far apart, the interactions between the particles are negligible. In such situations the average value of $\bar{\bar{S}}$ for a reference particle can be calculated assuming that the reference particle alone is present in an infinite matrix. Thus,

$$\langle \bar{\bar{\sigma}} \rangle = \langle \bar{\bar{\sigma}}_o \rangle + \frac{N}{V} \langle \bar{\bar{S}}^o \rangle \tag{12.5}$$

where $\langle \bar{\bar{S}}^o \rangle$ is the average value of the dipole strength of the reference particle present in an infinite matrix in which the strain tensor is $\langle \bar{\bar{e}} \rangle$ far from the particle. Note that $\langle \bar{\bar{e}} \rangle$ is the imposed bulk strain on the composite.

12.3 DIPOLE STRENGTH OF SPHERICAL PARTICLES

For a particulate composite with spherical particles, the orientation effects of the particle are irrelevant. Consequently, one can write Equation 12.5 as:

$$\langle \bar{\bar{\sigma}} \rangle = \langle \bar{\bar{\sigma}}_o \rangle + \left(\frac{3\phi}{4\pi R^3} \right) \bar{\bar{S}}^o \tag{12.6}$$

where R is the particle radius, ϕ is the volume fraction of particles, and $\bar{\bar{S}}^o$ is the dipole strength of a single spherical particle located in an infinite matrix, given by Equation 12.2.

Combining Equation 12.3 and Equation 12.6 leads to the following constitutive equation for dilute particulate composites with spherical particles:

$$\langle \bar{\bar{\sigma}} \rangle = \lambda_c \left(tr \langle \bar{\bar{e}} \rangle \right) \bar{\bar{\delta}} + 2G_c \langle \bar{\bar{e}} \rangle + \left(\frac{3\phi}{4\pi R^3} \right) \bar{\bar{S}}^o \tag{12.7}$$

The dipole strength of a single spherical particle located in an infinite matrix, given by Equation 12.2, can be determined from the knowledge of the displacement and stress fields in the particle and matrix phases.

Consider a spherical solid particle of radius R suspended in an infinite continuous phase (matrix). The individual phases (particle and matrix) are assumed to be isotropic and linearly elastic with stress–strain relations given by the generalized Hooke's law for isotropic materials:

$$\bar{\bar{\sigma}}_d = \lambda_d \left(tr \, \bar{\bar{e}}_d \right) \bar{\bar{\delta}} + 2 \, G_d \, \bar{\bar{e}}_d \tag{12.8}$$

$$\bar{\bar{\sigma}}_c = \lambda_c \left(tr\, \bar{\bar{e}}_c \right) \bar{\bar{\delta}} + 2\, G_c\, \bar{\bar{e}}_c \qquad (12.9)$$

Equation 12.8 represents the stress–strain relation for the dispersed (particulate) phase, and Equation 12.9 represents the stress–stress relation for the matrix phase. λ_d and G_d are Lamé's constants for the dispersed phase, and λ_c and G_c are Lamé's constants for the matrix. The coordinate system is fixed to the center of the particle, and the matrix is subjected to strain $\bar{\bar{e}}_\infty$ at infinity.

In the absence of body forces, the displacement fields inside the particle and in the matrix phase are governed by the following Navier–Cauchy equations (see Chapter 2):

$$G_d\, \nabla^2\, \vec{\xi}^* + \left(\lambda_d + G_d \right) \nabla \nabla \bullet \vec{\xi}^* = 0 \qquad \text{for} \qquad r < R \qquad (12.10)$$

$$G_c\, \nabla^2\, \vec{\xi} + \left(\lambda_c + G_c \right) \nabla \nabla \bullet \vec{\xi} = 0 \quad \text{for} \quad r > R \qquad (12.11)$$

where $\vec{\xi}^*$ is the displacement field inside the particle and $\vec{\xi}$ is the displacement field in the matrix phase.

The boundary conditions are:

1. Far away from the particle ($r \rightarrow \infty$):

$$\bar{\bar{e}} \rightarrow \bar{\bar{e}}_\infty \qquad (12.12)$$

2. At the surface of the particle ($r = R$):

$$\bar{\bar{\sigma}}_c \bullet \hat{n} = \bar{\bar{\sigma}}_d \bullet \hat{n} \qquad (12.13)$$

that is, the tractions are continuous across the particle–matrix interface.
3. At the surface of the particle ($r = R$):

$$\vec{\xi}^* = \vec{\xi} \qquad (12.14)$$

that is, the displacements are continuous across the particle–matrix interface.

The solutions of the Navier–Cauchy equations, Equation 12.10 and Equation 12.11, subject to the given boundary conditions (Equation 12.12 to Equation 12.14) give the displacement fields $\vec{\xi}$ and $\vec{\xi}^*$ in the matrix and dispersed phases. From knowledge of the displacement fields, the strain and stress fields ($\bar{\bar{e}}$ and $\bar{\bar{\sigma}}$) can be determined.

Chen and Acrivos [16] obtained the following expression for $\overline{\overline{S}}^o$ using Equation 12.2 and the solutions of the Navier–Cauchy equations:

$$\overline{\overline{S}}^o = \left(\frac{4}{3}\pi R^3\right)\left(K_c + \frac{4}{3}G_c\right)\alpha_1 \, (tr\,\overline{\overline{e}}_\infty)\,\overline{\overline{\delta}}$$

$$+ (40\,\pi\,R^3)\,G_c\,(1-v_c)\,\alpha_2\left[\overline{\overline{e}}_\infty - \frac{1}{3}tr\,(\overline{\overline{e}}_\infty)\,\overline{\overline{\delta}}\right] \tag{12.15}$$

where v_c is Poisson's ratio of the continuous phase (matrix), and α_1 and α_2 are given as:

$$\alpha_1 = \left[\frac{3\left(K_d - K_c\right)}{3K_d + 4G_c}\right] \tag{12.16}$$

$$\alpha_2 = \left[\frac{G_d - G_c}{2G_d\left(4 - 5\,v_c\right) + G_c\left(7 - 5\,v_c\right)}\right] \tag{12.17}$$

K_c and K_d are the bulk moduli of matrix and dispersed phase, respectively.

From Equation 12.7 and Equation 12.15, the following constitutive equation is obtained for dilute particulate composites with spherical particles:

$$\overline{\overline{\sigma}} = \lambda_c\left[1 + \frac{\phi}{\lambda_c}\left\{\left(K_c + \frac{4}{3}G_c\right)\alpha_1 - 10\,G_c(1 - v_c)\alpha_2\right\}\right]$$

$$\times (tr\,\overline{\overline{e}})\,\overline{\overline{\delta}} + 2G_c\left[1 + 15\phi(1 - v_c)\alpha_2\right]\overline{\overline{e}} \tag{12.18}$$

In obtaining Equation 12.18 from the single-particle solution, Equation 12.15, the applied strain at infinity $\overline{\overline{e}}_\infty$ was set equal to the imposed bulk strain on the composite $\langle\overline{\overline{e}}\rangle$. To simplify the notation, the angular brackets $\langle\,\rangle$ have been dropped in writing Equation 12.18.

12.4 EFFECTIVE ELASTIC MODULI OF PARTICULATE COMPOSITES WITH SPHERICAL PARTICLES

The form of the constitutive equation, Equation 12.18, indicates that a dilute particulate composite with spherical particles behaves as a Hookean elastic or linear elastic solid. The effective moduli of the composite can be defined by:

$$\overline{\overline{\sigma}} = \lambda\left(tr\,\overline{\overline{e}}\right)\overline{\overline{\delta}} + 2G\overline{\overline{e}} \tag{12.19}$$

Upon comparison of Equation 12.18 and Equation 12.19, the following relations are obtained for the effective elastic properties of dilute particulate composite with spherical particles:

$$\frac{G}{G_c} = 1 + 15\left(1 - v_c\right)\alpha_2 \, \phi \qquad (12.20)$$

$$\frac{\lambda}{\lambda_c} = 1 + \frac{1}{\lambda_c}\left[\left(K_c + \frac{4}{3}G_c\right)\alpha_1 - 10\,G_c\left(1 - v_c\right)\alpha_2\right]\phi \qquad (12.21)$$

Note that the Lamé constant λ is related to bulk and shear moduli as:

$$\lambda = K - \frac{2}{3}G \qquad (12.22)$$

Substitution of Equation 12.16 and Equation 12.17 into Equation 12.20 and Equation 12.21, and using Equation 12.22, the expressions for the moduli and their intrinsic values can be obtained. The shear modulus and its intrinsic value are:

$$\frac{G}{G_c} = 1 + \left[\frac{15\left(1 - v_c\right)\left(G_d - G_c\right)}{2G_d\left(4 - 5v_c\right) + G_c\left(7 - 5v_c\right)}\right]\phi \qquad (12.23a)$$

$$[G] = \underset{\phi \to 0}{Lim}\frac{G/G_c - 1}{\phi} = \frac{15(1 - v_c)(\lambda_G - 1)}{2\lambda_G(4 - 5v_c) + (7 - 5v_c)} \qquad (12.23b)$$

where $[G]$ is intrinsic shear modulus and λ_G is shear modulus ratio G_d/G_c. The bulk modulus is:

$$\frac{K}{K_c} = 1 + \left[\left(\frac{3K_c + 4G_c}{3K_c}\right)\left(\frac{3K_d - 3K_c}{3K_d + 4G_c}\right)\right]\phi \qquad (12.24)$$

Because $G/K = [3(1 - 2v)]/[2(1 + v)]$ for isotropic materials, Equation 12.24 can be rewritten as:

$$\frac{K}{K_c} = 1 - \left[\frac{3(1 - v_c)(K_c - K_d)}{2K_c(1 - 2v_c) + K_d(1 + v_c)}\right]\phi \qquad (12.25a)$$

The intrinsic bulk modulus is given as:

$$[K] = \lim_{\phi \to 0} \frac{K/K_c - 1}{\phi} = -\left[\frac{3(1-v_c)(1-\lambda_K)}{2(1-2v_c)+\lambda_K(1+v_c)}\right]\phi \qquad (12.25b)$$

where $[K]$ is the intrinsic bulk modulus and λ_K is the bulk modulus ratio K_d/K_c.

With the results obtained for the shear and bulk moduli (G and K), the other two elastic constants, Young's modulus (E) and Poisson's ratio (v), can be obtained from the following standard relations for isotropic materials [19]:

$$E = \frac{9\,K\,G}{3\,K + G} \qquad (12.26)$$

$$v = \frac{3K - 2G}{6K + 2G} = \frac{1}{2}\left(1 - \frac{E}{3K}\right) = \frac{E}{2G} - 1 \qquad (12.27)$$

It can readily be shown that Young's modulus of a dilute dispersion of spherical solid particles in a solid matrix is given by the following expression [20]:

$$\frac{E}{E_c} = 1 + \left\{\left[\frac{10(1-v_c^2)\,(1+v_c)\,E_d - 10(1-v_c^2)\,(1+v_d)\,E_c}{2(4-5v_c)\,(1+v_c)\,E_d + (7-5v_c)(1+v_d)\,E_c}\right]\right.$$
$$\left. + \left[\frac{(1-v_c)(1-2v_c)\,E_d - (1-v_c)(1-2v_d)\,E_c}{(1+v_c)\,E_d + 2E_c\,(1-2v_d)}\right]\right\}\phi \qquad (12.28a)$$

The intrinsic Young's modulus [E] is given by:

$$[E] = \lim_{\phi \to 0} \frac{E/E_c - 1}{\phi} = \left\{\left[\frac{10(1-v_c^2)(1+v_c)\lambda_E - 10(1-v_c^2)\,(1+v_d)}{2(4-5v_c)(1+v_c)\lambda_E + (7-5v_c)(1+v_d)}\right]\right.$$
$$\left. + \left[\frac{(1-v_c)(1-2v_c)\lambda_E - (1-v_c)(1-2v_d)}{(1+v_c)\lambda_E + 2(1-2v_d)}\right]\right\} \qquad (12.28b)$$

where λ_E is Young's modulus ratio E_d/E_c, E_d and E_c are Young's moduli of the dispersed phase and the matrix phase, respectively, and v_d and v_c are the Poisson ratios of the dispersed and matrix phases, respectively.

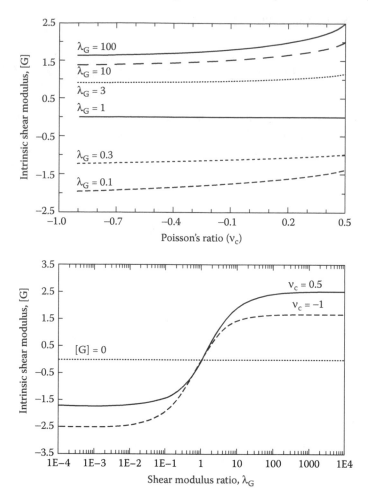

FIGURE 12.1 Intrinsic shear modulus [G] plots of particulate composite with spherical particles.

Figure 12.1 shows the intrinsic shear modulus [G] plots generated from Equation 12.23b. For a given shear modulus ratio λ_G, [G] generally increases with the increase in v_c (matrix Poisson's ratio). For a given v_c, [G] increases with the increase in λ_G. At $\lambda_G = 1$, [G] is zero regardless of the value of v_c. When $\lambda_G < 1$, [G] is negative and when $\lambda_G > 1$, [G] is positive.

Figure 12.2 shows intrinsic bulk modulus [K] plots generated from Equation 12.25b. For a given bulk modulus ratio λ_K, [K] generally decreases with the increase in v_c. For a given v_c, [K] increases with the increase in λ_K. At $\lambda_K = 1$, [K] is zero regardless of the value of v_c. When $\lambda_K < 1$, [K] is negative and when $\lambda_K > 1$, [K] is positive.

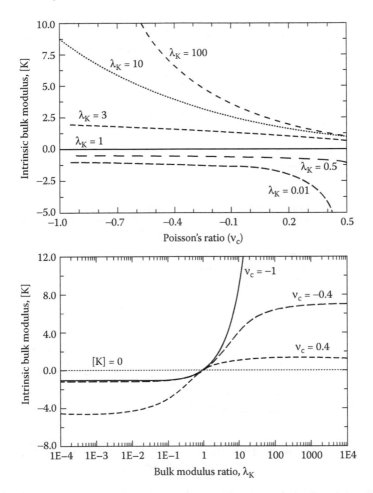

FIGURE 12.2 Intrinsic bulk modulus [*K*] plots of particulate composite with spherical particles.

Figure 12.3 shows the intrinsic Young's modulus [*E*] plots generated from Equation 12.28b. In the top graph of Figure 12.3, the matrix is taken to be incompressible ($v_c = 0.5$), and in the bottom graph of Figure 12.3, the particles are taken to be incompressible ($v_d = 0.5$). For a given v_d (particle Poisson's ratio), the intrinsic Young's modulus [*E*] of composite with incompressible matrix ($v_c = 0.5$) increases with the increase in λ_E, the Young's modulus ratio. Also, [*E*] decreases with the increase in v_d in the intermediate range of λ_E ($10^{-2} < \lambda_E < 100$). The crossover value of the modulus ratio λ_E where [*E*] crosses from the region of [*E*] > 0 to the region [*E*] < 0, increases with the increase in v_d. The crossover value of λ_E is unity only when v_d is near 0.5. For a given v_c, [*E*] increases with the increase λ_E when particles are incompressible ($v_d = 0.5$).

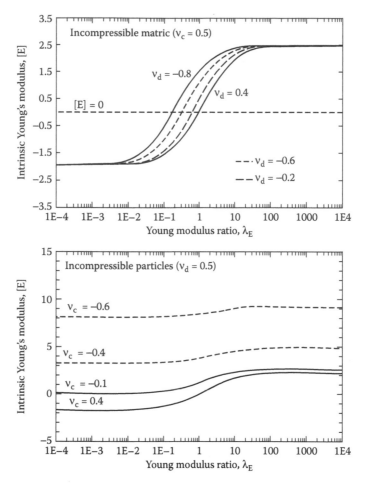

FIGURE 12.3 Intrinsic Young's modulus $[E]$ plots of particulate composite with spherical particles.

12.4.1 COMPOSITES WITH RIGID SPHERICAL PARTICLES

For particulate composite reinforced by rigid particles ($v_d \to 0.5$, $K_d \to \infty$, $G_d \to \infty$, $E_d \to \infty$), the expressions for the shear, bulk, and Young's moduli reduce to:

$$\frac{G}{G_c} = 1 + \left[\frac{15(1-v_c)}{2(4-5v_c)}\right]\phi \tag{12.29}$$

$$\frac{K}{K_c} = 1 + \left[\frac{3(1-v_c)}{(1+v_c)}\right]\phi \tag{12.30}$$

$$\frac{E}{E_c} = 1 + \left[\frac{3(1-v_c)(5v_c^2 - v_c + 3)}{(1+v_c)(4-5v_c)} \right] \phi \qquad (12.31)$$

If in addition to rigid particles, the composite matrix is incompressible ($v_c \to 0.5$, $K_c \to \infty$), the expressions for the shear and Young's moduli become the same, that is,

$$\frac{G}{G_c} = \frac{E}{E_c} = 1 + \frac{5}{2}\phi \qquad (12.32)$$

12.4.2 Composites with Incompressible Matrix

For particulate composites with incompressible matrix ($v_c = 0.5$), the expressions for the composite effective elastic properties G, K, and E are given as follows:

$$\frac{G}{G_c} = 1 + \left[\frac{5(G_d - G_c)}{2(G_d + 1.5G_c)} \right] \phi \qquad (12.33)$$

$$K = \frac{3K_d + 4G_c}{3\phi} \qquad (12.34)$$

$$\frac{E}{E_c} = 1 + \left[\left\{ \frac{15(E_d/E_c) - 10(1+v_d)}{6(E_d/E_c) + 6(1+v_d)} \right\} - \left\{ \frac{1 - 2v_d}{3(E_d/E_c) + 4(1-2v_d)} \right\} \right] \phi \qquad (12.35)$$

If both phases are incompressible, that is, $v_c = 0.5$ and $v_d = 0.5$, the expression for G remains the same as Equation 12.33, K becomes infinite, and E is given by:

$$\frac{E}{E_c} = 1 + \left[\frac{5(E_d - E_c)}{2(E_d + 1.5E_c)} \right] \phi \qquad (12.36)$$

12.4.3 Composites with Pores

Composites with pores (pore-solid composite materials) find many engineering applications ranging from acoustic absorption to heat shields in reentry vehicles. Many industrial materials such as wood, carbon, ceramic, foam, and brick are porous in nature. Porous ceramics are widely used as filters, catalyst supports, low dielectric substances, thermal insulators, and as lightweight structural components. Pore-solid composites are of interest in biomedical applications as well [21–23]. Even in nature many of the solids are porous.

For pore-solid composites, the expressions for the shear, bulk, and Young's moduli (Equation 12.23, Equation 12.24, and Equation 12.28) reduce to:

$$\frac{G}{G_c} = 1 - \left[\frac{15(1-\nu_c)}{7-5\nu_c}\right]\phi \tag{12.37}$$

$$\frac{K}{K_c} = 1 - \left[\frac{3(1-\nu_c)}{2(1-2\nu_c)}\right]\phi \tag{12.38}$$

$$\frac{E}{E_c} = 1 - \left[\frac{3(1-\nu_c)(9+5\nu_c)}{2(7-5\nu_c)}\right]\phi \tag{12.39}$$

The moduli of the dispersed phase (voids) are taken to be zero, that is, $G_d \to 0$, $E_d \to 0$, and $K_d \to 0$. It is interesting to note that whereas the effective shear and bulk moduli (G and K) of void containing materials are strongly dependent on the matrix-phase Poisson's ratio, ν_c, the effective Young's modulus E of pore-solid composites is insensitive to ν_c [24], especially when $0 \leq \nu_c \leq 0.5$.

12.5 EFFECTIVE ELASTIC MODULI OF PARTICULATE COMPOSITES WITH DISK-SHAPED PARTICLES

For dilute isotropic particulate composites with disk-shaped particles, the elastic properties K and G are given as follows [25]:

$$\frac{K}{K_c} = 1 + \phi\left[\frac{(K_d - K_c)(3K_c + 4G_d)}{K_c(3K_d + 4G_d)}\right] \tag{12.40}$$

$$\frac{G}{G_c} = 1 + \phi\left[\frac{(G_d - G_c)(G_c + G_g)}{G_c(G_g + G_d)}\right] \tag{12.41}$$

where

$$G_g = \frac{G_d(9K_d + 8G_d)}{6(K_d + 2G_d)} \tag{12.42}$$

Equation 12.40 and Equation 12.41 could also be written in the following forms:

$$\frac{K}{K_c} = 1 + \phi(\lambda_K - 1)\left[\frac{(1+\nu_d) + 2\lambda_K(1-2\nu_d)}{3\lambda_K(1-\nu_d)}\right] \quad (12.43)$$

$$\frac{G}{G_c} = 1 + \phi(\lambda_G - 1)\left[\frac{(8-10\nu_d) + \lambda_G(7-5\nu_d)}{15\lambda_G(1-\nu_d)}\right] \quad (12.44)$$

Figure 12.4 compares the relative shear modulus G_r (defined as G/G_c) of composites with disks and spheres as particles. When the shear modulus ratio

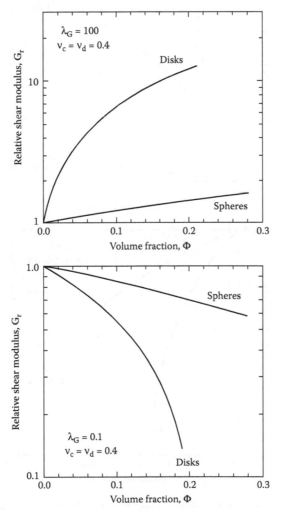

FIGURE 12.4 Relative shear modulus of composites with disks and spheres as particles.

λ_G is 100, the relative shear modulus is larger than one and G_r of composites with disk-shaped particles is much larger than the G_r of composites with spherical particles at the same volume fraction of particles (ϕ). When $\lambda_G = 0.1$, the opposite effect is observed. Now $G_r < 1$ and spherical particles result in a larger value of G_r

Figure 12.5 compares the relative bulk modulus K_r (defined as K/K_c) of composites with disks and spheres as particles. The relative bulk modulus

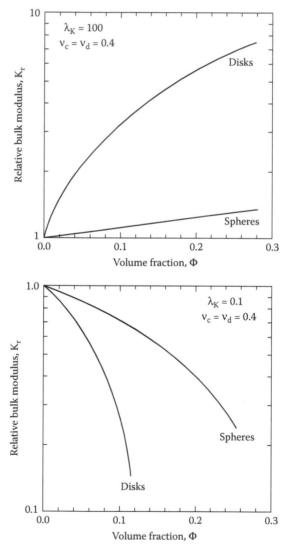

FIGURE 12.5 Relative bulk modulus of composites with disks and spheres as particles.

exhibits behavior similar to that of G_r. When λ_K is 100, $K_r > 1$ and K_r of composites with disk-shaped particles is much larger than the K_r of those with spherical particles. When $\lambda_K = 0.1$, $K_r < 1$ and spherical particles result in a larger value of K_r.

12.6 BOUNDS FOR THE EFFECTIVE ELASTIC PROPERTIES OF PARTICULATE COMPOSITES

The equations for the effective elastic properties of particulate composites given in the preceding section are valid for dilute composites with spherical or disk-shaped inclusions (particles). Interestingly, the elastic properties of isotropic two-phase composites (dilute as well as concentrated) are restricted by definite lower and upper bounds [16,26–39], regardless of the geometry of the inclusions (spherical or nonspherical). Such bounds depend only on the volume fractions of the components and the properties of the components. The effective elastic property of the composite lies somewhere in the interval between the lower and upper bounds.

A composite material composed of isotropic dispersed phase (unknown geometry) distributed randomly in an isotropic matrix has isotropic effective properties. Assuming that the composite is linearly elastic, it has two independent elastic constants (properties) such as bulk modulus (K) and shear modulus (G). In the literature, the bounds for bulk modulus and shear modulus are reported simultaneously.

The elementary bounds for bulk modulus (K) and shear modulus (G) of isotropic two-phase composites are known as Voigt–Reuss bounds [38,39]; the Voigt bounds represent the upper bounds on modulus (K or G) and the Reuss bounds represent the lower bounds on modulus. The Voigt (upper) bounds for the effective modulus of isotropic composite are [38]:

$$K \le \phi\, K_d + (1 - \phi)\, K_c \tag{12.45}$$

$$G \le \phi\, G_d + (1 - \phi)\, G_c \tag{12.46}$$

The Reuss (lower) bounds for the effective modulus of isotropic composite are given as [39]:

$$K \ge \dfrac{1}{\left(\dfrac{\phi}{K_d} + \dfrac{1-\phi}{K_c}\right)} \tag{12.47}$$

$$G \geq \frac{1}{\left(\dfrac{\phi}{G_d} + \dfrac{1-\phi}{G_c}\right)} \tag{12.48}$$

The actual effective modulus of the isotropic composite is expected to lie somewhere in the interval between the Voigt and Reuss bounds, no matter what the geometry (microstructure) may be. The upper and lower bounds given by Equation 12.45 to Equation 12.48 are generally not close enough to provide a good estimate of the effective modulus of the composite.

Hashin and Shtrikman [30] and Walpole [33] employed a variational theorem to obtain improved (and much tighter) upper and lower bounds for the effective bulk and shear moduli of isotropic composites. The Hashin–Shtrikman–Walpole (HSW) bounds are as follows [16,30,33]:

$$\left[\frac{(K_c+K_\ell)\phi}{K_c+K_\ell+(K_d-K_c)(1-\phi)}\right] \leq \left(\frac{K-K_c}{K_d-K_c}\right)$$
$$\leq \left[\frac{(K_c+K_g)\phi}{K_c+K_g+(K_d-K_c)(1-\phi)}\right] \tag{12.49}$$

$$\left[\frac{(G_c+G_\ell)\phi}{G_c+G_\ell+(G_d-G_c)(1-\phi)}\right] \leq \left(\frac{G-G_c}{G_d-G_c}\right)$$
$$\leq \left[\frac{(G_c+G_g)\phi}{G_c+G_g+(G_d-G_c)(1-\phi)}\right] \tag{12.50}$$

where

$$K_\ell = \frac{4}{3}G_c, \quad K_g = \frac{4}{3}G_d \tag{12.51}$$

$$G_\ell = \frac{3}{2}\left[\frac{1}{G_c}+\frac{10}{9K_c+8G_c}\right]^{-1}, \quad G_g = \frac{3}{2}\left[\frac{1}{G_d}+\frac{10}{9K_d+8G_d}\right]^{-1} \tag{12.52}$$

provided that $(G_d - G_c)(K_d - K_c) \geq 0$. If $(G_d - G_c)(K_d - K_c) \leq 0$, then

$$K_\ell = \frac{4}{3} G_d, \quad K_g = \frac{4}{3} G_c \tag{12.53}$$

$$G_\ell = \frac{3}{2} \left[\frac{1}{G_c} + \frac{10}{9 K_d + 8 G_c} \right]^{-1}, \quad G_g = \frac{3}{2} \left[\frac{1}{G_d} + \frac{10}{9 K_c + 8 G_d} \right]^{-1} \tag{12.54}$$

Note that when $G_c = G_d$ the lower and upper bounds for shear modulus coincide and composite $G = G_c$. The lower and upper bounds for bulk modulus also coincide when $G_c = G_d$ and composite K is given by:

$$K = K_c + \left[\frac{(K_d - K_c)(3K_c + 4G_c)\phi}{(3K_d + 4G_c) - (3K_d - 3K_c)\phi} \right] \tag{12.55}$$

Figure 12.6 to Figure 12.9 compare the Voigt–Reuss bounds with the HSW bounds under different conditions ($\lambda_K > 1$, $\lambda_G > 1$ in Figure 12.6; $\lambda_K < 1$, $\lambda_G < 1$ in Figure 12.7; $\lambda_K < 1$, $\lambda_G > 1$ in Figure 12.8; $\lambda_K > 1$, $\lambda_G < 1$ in Figure 12.9). Clearly, the HSW bounds are much tighter, and hence better, compared to the Voigt–Reuss bounds.

The HSW bounds, although better than the Voigt–Reuss bounds, yield satisfactory estimates for the effective moduli, in that the upper and lower bounds are close, only when the ratios between the corresponding moduli of the two phases are not too large. When the moduli of the two phases are very different, the bounds become too wide to be of any practical value.

It is interesting to note that for dilute particulate composites when $(G_d - G_c)(K_d - K_c) \geq 0$, the lower HSW bounds for effective bulk and shear moduli correspond to the formulas for composites of spherical particles and the upper HSW bounds correspond to the formulas for composites of disk-shaped particles. This indicates that when $(G_d - G_c)(K_d - K_c) \geq 0$, the spherical particles provide the minimum reinforcing effect, whereas the disk-shaped particles provide the maximum reinforcing effect [25].

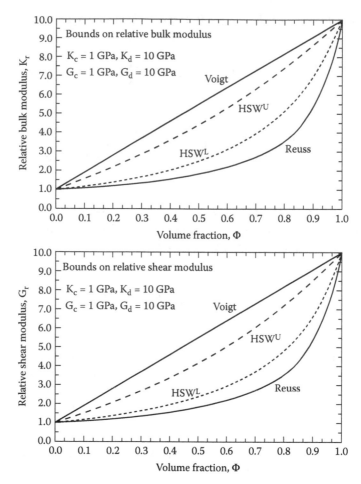

FIGURE 12.6 Voigt–Reuss and Hashin–Shtrikmann–Walpole (HSW) bounds for the relative bulk and shear moduli of isotropic composites under the conditions $\lambda_K > 1$ and $\lambda_G > 1$. (Note that $\lambda_K = K_d/K_c$ and $\lambda_G = G_d/G_c$).

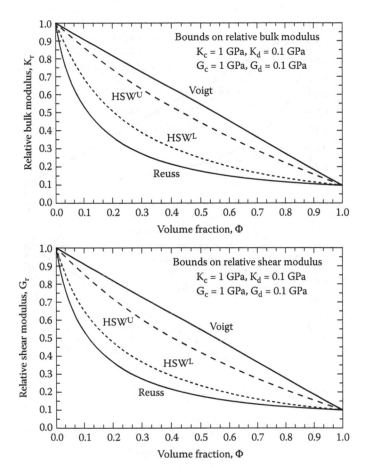

FIGURE 12.7 Voigt–Reuss and HSW bounds for the relative bulk and shear moduli of isotropic composites under the conditions $\lambda_K < 1$ and $\lambda_G < 1$. (where $\lambda_K = K_d/K_c$ and $\lambda_G = G_d/G_c$).

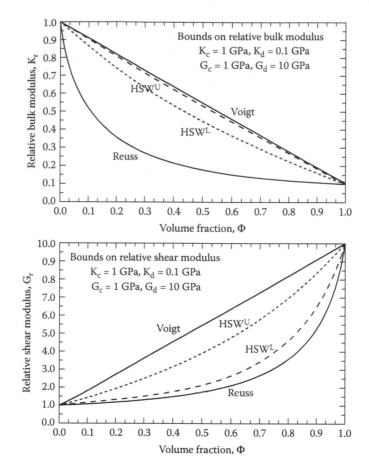

FIGURE 12.8 Voigt–Reuss and HSW bounds for the relative bulk and shear moduli of isotropic composites under the conditions $\lambda_K < 1$ and $\lambda_G > 1$. (where $\lambda_K = K_d/K_c$ and $\lambda_G = G_d/G_c$).

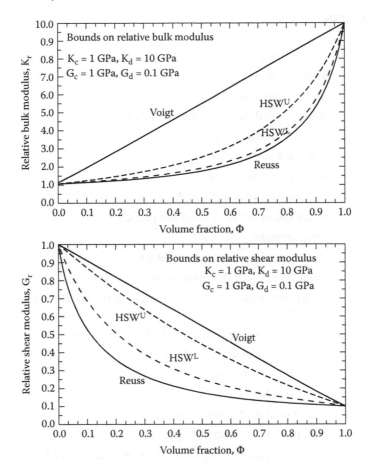

FIGURE 12.9 Voigt–Reuss and HSW bounds for the relative bulk and shear moduli of isotropic composites under the conditions $\lambda_K > 1$ and $\lambda_G < 1$. (where $\lambda_K = K_d/K_c$ and $\lambda_G = G_d/G_c$).

NOTATION

A	surface area
A_0	surface area of a particle
E	Young's modulus
E_c	Young's modulus of continuous phase
$[E]$	intrinsic Young's modulus
$\overline{\overline{e}}$	strain tensor
$\overline{\overline{e}}_c$	strain tensor in matrix
$\overline{\overline{e}}_d$	strain tensor in dispersed phase
$\overline{\overline{e}}_\infty$	strain tensor far away from the particle
G	shear modulus
G_c	shear modulus of continuous phase
G_d	shear modulus of dispersed phase
G_g	parameter given by Equation 12.42, Equation 12.52, or Equation 12.54
G_l	parameter given by Equation 12.52 or Equation 12.54
G_r	relative shear modulus, defined as G/G_c
$[G]$	intrinsic shear modulus
K	bulk modulus
K_c	bulk modulus of continuous phase
K_d	bulk modulus of dispersed phase
K_g	parameter given by Equation 12.51 or Equation 12.53
K_l	parameter given by Equation 12.51 or Equation 12.53
K_r	relative bulk modulus, defined as K/K_c
$[K]$	intrinsic bulk modulus
$\hat{n}$	outward unit normal vector
N	number of particles
R	particle radius
$\overline{\overline{S}}$	dipole strength of a particle
$\overline{\overline{S}}^o$	dipole strength of a single particle located in an infinite matrix
$\overline{\overline{S}}_i$	dipole strength of ith particle
tr	trace of a tensor
V	volume
$\vec{x}$	position vector

Greek

α_1	parameter given by Equation 12.16
α_2	parameter given by Equation 12.17
$\bar{\bar{\delta}}$	unit tensor
∇^2	Laplacian
λ	Lamé constant
λ_c	matrix Lamé constant
λ_d	Lamé constant of dispersed phase (particles)
λ_E	Young's modulus ratio (E_d / E_c)
λ_G	shear modulus ratio (G_d / G_c)
λ_K	bulk modulus ratio (K_d / K_c)
ν	Poisson's ratio
ν_c	Poisson's ratio of continuous phase
$\vec{\xi}$	displacement field
$\vec{\xi}^*$	displacement field inside the particle
$\bar{\bar{\sigma}}$	stress tensor
$\bar{\bar{\sigma}}_o$	stress tensor in pure matrix material under condition corresponding to the imposed bulk strain on the composite
$\bar{\bar{\sigma}}_c$	stress tensor in matrix phase
$\bar{\bar{\sigma}}_d$	stress tensor in dispersed phase (particle)
ϕ	volume fraction of particles
$\langle X \rangle$	volume average of quantity X

REFERENCES

1. F. Chabrier, C.H. Lloyd, S.N. Scrimgeour, *Dent. Mater.*, 15: 33–38, 1999.
2. Y. Abe, P. Lambrechts, S. Inone, M.J. Braem, M. Takeuch, G. Vanherle, B. Van Meerbeek, *Dent. Mater.*, 17: 520–525, 2001.
3. I.B. Lee, H.H. Son, C.M. Um, *Dent. Mater.*, 19: 298–307, 2003.
4. M. Braem, V.E. Van Doren, P. Lambrechts, G. Vanherle, *J. Mater. Sci.*, 22: 2037–2042, 1987.
5. M. Braem, *J. Dent. Res.*, 65: 648–653, 1986.
6. G. Willems, P. Lambrechts, M. Braem, J.P. Celis, G. Vanherle, *Dent. Mater.*, 8: 310–319, 1992.
7. K.F. Leinfelder, S.C. Bayne, E.J. Swift, *J. Esthet. Dentistry*, 11: 234–249, 1999.

8.　M. Braem, W. Finger, V.E. Van Doren, P. Lambrechts, G. Vanherle, *Dent. Mater.*, 5: 346–349, 1989.
9.　T.L. Smith, *Ind. Eng. Chem.*, 52: 776–780, 1960.
10.　P.J. Blatz, *Ind. Eng. Chem.*, 48: 727–729, 1956.
11.　R.F. Landel, T.L. Smith, *ARS J.*, 31: 599–608, 1961.
12.　A.K. Kaw, *Mechanics of Composite Materials*, CRC Press, Boca Raton, FL, 1997.
13.　L.J. Cohen, O. Ishai, *J. Comp. Mater.*, 1: 390–403, 1967.
14.　E. J. Garboczi, J.G. Berryman, *Mech. Mater.*, 33: 455–470, 2001.
15.　W.B. Russel, A. Acrivos, *J. Appl. Maths Phys.* (ZAMP) 23: 435–464, 1972.
16.　H.S. Chen, A. Acrivos, *Int. J. Solids Struct.*, 14: 349–364, 1978.
17.　H.S. Chen, A. Acrivos, *Int. J. Solids Struct.*, 14: 331–348, 1978.
18.　G.E. Mase, *Continuum Mechanics*, Schaum's outline series, McGraw-Hill, New York, 1970, chap. 6.
19.　P. Sherman, *Industrial Rheology.* Academic Press, London, 1970.
20.　R. Pal, *Comp. Part B: Eng.*, 36: 513–523, 2005.
21.　K. Zhang, Y. Wang, M.A. Hillmyer, L.F. Francis, *Biomaterials*, 25: 2489–2500, 2004.
22.　K. Zhang, Y. Ma, L.F. Francis, *J. Biomed Mater. Res.*, 61: 551–563, 2002.
23.　M. Milosevski, J. Bossert, D. Milosevski, N. Gruevska, *Ceram. Int.*, 25: 693–696, 1999.
24.　R.M. Christensen, *Proc. R. Soc. Lond. A*, 440: 461–473, 1993.
25.　S. Boucher, *J. Comp. Mater.*, 8: 82–89, 1974.
26.　Z. Hashin, in *Mechanics of Composite Materials*, F.W. Wendt, H. Liebowitz, N. Perrone, Eds., Pergamon Press, Oxford, 1970, pp. 201–242.
27.　R.M. Christensen, *Mechanics of Composite Materials*, John Wiley & Sons, New York, 1979.
28.　R.M. Christensen, in *Mechanics of Composite Materials — Recent Advances*, Z. Hashin, C.T. Herakovich, Eds., Pergamon Press, New York, 1982, pp. 1–16.
29.　Z. Hashin, *J. Appl. Mech.*, 29: 143–150, 1962.
30.　Z. Hashin, S. Shtrikman, *J. Mech. Phys. Solids*, 11: 127–140, 1963.
31.　R. Hill, *J. Mech. Phys. Solids*, 11: 357–372, 1963.
32.　Z. Hashin, *Appl. Mech. Rev.*, 17: 1–9, 1964.
33.　L.J. Walpole, *J. Mech. Phys. Solids*, 14: 151–162, 1966.
34.　R. Roscoe, *Rheol. Acta*, 12: 404–411, 1973.
35.　C.W. Milton, N. Phan-Thien, *Proc. R. Soc. Lond. A*, 380: 305–331, 1982.
36.　P.D. Chinh, *Mech. Mater.*, 27: 249–260, 1998.
37.　L.V. Gibiansky, O. Sigmund, *J. Mech. Phys. Solids*, 48: 461–498, 2000.
38.　W. Voigt, *Lehrbuch der Kristallphysik*, Teubner-Verlag, Leipzig, 1928.
39.　A. Reuss, *Z. Angew. Math. Mech.*, 9: 49–58, 1929.

13 Fiber-Reinforced Composites

13.1 INTRODUCTION

Fiber-reinforced composites are the most widely used composite materials. For instance, fiber-reinforced polymers are extensively used in aerospace applications where materials with low density but high strength and stiffness are required. Fiber-reinforced composites find applications in many other industries, such as automotive, sports, and marine [1].

Fiber-reinforced composites can be classified into two broad groups: *Continuous-fiber composites* and *discontinuous-* or *short-fiber composites* [2]. Continuous-fiber composites consist of long continuous fibers, whereas discontinuous-fiber composites consist of short fibers or whiskers as the reinforcing phase. Continuous- and short-fiber composites can be further classified as either aligned or randomly distributed fiber composites. In the case of aligned or unidirectional fiber composites, all fibers are oriented in the same direction. In the random case, the spatial orientation of the fibers is random. Figure 13.1 schematically illustrates various types of fiber composites.

The elastic properties of fiber-reinforced composites are of importance in many applications involving the use of these materials [1–14].

13.2 ELASTIC BEHAVIOR OF CONTINUOUS-FIBER COMPOSITES

13.2.1 UNIDIRECTIONAL CONTINUOUS-FIBER COMPOSITES

A unidirectional fiber composite in which all the fibers are aligned in one direction (Figure 13.2) is transversely isotropic so long as the fibers are positioned uniformly or randomly in space. Such composites are fully characterized by five independent elastic constants [7–8,11–12,14] such as:

E_{11} – longitudinal Young's modulus (Young's modulus in the fiber direction)
E_{22} – transverse Young's modulus (Young's modulus in the direction transverse to the fibers)
G_{12} – in plane or longitudinal shear modulus
G_{23} – out-of-plane or transverse shear modulus
v_{12} – in-plane or longitudinal Poisson's ratio

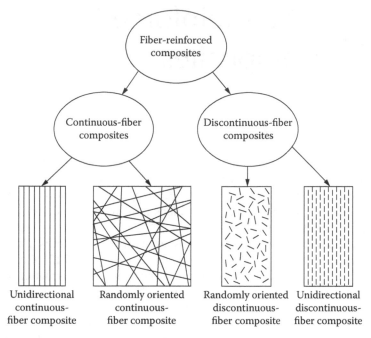

FIGURE 13.1 Classification of fiber-reinforced composites.

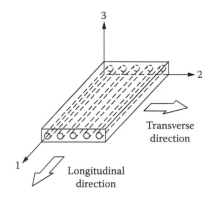

FIGURE 13.2 Unidirectional fiber composite where all fibers are aligned in one direction.

Figure 13.3 illustrates various stresses related to the definitions of these elastic constants:

$$E_{11} = \frac{\sigma_{11}}{e_{11}} \qquad (13.1)$$

$$E_{22} = \frac{\sigma_{22}}{e_{22}} \qquad (13.2)$$

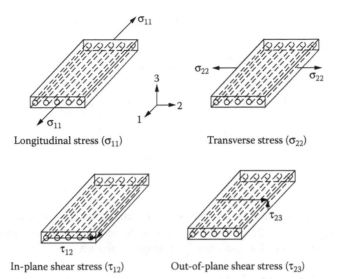

FIGURE 13.3 Illustration of various stresses in a unidirectional fiber-reinforced composite.

$$G_{12} = \frac{\tau_{12}}{e_{12}} \tag{13.3}$$

$$G_{23} = \frac{\tau_{23}}{e_{23}} \tag{13.4}$$

$$\nu_{12} = -\frac{e_{22}}{e_{11}} \tag{13.5}$$

where σ_{11} and σ_{22} are normal stresses in directions 1 and 2, respectively, τ_{12} and τ_{23} are shear stresses, e_{11} and e_{22} are normal strains, and e_{12} and e_{23} are shear strains. Note that for transversely isotropic material with axis of symmetry coinciding with the fiber direction:

$$E_{33} = E_{22} \tag{13.6}$$

$$G_{13} = G_{12} \tag{13.7}$$

$$\nu_{23} = \frac{E_{22}}{2\,G_{23}} - 1 \tag{13.8}$$

$$v_{12} = v_{13} \tag{13.9}$$

$$v_{31} = v_{21} \tag{13.10}$$

$$v_{32} = v_{23} \tag{13.11}$$

$$\frac{v_{12}}{E_{11}} = \frac{v_{21}}{E_{22}} \tag{13.12}$$

Consequently, only five of the twelve elastic constants (E_{11}, E_{22}, E_{33}, G_{12}, G_{23}, G_{13}, v_{12}, v_{13}, v_{23}, v_{21}, v_{31}, and v_{32}) present in the stress–strain relation (Hooke's law) are independent for transversely isotropic materials.

13.2.1.1 Elementary Models

The simplest approach to model the longitudinal behavior of unidirectional fiber composites is to treat fibers (assumed anisotropic) and matrix (assumed isotropic) as springs connected in parallel [2–5,8,11–12,14] (Figure 13.4a). When a normal stress σ_{11} is applied on the composite in the fiber direction, each component of the composite (fibers and matrix) will undergo the same longitudinal strain provided the two components adhere perfectly and have the same Poisson's ratio. As the two components (fiber and matrix phases) act in parallel, the total load sustained by the composite F is equal to the load carried by the fiber phase F_f plus the load carried by the continuous phase F_c. Thus,

$$F = F_f + F_c \tag{13.13}$$

Because F = stress × cross-sectional area, one can write:

$$\sigma_{11} A = \sigma_{f1} A_f + \sigma_{c1} A_c \tag{13.14}$$

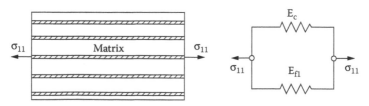

Figure 13.4a Parallel arrangement of matrix and fibers. The composite is subjected to longitudinal stress.

where A, A_f, and A_c are cross-sectional areas of composite, fiber phase, and matrix, respectively, σ_{f1} and σ_{c1} are normal stresses (in direction 1) in fiber phase and matrix, respectively. If the composite, fibers, and matrix phase lengths are all equal, then the volume fraction of fibers (ϕ) is:

$$\phi = \frac{A_f}{A} = 1 - \frac{A_c}{A} \tag{13.15}$$

Combining Equation 13.14 and Equation 13.15 gives:

$$\sigma_{11} = \phi\,\sigma_{f1} + (1-\phi)\,\sigma_{c1} \tag{13.16}$$

Because the strains in the composite, fibers, and matrix are equal (an isostrain situation), Equation 13.16 can be rewritten as:

$$\frac{\sigma_{11}}{e_L} = \phi\,\frac{\sigma_{f1}}{e_{Lf}} + (1-\phi)\,\frac{\sigma_{c1}}{e_{Lc}} \tag{13.17}$$

where e_L, e_{Lf}, and e_{Lc} are the longitudinal strains in the composite, fibers, and matrix, respectively (Note that $e_L = e_{Lf} = e_{Lc}$). Equation 13.17 indicates that:

$$E_{11} = \phi\,E_{f1} + (1-\phi)E_c \tag{13.18}$$

where E_{f1} is the longitudinal Young's modulus of fiber and E_c is the matrix Young's modulus (the matrix is assumed to be isotropic). Equation 13.18 is known as the Voigt estimate of longitudinal Young's modulus of fiber-reinforced composite. It is also referred to as rule-of-mixtures (ROM) for Young's modulus in the fiber direction.

To model the transverse behavior of unidirectional fiber composite, fiber and continuous (matrix) phases are assumed to be connected in series [2–5,8,11–12,14] (Figure 13.4b). When a normal stress σ_{22} is applied on the

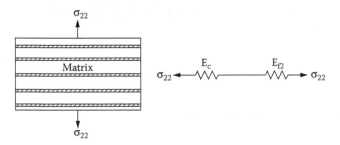

Figure 13.4b Series arrangement of matrix and fibers. The composite is subjected to transverse stress.

composite in the transverse direction, each component of the composite (fibers and matrix) will experience the same stress provided the two components adhere perfectly and have the same Poisson's ratio. The total displacement of the composite (L) in the transverse direction is the sum of displacements of the components, that is,

$$L = L_f + L_c \tag{13.19}$$

Let the thickness of the composite in the transverse direction be t, the cumulative thickness of the fibers be t_f and the cumulative thickness of the matrix phase be t_c. Thus,

$$L = e_T t \tag{13.20}$$

$$L_f = e_{Tf} t_f \tag{13.21}$$

$$L_c = e_{Tc} t_c \tag{13.22}$$

where e_T, e_{Tf}, and e_{Tc} are transverse strains in the composite, fibers, and matrix, respectively. Substituting Equation 13.20 to Equation 13.22 in Equation 13.19 gives:

$$e_T = e_{Tf}\left(\frac{t_f}{t}\right) + e_{Tc}\left(\frac{t_c}{t}\right) \tag{13.23}$$

As fibers and matrix are linear elastic materials, Equation 13.23 can be rewritten as:

$$\frac{\sigma_{22}}{E_{22}} = \frac{\sigma_{f2}}{E_{f2}}\left(\frac{t_f}{t}\right) + \frac{\sigma_{c2}}{E_c}\left(\frac{t_c}{t}\right) \tag{13.24}$$

where E_{f2} is the transverse modulus of the fiber, and σ_{f2} and σ_{c2} are normal stresses (in direction 2) in fiber phase and matrix, respectively.

Assuming that:

$$\phi = \frac{t_f}{t} = 1 - \frac{t_c}{t} \tag{13.25}$$

Equation 13.24 can be expressed as:

$$\frac{\sigma_{22}}{E_{22}} = \phi\left(\frac{\sigma_{f2}}{E_{f2}}\right) + (1-\phi)\left(\frac{\sigma_{c2}}{E_c}\right) \tag{13.26}$$

Because the stresses in the composite, fibers, and matrix are equal (an iso-stress situation), Equation 13.26 gives:

$$\frac{1}{E_{22}} = \frac{\phi}{E_{f2}} + \frac{1-\phi}{E_c} \tag{13.27}$$

Equation 13.27 is known as the *Reuss* estimate of transverse Young's modulus of fiber-reinforced composite. It is also referred to as ROM for Young's modulus in a direction transverse to the fibers.

Figure 13.5 shows the plots of E_{11}/E_c and E_{22}/E_c as functions of fiber volume fraction, calculated from Equation 13.18 and Equation 13.27. Interestingly, fibers do not contribute appreciably to the stiffness in the transverse direction (increase in E_{22}/E_c with ϕ is small). However, the stiffness (modulus) in the longitudinal direction is greatly enhanced by the fibers (increase in E_{11}/E_c with ϕ is large).

The expressions for the in-plane or longitudinal shear modulus (G_{12}) and out-of-plane or transverse shear modulus (G_{23}) can be derived using the simple isostress model; that is, the shear stresses in the composite, the fibers, and the matrix are equal. The isostress or Reuss model leads to the following expressions for G_{12} and G_{23}:

$$\frac{1}{G_{12}} = \frac{\phi}{G_{f12}} + \frac{1-\phi}{G_c} \tag{13.28}$$

$$\frac{1}{G_{23}} = \frac{\phi}{G_{f23}} + \frac{1-\phi}{G_c} \tag{13.29}$$

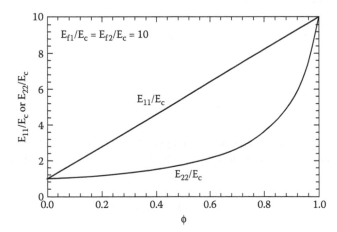

FIGURE 13.5 Variation of relative longitudinal Young's modulus (E_{11}/E_c) and relative transverse Young's modulus (E_{22}/E_c) with volume fraction of fibers in a fiber-reinforced composite.

where G_{f12} and G_{f23} are longitudinal and transverse shear moduli of the fibers.

The in-plane and out-of-plane shear moduli of fiber-reinforced composites are matrix-dominated properties. The matrix contributes more to the development of G_{12} and G_{23} than the fibers.

The expression for the in-plane or longitudinal Poisson's ratio (v_{12}) can be derived as follows. When the composite is subjected to longitudinal stress σ_{11}, the transverse strains in the composite, fibers, and matrix can be expressed in terms of longitudinal strains (e_L) and the Poisson's ratios as:

$$e_T = -v_{12}\, e_L \tag{13.30}$$

$$e_{Tf} = -v_{f12}\, e_{Lf} \tag{13.31}$$

$$e_{Tc} = -v_c\, e_{Lc} \tag{13.32}$$

where v_{f12} and v_c are longitudinal Poisson's ratio of fiber and Poison's ratio of matrix, respectively. The displacements in the transverse direction can be written as:

$$L = t\, e_T = -tv_{12}\, e_L \tag{13.33}$$

$$L_f = t_f e_{Tf} = -t_f v_{f12}\, e_{Lf} \tag{13.34}$$

$$L_c = t_c\, e_{Tc} = -t_c\, v_c\, e_{Lc} \tag{13.35}$$

From Equation 13.33 to Equation 13.35 and

$$L = L_f + L_c \tag{13.36}$$

one can write:

$$-t\, v_{12} e_L = -t_f\, v_{f12}\, e_{Lf} - t_c\, v_c\, e_{Lc} \tag{13.37}$$

The longitudinal strains in the composite, fibers, and matrix are the same when the composite is subjected to longitudinal stress. Thus, Equation 13.37 reduces to:

$$t\, v_{12} = t_f\, v_{f12} + t_c\, v_c \tag{13.38}$$

or

$$v_{12} = \left(\frac{t_f}{t}\right) v_{f12} + \left(\frac{t_c}{t}\right) v_c \tag{13.39}$$

Using Equation 13.25, Equation 13.39 can be rewritten as:

$$v_{12} = \phi\, v_{f12} + (1 - \phi)\, v_c \tag{13.40}$$

This is the ROM or Voigt estimate for the in-plane or longitudinal Poisson's ratio of a unidirectional composite.

The ROM formulas for the elastic constants of unidirectional fiber-reinforced composites (Equation 13.18, Equation 13.27, Equation 13.28, Equation 13.29, and Equation 13.40) provide only approximate values. The values of longitudinal properties such as E_{11} and v_{12} predicted by the ROM formulas generally agree well with the experimental data. However, the values of the transverse properties such as E_{22} and G_{23} predicted by the ROM formulas are generally much lower than the experimental values. Thus, there is a need for more accurate models to predict the elastic properties of fiber-reinforced composites.

13.2.1.2 Improved Models

The composite cylinders model was introduced by Hashin and Rosen [15] to determine the elastic properties of unidirectional fiber-reinforced composites. According to this model, a unidirectional continuous-fiber composite can be visualized as an assemblage of infinitely long concentric cylinders, each consisting of a fiber core surrounded by a matrix phase annulus as shown in Figure 13.6. To fill the entire volume of the composite with concentric cylinders, the size of the cylinders (radii "R" and "a") are allowed to vary with each composite cylinder. However, the volume fraction of fiber in each composite cylinder, and hence the ratio of radii R/a, is required to be constant.

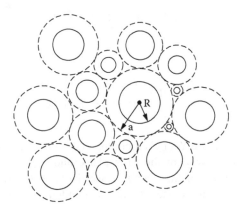

FIGURE 13.6 Composite cylinder model of unidirectional fiber-reinforced composite.

The elastic properties of a fiber-reinforced composite are determined from the analysis of an individual concentric cylinder [7,16–19] composed of isotropic linearly elastic fiber and matrix phases. The analysis of the response of a single concentric cylinder to a given applied strain allows the determination of four of the five independent elastic constants of a transversely isotropic composite. The expressions for the four elastic properties determined from the composite cylinder model are as follows:

$$E_{11} = \phi E_f + (1-\phi) E_c + 4\phi (1-\phi) G_c \left[\frac{(\nu_f - \nu_c)^2}{\dfrac{(1-\phi)G_c}{K_f + \dfrac{G_f}{3}} + \dfrac{\phi G_c}{K_c + \dfrac{G_c}{3}} + 1} \right] \quad (13.41)$$

$$\nu_{12} = \phi \, \nu_f + (1-\phi) \nu_c + \frac{\phi(1-\phi)(\nu_f - \nu_c) \left[\dfrac{G_c}{K_c + \dfrac{G_c}{3}} - \dfrac{G_c}{K_f + \dfrac{G_f}{3}} \right]}{\left[\dfrac{(1-\phi)G_c}{\left(K_f + \dfrac{G_f}{3}\right)} + \dfrac{\phi G_c}{\left(K_c + \dfrac{G_c}{3}\right)} + 1 \right]} \quad (13.42)$$

$$K_{23} = K_c + \frac{G_c}{3} + \frac{\phi}{\left[\dfrac{1}{K_f - K_c + \dfrac{1}{3}(G_f - G_c)} + \dfrac{1-\phi}{K_c + \dfrac{4}{3}G_c} \right]} \quad (13.43)$$

$$\frac{G_{12}}{G_c} = \left[\frac{G_f(1+\phi) + G_c (1-\phi)}{G_f (1-\phi) + G_c (1+\phi)} \right] \quad (13.44)$$

where E_f, K_f, G_f, and ν_f are Young's modulus, bulk modulus, shear modulus, and Poisson's ratio, respectively, of the fiber phase and E_c, K_c, G_c, and ν_c are the corresponding properties of the matrix phase. Note that here K_{23} (plane strain bulk modulus) is chosen as one of the independent elastic properties of the composite instead of the transverse Young's modulus E_{22}. The transverse Young's modulus is related to K_{23} as [7]:

$$E_{22} = \frac{4\,G_{23}\,K_{23}}{K_{23} + G_{23} + 4v_{12}^2\,G_{23}\,K_{23}/E_{11}} \tag{13.45}$$

It is not possible to determine the expression for the transverse shear modulus (out-of-plane shear modulus) G_{23} using the composite cylinder model [7,16–19]. Christensen and Lo [16] have determined this fifth independent elastic property (G_{23}) of a unidirectional continuous fiber composite using the generalized self-consistent model, also referred to as a three-phase model. According to the generalized self-consistent model, the composite (unidirectional continuous fiber composite in the present case) is first treated as an equivalent "effective medium" that is homogeneous and has the same macroscopic elastic properties as that of the composite. Then a small portion of the effective homogenous medium is replaced by the actual components of the composite. This is illustrated in Figure 13.7, where a cylindrical fiber is embedded in a concentric cylindrical annulus of the matrix phase, which in turn is embedded in an infinite effective homogenous medium (whose properties are to be determined). The radius a is chosen to give the correct volume fraction of the fibers in the composite, that is $a = R/(\phi)^{1/3}$. The properties of the effective medium are then determined by supposing that if a small portion of the effective homogeneous medium were replaced by the actual components of the composite, no difference in elastic behavior could be detected by macroscopic observations.

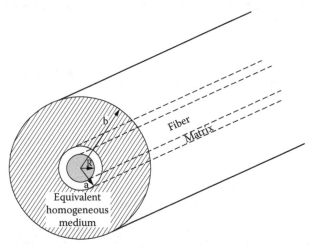

FIGURE 13.7 Generalized self-consistent model (three-phase cylinder model) of fiber-reinforced composite.

Interestingly, the generalized self-consistent model gives the same expressions for the four elastic properties (E_{11}, v_{12}, K_{23}, and G_{12}) as those obtained by the composite cylinders model of Hashin and Rosen [15]. The generalized self-consistent model has the additional advantage in that it gives the following expression for the transverse shear modulus G_{23} at small fiber volume fractions [7]:

$$\frac{G_{23}}{G_c} = 1 + \frac{\phi}{\left(\dfrac{G_c}{G_f - G_c}\right) + \dfrac{\left(K_c + \dfrac{7}{3}G_c\right)}{\left(2K_c + \dfrac{8}{3}G_c\right)}(1-\phi)} \tag{13.46}$$

In summary, Equation 13.41 to Equation 13.44 and Equation 13.46 provide reasonable predictions of the five independent elastic properties of a unidirectional continuous-fiber composite composed of isotropic fiber and matrix phases.

13.2.1.3 Bounds on the Elastic Properties of Unidirectional Continuous-Fiber Composites

Hashin [20] and Hill [21] have developed the upper and lower bounds on the effective elastic properties of fiber-reinforced composites that are transversely isotropic and consist of fibers of irregular shapes. The bounds on the five elastic constants, under the conditions that $G_f > G_c$ and $\mu_f > \mu_c$, are given by the following expressions [22]:

$$E_f\phi + E_c(1-\phi) + \frac{4\phi(1-\phi)(v_c - v_f)^2}{[(1-\phi)/\mu_f] + [\phi/\mu_c] + [1/G_c]} \leq E_{11} \leq E_f\phi + E_c(1-\phi)$$
$$\tag{13.47}$$
$$+ \frac{4\phi(1-\phi)(v_c - v_f)^2}{[(1-\phi)/\mu_f] + [\phi/\mu_c] + [1/G_f]}$$

$$v_c(1-\phi) + v_f\phi + \frac{\phi(1-\phi)(v_f - v_c)[(1/\mu_c) - (1/\mu_f)]}{[(1-\phi)/\mu_f] + [\phi/\mu_c] + [1/G_c]}$$
$$\tag{13.48}$$
$$\leq v_{12} \leq v_c(1-\phi) + v_f\phi + \frac{\phi(1-\phi)(v_f - v_c)[(1/\mu_c) - (1/\mu_f)]}{[(1-\phi)/\mu_f] + [\phi/\mu_c] + [1/G_f]}$$

$$\mu_c + \frac{\phi}{[1/(\mu_f - \mu_c)] + [(1-\phi)/(\mu_c + G_c)]}$$

$$\leq \mu_{23} \leq \mu_f + \frac{(1-\phi)}{[1/(\mu_c - \mu_f)] + [(1-\phi)/(\mu_f + G_f)]}$$

(13.49)

$$G_c + \frac{\phi}{[1/(G_f - G_c)] + [(1-\phi)/(2G_c)]}$$

$$\leq G_{12} \leq G_f + \frac{(1-\phi)}{[1/(G_c - G_f)] + [(\phi/(2G_f)]}$$

(13.50)

$$G_c + \frac{\phi}{[1/(G_f - G_c)] + [(1-\phi)(\mu_c + 2G_c)/\{2G_c(\mu_c + G_c)\}]}$$

$$\leq G_{23} \leq G_f + \frac{(1-\phi)}{[1/(G_c - G_f)] + [\phi(\mu_f + 2G_f)/\{2G_f(\mu_f + G_f)\}]}$$

(13.51)

where $\mu = K + (G/3)$.

13.2.2 RANDOMLY ORIENTED CONTINUOUS-FIBER COMPOSITES

Christensen and Waals [23] developed the expressions for the effective elastic properties of randomly oriented continuous-fiber composites using the results of a unidirectional continuous-fiber composite with appropriate averaging for angular dependence. Because the fibers are randomly oriented, the composite is isotropic with only two independent elastic properties. Christensen and Waals' analysis leads to the following expressions for the Young's modulus and Poisson's ratio of randomly oriented continuous-fiber composites:

$$E = \frac{\left[E_{11} + 4(1+v_{12})^2 K_{23} \right]\left[E_{11} + (1-2v_{12})^2 K_{23} + 6(G_{12} + G_{23}) \right]}{3\left[2E_{11} + (8v_{12}^2 + 12v_{12} + 7)K_{23} + 2(G_{12} + G_{23}) \right]}$$

(13.52)

$$v = \frac{\left[E_{11} + 2(2v_{12}^2 + 8v_{12} + 3) K_{23} - 4(G_{12} + G_{23}) \right]}{2\left[2E_{11} + (8v_{12}^2 + 12v_{12} + 7)K_{23} + 2(G_{12} + G_{23}) \right]}$$

(13.53)

Thus, Equation 13.52 and Equation 13.53 along with the equations for the aligned system (Equation 13.41 to Equation 13.46) can be used to predict the elastic properties (E and v) of isotropic randomly oriented continuous-fiber composites.

13.3 ELASTIC BEHAVIOR OF DISCONTINUOUS SHORT-FIBER COMPOSITES

13.3.1 UNIDIRECTIONAL SHORT-FIBER COMPOSITES

Russel [24] and Russel and Acrivos [25] have developed expressions for the elastic properties of aligned (unidirectional) short-fiber composites at dilute fiber concentrations by considering the response of a single fiber (embedded in the matrix phase) to a given applied strain. The "slender-body" approximation was applied to fiber, that is, the slenderness ratio (κ) (defined as fiber diameter divided by the fiber length) was considered to be small ($\kappa \ll 1$).

For unidirectional composite of slender prolate spheroids at dilute concentrations, the expressions for the five independent elastic properties are [24]:

$$\frac{E_{11}}{E_c} = 1 + \phi \left[\frac{\dfrac{\Delta G}{2(1-v_c)}(3\Delta\lambda + 2\Delta G) + \dfrac{E_c(1-2v_c)}{2(1+v_c)}\Delta\lambda}{} \right.$$
$$\left. + \frac{\dfrac{E_c(1+2v_c^2)\Delta G}{(1+v_c)(1-2v_c)}}{\Delta G(3\Delta\lambda + 2\Delta G)\left(\dfrac{1+v_c}{1-v_c}\right)\kappa^2\left(\ell n\dfrac{2}{\kappa} - \dfrac{5-4v_c}{4(1-v_c)}\right)} \right.$$
$$\left. + \frac{E_c(\Delta\lambda + \Delta G)}{2(1-v_c)} + G_c(3\lambda_c + 2G_c) \right] \tag{13.54}$$

$$v_{12} = v_c + \frac{1}{2}\phi(1+v_c)(1-2v_c)$$

$$\times \left[\frac{\Delta G(3\Delta\lambda + 2\Delta G)\left(\dfrac{1-2v_c}{1-v_c}\right)\kappa^2\left(\ell n\dfrac{2}{\kappa} - \dfrac{3-4v_c}{2(1-2v_c)}\right)}{} \right.$$
$$\left. + \frac{\dfrac{E_c E_f(v_f - v_c)}{(1+v_c)(1+v_f)(1-2v_c)(1-2v_f)}}{\Delta G(3\Delta\lambda + 2\Delta G)\left(\dfrac{1+v_c}{1-v_c}\right)\kappa^2\left(\ell n\dfrac{2}{\kappa} - \dfrac{5-4v_c}{4(1-v_c)}\right)} \right.$$
$$\left. + \frac{E_c(\Delta\lambda + \Delta G)}{2(1-v_c)} + G_c(3\lambda_c + 2G_c) \right] \tag{13.55}$$

$$K_{23} = \frac{E_c}{2(1+v_c)(1-2v_c)}$$

$$+\phi \begin{bmatrix} \Delta G(3\Delta\lambda+2\Delta G)\dfrac{E_c}{1-2v_c}\kappa^2\left(\ell n\dfrac{2}{\kappa}-\dfrac{5-4v_c}{4(1-v_c)}\right) \\[2ex] +\dfrac{\dfrac{E_c^2(\Delta\lambda+\Delta G)}{2(1+v_c)(1-2v_c)}}{\Delta G(3\Delta\lambda+2\Delta G)\left(\dfrac{1+v_c}{1-v_c}\right)\kappa^2\left(\ell n\dfrac{2}{\kappa}-\dfrac{5-4v_c}{4(1-v_c)}\right)} \\[4ex] +\dfrac{E_c}{2(1-v_c)}(\Delta\lambda+\Delta G)+G_c(3\lambda_c+2G_c) \end{bmatrix} \quad (13.56)$$

$$G_{12} = \frac{E_c}{2(1+v_c)}\left[1+2\phi\left(\frac{G_f-G_c}{G_f+G_c}\right)\right] \quad (13.57)$$

$$G_{23} = \frac{E_c}{2(1+v_c)}\left[1+4\phi(1-v_c)\frac{(G_f-G_c)}{G_c+(3-4v_c)G_f}\right] \quad (13.58)$$

where $\Delta\lambda = \lambda_f - \lambda_c$, $\Delta G = G_f - G_c$, and λ is the Lamé constant. When the fibers are nearly rigid such that $(E_f/E_c)\kappa^2\ell n(2/\kappa) \gg 1$, the expressions for the elastic properties E_{11}, v_{12}, and K_{23} of unidirectional short-fiber composites reduce to:

$$\frac{E_{11}}{E_c} = 1+\phi\left\{\frac{1/\left[2(1+v_c)\right]}{\kappa^2\left[\ell n\left(\dfrac{2}{\kappa}\right)-(5-4v_c)/(4-4v_c)\right]}\right\} \quad (13.59)$$

$$v_{12} = v_c+\phi\left\{\frac{(1-2v_c)^2\,\ell n\left(\dfrac{2}{\kappa}\right)-\dfrac{1}{2}(3-4v_c)(1-2v_c)}{2\,\ell n\left(\dfrac{2}{\kappa}\right)-\dfrac{2(5-4v_c)}{4(1-v_c)}}\right\} \quad (13.60)$$

$$K_{23} = \frac{E_c}{2(1+v_c)(1-2v_c)}\left[1+2\phi(1-v_c)\right] \quad (13.61)$$

Because for rigid particles $G_f \gg G_c$, the expressions for the longitudinal and transverse shear moduli reduce to:

$$G_{12} = \frac{E_c}{2(1+v_c)} \left[1 + 2\phi\right] \tag{13.62}$$

$$G_{23} = \frac{E_c}{2(1+v_c)} \left[1 + 4\left(\frac{1-v_c}{3-4v_c}\right)\phi\right] \tag{13.63}$$

When the fibers are nonrigid such that $(E_f/E_c)\kappa^2 \ln(2/\kappa) \ll 1$, the expressions for the elastic properties E_{11}, v_{12}, and K_{23} of unidirectional short-fiber composites reduce to:

$$\frac{E_{11}}{E_c} = 1 + \phi \left[\frac{E_f}{E_c} - 1 + \frac{2(v_f - v_c)^2}{(1+v_c) + \left(\dfrac{E_c}{E_f}\right)(1+v_f)(1-2v_f)}\right] \tag{13.64}$$

$$v_{12} = v_c + \phi \left[\frac{2(v_f - v_c)(1-v_c)(1+v_c)}{(1+v_c) + \left(\dfrac{E_c}{E_f}\right)(1-v_f)(1-2v_f)}\right] \tag{13.65}$$

$$K_{23} = \frac{E_c}{2(1+v_c)(1-2v_c)} \left[1 + 2\phi(1-v_c)\left\{\frac{1 - \dfrac{E_c(1+v_f)(1-2v_f)}{E_f(1+v_c)(1-2v_c)}}{1 + \dfrac{E_c(1+v_f)(1-2v_f)}{E_f(1+v_c)}}\right\}\right] \tag{13.66}$$

If $G_f \ll G_c$, the expressions for the longitudinal and transverse shear moduli reduce to:

$$G_{12} = \frac{E_c}{2(1+v_c)} \left[1 - 2\phi\right] \tag{13.67}$$

$$G_{23} = \frac{E_c}{2(1+v_c)}\left[1-4\left(\frac{1-v_c}{3-4v_c}\right)\phi\right] \qquad (13.68)$$

Figure 13.8 to Figure 13.12 show the plots of intrinsic longitudinal Young's modulus [defined as $(E_{11}/E_c - 1)/\phi$], longitudinal Poisson's ratio (v_{12}), transverse

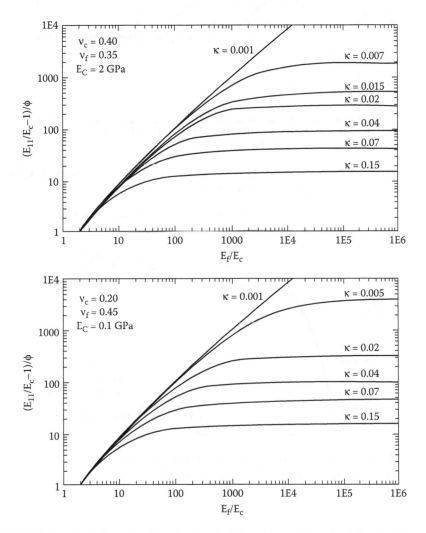

FIGURE 13.8 Variation of intrinsic longitudinal Young's modulus of unidirectional composite of slender prolate spheroids with Young's modulus ratio E_f/E_c (E_f is the fiber-phase Young's modulus, and E_c is the matrix-phase Young's modulus).

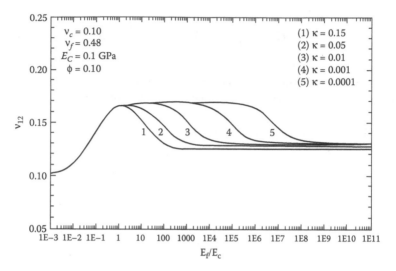

FIGURE 13.9 Variation of longitudinal Poisson's ratio (ν_{12}) of unidirectional composite of slender prolate spheroids with Young's modulus ratio E_f/E_c.

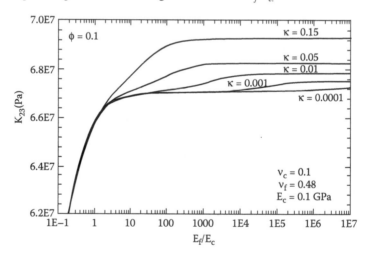

FIGURE 13.10 Variation of transverse bulk modulus (K_{23}) of unidirectional composite of slender prolate spheroids with Young's modulus ratio E_f/E_c.

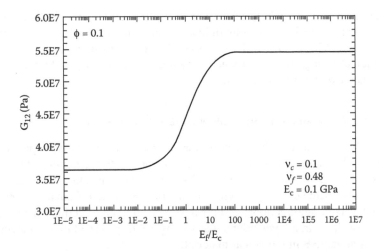

FIGURE 13.11 Variation of longitudinal shear modulus (G_{12}) of unidirectional composite of slender prolate spheroids with Young's modulus ratio E_f/E_c.

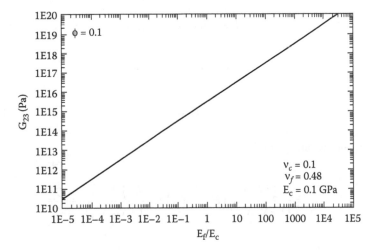

FIGURE 13.12 Variation of transverse shear modulus (G_{23}) of unidirectional composite of slender prolate spheroids with Young's modulus ratio E_f/E_c.

bulk modulus (K_{23}), longitudinal shear modulus (G_{12}), and transverse shear modulus (G_{23}) for unidirectional composite of slender prolate spheroids. The plots are generated from Equation 13.54 to Equation 13.58. The elastic constants are plotted as functions of Young's modulus ratio E_f/E_c, where E_f is the fiber-phase Young's modulus and E_c is the continuous-phase (matrix) Young's modulus. Of the five independent elastic properties of unidirectional short-fiber composites, two (G_{12} and G_{23}) are independent of the fiber slenderness ratio κ. The longitudinal Young's modulus E_{11} is most sensitive to the fiber slenderness ratio κ. The intrinsic longitudinal Young's modulus (Figure 13.8) shows a strong dependence on the slenderness ratio κ, especially when $E_f/E_c > 10$. The reinforcement effect of fibers, as reflected in the intrinsic longitudinal Young's modulus, decreases with the increase in κ. When $\kappa \to 0$, which corresponds to infinitely long fibers, the intrinsic longitudinal Young's modulus (hence, longitudinal modulus E_{11} and reinforcement effect) is maximum for given E_f/E_c.

13.3.2 RANDOMLY ORIENTED SHORT-FIBER COMPOSITES

Randomly oriented short-fiber composites are macroscopically isotropic and can be modeled as laminated systems [26–28]. According to the laminated model of randomly, or nearly randomly, oriented short-fiber composites, the actual composite is considered to be equivalent to a material constructed from a large number of thin layers of unidirectional composites oriented in all directions. The volume fraction of fibers in a layer oriented at an angle θ is governed by the percentage of fibers at angle θ in the actual material.

The elastic properties of randomly oriented short-fiber composites, determined from the laminated model, can be expressed as:

$$E = 4\, w_2\, (w_1 - w_2)/w_1 \tag{13.69}$$

$$\nu = (w_1 - 2\, w_2)/w_1 \tag{13.70}$$

$$G = w_2 \tag{13.71}$$

where w_1 and w_2 are given by:

$$w_1 = \frac{(3/8)}{\left(1 - \nu_{12}\nu_{21}\right)}\left(E_{11} + E_{22}\right) + \frac{(1/4)\,\nu_{12}\,E_{22}}{\left(1 - \nu_{12}\,\nu_{21}\right)} + \frac{G_{12}}{2} \tag{13.72}$$

$$w_2 = \frac{(1/8)}{\left(1 - \nu_{12}\nu_{21}\right)}\left(E_{11} + E_{22}\right) - \frac{(1/4)\,\nu_{12}\,E_{22}}{\left(1 - \nu_{12}\,\nu_{21}\right)} + \frac{G_{12}}{2} \tag{13.73}$$

where ν_{12}, ν_{21}, E_{11}, E_{22}, and G_{12} are the elastic properties of unidirectional short-fiber composites. Thus, the evaluation of the elastic properties of a randomly oriented short-fiber composite requires the knowledge of elastic properties of unidirectional short-fiber composites discussed in Section 13.3.1.

NOTATION

A	cross-sectional area of composite
A_c	cross-sectional area of matrix
A_f	cross-sectional area of fiber
a	radius defined as $R/(\phi)^{1/3}$, where R is fiber radius and ϕ is volume fraction of fibers in the composite
b	radius of a large cylindrical region filled with effective homogeneous medium whose macroscopic properties are to be calculated
E	Young's modulus
E_{11}	longitudinal Young's modulus of composite in direction 1
E_{22}	transverse Young's modulus of composite in direction 2
E_{33}	transverse Young's modulus of composite in direction 3
E_c	Young's modulus of matrix (continuous phase)
E_f	fiber Young's modulus
E_{f1}	longitudinal Young's modulus of fiber
E_{f2}	transverse Young's modulus of fiber
e_L	longitudinal strain in composite
e_{Lc}	longitudinal strain in matrix
e_{Lf}	longitudinal strain in fiber
e_T	transverse strain in composite
e_{Tc}	transverse strain in matrix
e_{Tf}	transverse strain in fiber phase
e_{11}, e_{22}	normal strains
e_{12}, e_{23}	shear strains
F	total load on the composite
F_c	load carried by the matrix
F_f	load carried by fibers
G	shear modulus
G_c	shear modulus of matrix
G_{12}	longitudinal shear modulus of composite in the plane that contains the fibers (shear stress loading in the 1-2 plane)
G_{13}	longitudinal shear modulus of the composite in the plane that contains the fibers (shear stress loading in the 1-3 plane)
G_{23}	out-of-plane or transverse shear modulus of composite
G_f	shear modulus of fiber
G_{f12}	longitudinal shear modulus of fiber
G_{f23}	transverse shear modulus of fiber
K_{23}	plane strain bulk modulus of composite
K_f	bulk modulus of fiber
L	total displacement of the composite in the transverse direction

L_c	displacement of the matrix phase in the transverse direction
L_f	displacement of the fiber phase in the transverse direction
R	radius of fiber
t	cumulative thickness of the composite in transverse direction
t_c	cumulative thickness of the matrix phase in transverse direction
t_f	cumulative thickness of the fibers in transverse direction
w_1	parameter given by Equation 13.72
w_2	parameter given by Equation 13.73

GREEK

ΔG	$= G_f - G_c$, where G_f is the shear modulus of fiber and G_c is shear modulus of matrix (continuous phase)
$\Delta\lambda$	$= \lambda_f - \lambda_c$, where λ_f is the Lamé constant of fiber and λ_c is Lamé constant of matrix
κ	slenderness ratio defined as fiber diameter divided by fiber length
λ	Lamé constant
μ	$= K + (G/3)$, where K is the bulk modulus and G is shear modulus
μ_c	$= K_c + (G_c/3)$, where subscript c refers to matrix (continuous phase)
μ_f	$= K_f + (G_f/3)$, where subscript f refers to fiber
ν_c	Poisson's ratio of matrix
ν_f	Poisson's ratio of fiber
ν_{12}	in-plane or longitudinal Poisson's ratio of composite (subscript 1 refers to the direction of the applied stress and the second subscript 2 corresponds to the direction of associated lateral stress)
ν_{21}	Poisson's ratio with applied stress in direction 1 and lateral strain in direction 2
ν_{13}	Poisson's ratio with applied stress in direction 1 and lateral strain in direction 3
ν_{31}	Poisson's ratio with applied stress in direction 3 and lateral strain in direction 1
ν_{23}	Poisson's ratio with applied stress in direction 2 and lateral strain in direction 3
ν_{32}	Poisson's ratio with applied stress in direction 3 and lateral strain in direction 2
ν_{f12}	longitudinal Poisson's ratio of fiber
σ_{c1}	normal stress in matrix (in direction 1)
σ_{c2}	normal stress in matrix (in direction 2)
σ_{f1}	normal stress in fiber (in direction 1)
σ_{f2}	normal stress in fiber (in direction 2)
σ_{11}, σ_{22}	normal stresses

τ_{12}, τ_{13} shear stresses
ϕ volume fraction of fibers

REFERENCES

1. P.K. Mallick, *Fiber-Reinforced Composites*, Marcel Dekker, New York, 1988.
2. I.M. Daniel, O. Ishai, *Engineering Mechanics of Composite Materials*, Oxford University Press, New York, 1994.
3. D. Hull, *An Introduction to Composite Materials*, Cambridge University Press, Cambridge, 1981.
4. R.F. Gibson, *Principles of Composite Material Mechanics*, McGraw-Hill, New York, 1994.
5. B.D. Agarwal, L.J. Broutman, *Analysis and Performance of Fiber Composites*, John Wiley & Sons, New York, 1990.
6. L. Nicolais, *Polym. Eng. Sci.*, 15: 137–149, 1975.
7. R.M. Christensen, *Mechanics of Composite Materials*, John Wiley & Sons, New York, 1979.
8. C.T. Herakovich, *Mechanics of Fibrous Composites*, John Wiley & Sons, New York, 1998.
9. F.L. Matthews, R.D. Rawlings, *Composite Materials: Engineering and Science*, Chapman and Hall, London, 1994.
10. K.K.U. Stellbrink, *Micromechanics of Composites*, Hanser Publishers, Munich, 1996.
11. L.P. Kellar, G.S. Springer, *Mechanics of Composite Structures*, Cambridge University Press, Cambridge, 2003.
12. E.J. Barbero, *Introduction to Composite Materials Design*, Taylor and Francis, Philadelphia, 1998.
13. A.K. Kaw, *Mechanics of Composite Materials*, CRC Press, Boca Raton, FL, 1997.
14. K.K. Chamela, *Composite Materials: Science and Engineering*, Springer-Verlag, New York, 1998.
15. Z. Hashin, B.W. Rosen, *J. Appl. Mech.*, 31: 223–232, 1964.
16. R.M. Christensen, K.H. Lo, *J. Mech. Phys. Solids*, 27: 315–330, 1979.
17. R.M. Christensen, in *Mechanics of Composite Materials — Recent Advances*, Z. Hashin, C.T. Herakovich, Eds., Pergamon Press, New York, 1982, pp. 1–16.
18. R.M. Christensen, *J. Mech. Phys. Solids*, 38: 379–404, 1990.
19. R.M. Christensen, *Int. J. Solids Struct.*, 12: 537–544, 1976.
20. Z. Hashin, *J. Mech. Phys. Solids*, 13: 119, 1965.
21. R. Hill, *J. Mech. Phys. Solids*, 12: 199–212, 1964.
22. Z. Hashin, in *Mechanics of Composite Materials*, F.W. Wendt, H. Liebowitz, N. Perrone, Eds., Pergamon Press, New York, 1970, pp. 201–242.
23. R.M. Christensen, F.M. Waals, *J. Comp. Mater.*, 6: 518–532, 1972.
24. W.B. Russel, *J. Appl. Maths Phys.* (ZAMP), 24: 581–600.
25. W.B. Russel, A. Acrivos, *J. Appl. Maths Phys.* (ZAMP), 23: 434–464.
26. J.C. Halpin, N.J. Pagano, *J. Comp. Mater.*, 3: 720–724, 1969.
27. J.C. Halpin, *J. Comp. Mater.*, 3: 732–734, 1969.
28. M. Manera, *J. Comp. Mater.*, 11: 235–247, 1977.

Part V

Linear Viscoelasticity of Particulate Dispersions and Composites

Part V, consisting of Chapter 14 to Chapter 16, deals with the linear viscoelastic behavior of particulate dispersions and composites.

Chapter 14 covers the basic background of linear viscoelasticity. The definitions and concepts related to the oscillatory strain behavior of materials are discussed. The Maxwell model describing the linear viscoelastic behavior of materials is presented.

Chapter 15 is devoted to the dynamic viscoelastic behavior of particulate dispersions. Complex shear modulus expressions are presented for dispersions of a variety of different types of particles, such as viscoelastic soft-solid particles, liquid droplets with pure interface, and liquid droplets coated with additive. In the latter case, the additive forms a thin film on the droplet surface. Depending on the nature of the additive, the film can be purely viscous, purely elastic, or viscoelastic. The film may also possess a nonzero bending rigidity.

Chapter 16 describes the dynamic behavior of viscoelastic composites in terms of the effective complex moduli. Using the elastic-viscoelastic correspondence principle, the expressions for the elastic moduli of composites are generalized to the corresponding expressions for complex moduli. The complex moduli expressions for both particulate composites and fiber-reinforced composites are presented. The predictions of the models are discussed.

14 Introduction to Linear Viscoelasticity

A purely elastic material (e.g., a Hookean solid) differs widely in its deformational characteristics from a purely viscous material (e.g., Newtonian fluid). The stress in a purely elastic material is a function of the instantaneous magnitude of deformation (strain) only, whereas the stress is a function of the instantaneous rate of deformation only, in a purely viscous material. Also, purely elastic materials return to their natural or undeformed state upon removal of applied loads, whereas purely viscous materials possess no tendency at all for deformational recovery. The term *viscoelastic* implies the simultaneous existence of viscous and elastic characteristics in a material. Thus, material behavior that incorporates a blend of both viscous and elastic characteristics is called *viscoelastic behavior.*

Many substances, including dispersions and composites, exhibit viscoelastic behavior. They flow under the influence of applied stresses, unlike purely elastic materials, which exhibit a constant strain and no flow. Upon removal of the applied stress, some of their deformation is gradually recovered and they exhibit elastic recovery, unlike purely viscous materials, which exhibit no recovery at all. Both steady and unsteady shear measurements are carried out to evaluate viscoelasticity of materials. In steady shearing motion, elasticity causes normal stress effects.

14.1 LINEAR VISCOELASTICITY

The term *linear viscoelasticity* implies the study of viscoelastic effects in a small strain region where strain varies linearly with stress (doubling the stress will double the strain). This implies that stress, strain, and their time derivatives are related by a linear differential equation with constant coefficients. The coefficients of the time differentials are material parameters that are independent of strain, stress, or strain rate. Also, the time derivatives are ordinary partial derivatives [1–3]. A general differential equation describing linear viscoelasticity has the form:

$$(1+P)\bar{\bar{\tau}} = Q\bar{\bar{e}} \tag{14.1}$$

where $\bar{\bar{\tau}}$ is the deviatoric stress tensor, $\bar{\bar{e}}$ is the strain tensor, and P and Q are linear differential operators with respect to time:

$$P = \sum_{k=1}^{k=m} \alpha_k \frac{\partial^k}{\partial t^k}, \quad Q = \sum_{k=0}^{k=n} \beta_k \frac{\partial^k}{\partial t^k} \tag{14.2}$$

Thus,

$$\left[1 + \sum_{k=1}^{k=m} \alpha_k \frac{\partial^k}{\partial t^k} \right] \overline{\overline{\tau}} = \sum_{k=0}^{k=n} \beta_k \frac{\partial^k \overline{\overline{e}}}{\partial t^k} \tag{14.3}$$

where α_k and β_k are materials parameters.

Among the various techniques available to study the linear viscoelastic behavior of materials, oscillatory testing at small strain amplitudes is very popular. The material is subjected to oscillatory strain and its stress response is monitored. For purely elastic materials, stress is in phase with the strain. For viscoelastic materials, there is a lag between the stress and strain. Although the material could be subjected to either oscillatory normal strain or oscillatory shear strain, only oscillatory shear strain is considered in the following analysis.

14.2 OSCILLATORY SHEAR

In the oscillatory shear experiment, the material is subjected to a sinusoidal shearing strain, which may be represented by either a sine or a cosine function. Thus,

$$e(t) = e_0 \sin(\omega t) \tag{14.4}$$

where e_0 is the strain amplitude and ω is the frequency of strain oscillation. If the strain amplitude is sufficiently small so that the material behaves linearly, it can be shown that the resulting steady-state stress will also oscillate sinusoidally at the same frequency but will be shifted by a phase angle δ with respect to the strain wave, that is,

$$\tau = \tau_0 \sin(\omega t + \delta) \tag{14.5}$$

The stress leads the strain by phase angle δ. For purely elastic materials, $\delta = 0°$ (stress is in phase with strain). For purely viscous materials, $\delta = 90°$ (stress leads the strain by $90°$ although the stress is in phase with the strain rate). For viscoelastic materials, $0 < \delta < 90°$. Figure 14.1 illustrates these points.

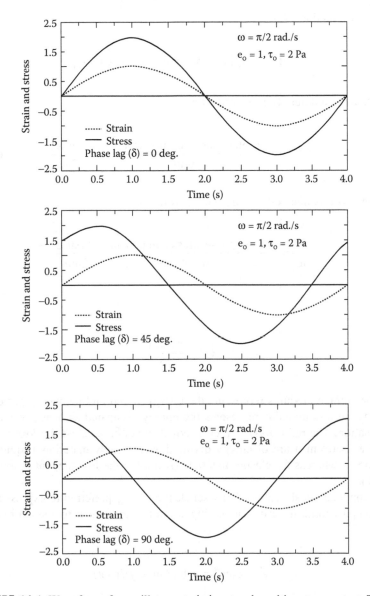

FIGURE 14.1 Wave forms for oscillatory strain input and resulting stress output. The top plot indicates that stress is in phase with strain ($\delta = 0°$ and the material is purely elastic). The middle plot indicates that stress leads the strain by angle $\delta = 45°$ (the material is viscoelastic). The bottom plot indicates that stress is $90°$ ahead of strain (the material is purely viscous).

From Equation 14.5:

$$\tau = \tau_0 \left[(\cos \delta) \sin \omega t + (\sin \delta) \cos \omega t \right] \tag{14.6}$$

The storage modulus (G') is defined as:

$$G' = \frac{\text{In-phase stress component}}{2 \times \text{strain}} = \frac{\tau_0 \cos \delta}{2e_0} \tag{14.7}$$

and the loss modulus (G'') is defined as follows:

$$\frac{G''}{\omega} = \frac{\text{Out-of-phase stress component}}{2 \times \text{strain rate}} = \frac{\tau_0 \sin \delta}{2\omega e_0} \tag{14.8}$$

that is,

$$G'' = \frac{\tau_0 \sin \delta}{2e_0} \tag{14.9}$$

The storage modulus represents the energy stored and recovered per cycle, whereas the loss modulus represents the energy dissipated as heat per cycle of deformation. The ratio G''/G' which is equal to tan δ, is called the *loss tangent* and is a useful measure of energy dissipation. For lossy materials (liquidlike), tan $\delta \gg 1$, whereas for elastic materials (solidlike) the loss tangent is low (tan $\delta \ll 1$).

In the theoretical treatment of viscoelasticity, it is preferred to express strain in a complex form, instead of assuming strain to vary similar to a sine or cosine of time. Thus,

$$e^* = e_0 e^{j\omega t} = e_0 (\cos \omega t + j \sin \omega t) \tag{14.10}$$

In this expression, the real and imaginary parts each represent two oscillatory strains of frequency ω. The resulting stress is also complex and is related to strain as:

$$\tau^* = 2G^* e^* \tag{14.11}$$

where G^* is the complex modulus. The complex modulus can be written in terms of storage and loss moduli as:

$$G^* = G' + jG'' \tag{14.12}$$

Equation 14.10, Equation 14.11, and Equation 14.12 give the following expression for complex stress:

$$\tau^* = 2e_0 \left(G' \cos \omega t - G'' \sin \omega t \right) + j(2e_0)(G' \sin \omega t + G'' \cos \omega t) \tag{14.13}$$

The real part of τ^* corresponds to the real part of e^*, and the imaginary parts of τ^* and e^* correspond to each other. Thus, the solutions to two closely related problems are obtained simultaneously using complex analysis. When $e = e_0 \cos \omega t$, stress response is $\tau = 2e_0(G' \cos \omega t - G'' \sin \omega t)$. When $e = e_0 \sin \omega t$, stress response is $\tau = 2e_0(G' \sin \omega t + G'' \cos \omega t)$. In both cases, stress is a mixture of cosine and sine functions indicating a phase shift between stress and strain.

The oscillatory shear behavior of a material can also be characterized in terms of *complex viscosity* (rather than complex modulus). The complex viscosity η^* of a material is related to the complex shear modulus G^* as follows:

$$\eta^* = G^*/j\omega = \eta' - j\eta'' \tag{14.14}$$

where $\eta'(=G''/\omega)$ is the dynamic viscosity and $\eta''(=G'/\omega)$ is the imaginary part of the complex viscosity.

14.3 MAXWELL MODEL

Maxwell [4] was probably the first to investigate the viscoelastic behavior of materials. His equation, which follows, is the simplest model of describing the linear viscoelastic behavior of materials:

$$\bar{\bar{\tau}} + t_o \frac{\partial \bar{\bar{\tau}}}{\partial t} = 2\eta_0 \frac{\partial \bar{\bar{e}}}{\partial t} \tag{14.15}$$

where t_o is the relaxation time and η_0 is the zero-shear viscosity of the material. Note that this equation is a special case of Equation 14.3 when $m = 1$, $n = 1$, and $\alpha_1 = t_o$, $\beta_0 = 0$, and $\beta_1 = 2\eta_0$.

When a Maxwellian fluid is subjected to a sinusoidal shearing strain given by:

$$e(t) = e_0 \cos(\omega t) \tag{14.16}$$

the shear stress is governed by the following equation:

$$\tau + t_o \frac{d\tau}{dt} = -2\eta_0 e_0 \omega \sin \omega t \tag{14.17}$$

The solution of this first-order linear differential equation is:

$$\tau = \frac{2\eta_0 \omega e_0}{1+\omega^2 t_o^2}\left(\omega t_o \cos \omega t - \sin \omega t\right) \tag{14.18}$$

Therefore, the storage modulus G' of a Maxwellian fluid is:

$$G' = \frac{\text{In-phase stress component}}{2 \times \text{strain}} = \frac{2\eta_0 \omega e_0 \omega t_o \cos \omega t}{(1+\omega^2 t_o^2) 2 e_0 \cos \omega t}$$
$$= \frac{\eta_0 t_o \omega^2}{1+\omega^2 t_o^2} \tag{14.19}$$

and the loss modulus G'' is:

$$G'' = \omega\left[\frac{\text{Out-of-phase-stress-component}}{2 \times \text{strain-rate}}\right]$$
$$= \omega\left(\frac{-2\eta_0 \omega e_0 \sin \omega t}{1+\omega^2 t_o^2}\right)\left(\frac{1}{-2e_0\omega\sin\omega t}\right) = \frac{\eta_0\omega}{1+\omega^2 t_o^2} \tag{20}$$

Note that the preceding expressions for storage and loss moduli could also be expressed as follows:

$$G' = \frac{G\omega^2 t_o^2}{1+\omega^2 t_o^2} \tag{14.21}$$

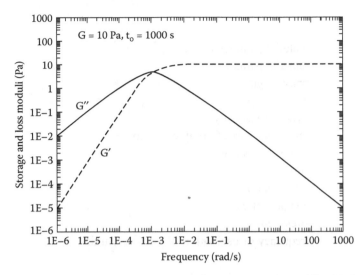

FIGURE 14.2 Variation of storage modulus (G') and loss modulus (G'') with frequency for a Maxwell material.

$$G'' = \frac{G\omega t_o}{1+\omega^2 t_o^2}$$ (14.22)

where $G(=\eta_0/t_o)$ is the high-frequency storage modulus of the material.

Figure 14.2 shows how the storage modulus (G') and loss modulus (G'') vary with frequency for a Maxwell material.

NOTATION

e	strain
e_0	strain amplitude
e^*	complex strain
$\bar{\bar{e}}$	strain tensor
G	high-frequency storage modulus
G^*	complex modulus
G'	storage modulus
G''	loss modulus
j	imaginary number
P	linear differential operator; see Equation 14.2
Q	linear differential operator; see Equation 14.2
t	time
t_o	relaxation time

GREEK

α_k	material parameter
β_k	material parameter
δ	phase angle
η_0	zero-shear viscosity
η'	dynamic viscosity
η''	imaginary part of complex viscosity
η^*	complex viscosity
τ	stress
τ_0	stress amplitude
τ^*	complex stress
$\bar{\bar{\tau}}$	stress tensor
ω	frequency of oscillation of stress or strain

REFERENCES

1. W. Flugge, *Viscoelasticity*, Blaisdell Publishing Co., Waltham, MA, 1967.
2. R. Darby, *Viscoelastic Fluids*, Marcel Dekker, New York, 1976.
3. H.A. Barnes, J.F. Hutton, K. Walters, *An Introduction to Rheology*, Elsevier, Amsterdam, 1989.
4. J.C. Maxwell, *Philos. Trans. R. Soc. Lond. A*, 157: 49–88, 1867.

15 Dynamic Viscoelastic Behavior of Particulate Dispersions

A particulate dispersion is expected to exhibit viscoelastic behavior if one or more of the following conditions are met:

1. The particles are composed of some soft rubberlike solid material (elastic or viscoelastic). The continuous-phase liquid is purely viscous (inelastic).
2. The particles are deformable liquid droplets with pure interface (no interfacial additives). The dispersed-phase and continuous-phase liquids are purely viscous. The mechanism for shape recovery of droplets is interfacial tension.
3. The particles are deformable liquid droplets coated with a thin film of some material (surfactants, polymers, etc.). The dispersed-phase and continuous-phase liquids are purely viscous. The mechanism of shape recovery of droplets is film elasticity or film-bending rigidity.
4. The particles are deformable liquid droplets composed of viscoelastic liquid.
5. The continuous-phase liquid of the dispersion is viscoelastic.

15.1 DISPERSIONS OF SOFT RUBBERLIKE SOLID PARTICLES

Frohlich and Sack [1] utilized a self-consistent, effective medium approach to derive a constitutive equation governing the viscoelastic behavior of a dispersion of soft elastic particles in a Newtonian (purely viscous) liquid. According to their approach, the dispersion is first treated as an equivalent "effective medium" that is homogeneous and has the same macroscopic rheological properties as those of the dispersion. Then, a small portion of the effective homogeneous medium is replaced by the actual components of the dispersion. The properties of the effective medium are then determined by supposing that if a small portion of the effective homogeneous medium is replaced by the actual components of the dispersion, no difference in rheological behavior would be detected by macroscopic observations.

In essence, Frohlich and Sack [1] consider two systems: system A and system B, as shown in Figure 15.1. System A consists of a large spherical region of radius b filled with the effective homogeneous medium (dispersion) whose

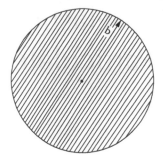 

System A - a large spherical region filled with the effective homogeneous medium (dispersion) whose macroscopic properties are to be determined

System B - a single particle of radius "R" is surrounded by an envelope of matrix fluid up to radius "a". The region a < r < b is filled with the effective homogeneous medium (dispersion)

FIGURE 15.1 Self-consistent effective medium model of dispersion.

macroscopic properties we wish to calculate. In system B, a single spherical particle of radius R, made up of elastic material (incompressible and isotropic) with shear modulus G_d, is surrounded by an envelope of continuous-phase fluid of the dispersion up to a certain radius a, and the remaining space $a \leq r \leq b$ is filled with an effective homogeneous medium (dispersion) whose properties we wish to calculate. The continuous-phase fluid of the dispersion is assumed to be incompressible and Newtonian with viscosity η_c. The radius a is chosen to give the correct volume fraction of the dispersed phase, that is, $a = R/(\phi)^{1/3}$.

The two systems are subjected to the same time-dependent stress at the boundary $r = b$, that is,

$$\left. \overline{\overline{\sigma}}_A(t) \right|_{r=b} = \left. \overline{\overline{\sigma}}_B(t) \right|_{r=b} \tag{15.1}$$

Subject to the stress specified at the boundary $r = b$, the velocity distributions in the two systems are determined. The governing equations are:

System A:

$$\eta \nabla^2 \vec{u} = \nabla p, \quad \nabla \bullet \vec{u} = 0 \quad \text{for} \quad 0 \leq r \leq b \tag{15.2}$$

where η is the dispersion viscosity.

System B (within the particle):

$$\nabla \bullet \overline{\overline{\sigma}}_d = 0, \quad \nabla \bullet \vec{u}_d = 0, \quad \overline{\overline{\tau}}_d = 2 G_d \overline{\overline{e}}_d \quad \text{for} \quad 0 \leq r \leq R \tag{15.3}$$

where $\overline{\overline{e}}_d$ is strain tensor of the particle.

System B (outside the particle):

$$\eta_c \nabla^2 \vec{u}_c = \nabla P_c, \quad \nabla \bullet \vec{u}_c = 0 \quad \text{for} \quad R \leq r \leq a \tag{15.4}$$

$$\eta \nabla^2 \vec{u} = \nabla P, \quad \nabla \bullet \vec{u} = 0 \quad \text{for} \quad a \leq r \leq b \tag{15.5}$$

The boundary conditions to be satisfied by system B are as follows:

1. Continuity of velocity at the surface of the particle, that is,

$$\vec{u}_c = \vec{u}_d = \frac{\partial \vec{r}}{\partial t} \quad \text{at} \quad r = R \tag{15.6}$$

2. Continuity of stress at the surface of the particle, that is,

$$\bar{\bar{\sigma}}_c \bullet \hat{n} = \bar{\bar{\sigma}}_d \bullet \hat{n} \quad \text{at} \quad r = R \tag{15.7}$$

3. Continuity of velocity at $r = a$, that is,

$$\vec{u}_c = \vec{u} \quad \text{at} \quad r = a \tag{15.8}$$

4. Continuity of stress at $r = a$, that is,

$$\bar{\bar{\sigma}}_c \bullet \hat{n} = \bar{\bar{\sigma}} \bullet \hat{n} \quad \text{at} \quad r = a \tag{15.9}$$

Because the two systems (A and B) are supposed to be indistinguishable when observations are made at a distance $r = b$ sufficiently large compared to a, the following condition has to be fulfilled:

$$\vec{u}_A \Big|_{r=b} = \vec{u}_B \Big|_{r=b} \tag{15.10}$$

Frohlich and Sack's analysis leads to the following constitutive equation, valid at small variable rates of strain, for dispersions of elastic spheres in Newtonian liquids:

$$\left(1 + \tau_1 \frac{\partial}{\partial t}\right) \bar{\bar{\tau}} = 2\eta_0 \left(1 + \tau_2 \frac{\partial}{\partial t}\right) \bar{\bar{E}} \tag{15.11}$$

where τ_1 is the relaxation time, τ_2 is the retardation time, η_0 is the zero shear viscosity of the dispersion, $\bar{\bar{\tau}}$ is the deviatoric stress tensor, and $\bar{\bar{E}}$ is the rate of strain tensor. Expressions for η_0, τ_1, and τ_2 are as follows:

$$\eta_0 = \eta_c \left[\frac{2+3\phi}{2-2\phi} \right] = \eta_c \left[1 + \frac{5}{2}\phi + O(\phi^2) \right] \tag{15.12}$$

$$\tau_1 = \frac{\eta_c}{2G_d} \left[\frac{3+2\phi}{1-\phi} \right] = \frac{3\eta_c}{2G_d} \left[1 + \frac{5}{3}\phi + O(\phi^2) \right] \tag{15.13}$$

$$\tau_2 = \frac{3\eta_c}{G_d} \left[\frac{1-\phi}{2+3\phi} \right] = \frac{3\eta_c}{2G_d} \left[1 - \frac{5}{2}\phi + O(\phi^2) \right] \tag{15.14}$$

Note that the Frohlich and Sack constitutive equation is a special case of the general differential equation describing linear viscoelasticity (Equation 14.3). When $m = 1$ and $n = 2$, the general equation reduces to:

$$\left(1 + \alpha_1 \frac{\partial}{\partial t} \right) \bar{\bar{\tau}} = \left(\beta_0 + \beta_1 \frac{\partial}{\partial t} + \beta_2 \frac{\partial^2}{\partial t^2} \right) \bar{\bar{e}} \tag{15.15}$$

This equation becomes Equation 15.11 if $\alpha_1 = \tau_1$, $\beta_0 = 0$, $\beta_1 = 2\eta_0$, and $\beta_2 = 2\eta_0\tau_2$.

When the dispersion is subjected to harmonic strain (complex notation $e^* = e_0 e^{j\omega t}$), the Frohlich and Sack equation in conjunction with $\tau^* = 2G^*e^*$ gives:

$$[1 + \tau_1 j\omega](2e^*G^*) = 2\eta_0[1 + \tau_2 j\omega](e^* j\omega) \tag{15.16}$$

From Equation 15.16:

$$G^* = j\omega\eta_0 \left[\frac{1 + j\omega\tau_2}{1 + j\omega\tau_1} \right] \tag{15.17}$$

Because $G^* = G' + jG''$, Equation 15.17 gives the following expressions for storage and loss moduli:

$$G' = \frac{\eta_0\omega^2(\tau_1 - \tau_2)}{1 + \omega^2\tau_1^2} \tag{15.18}$$

$$G'' = \frac{\eta_0 \omega (1 + \omega^2 \tau_1 \tau_2)}{1 + \omega^2 \tau_1^2} \tag{15.19}$$

In the limit $G_d \to \infty$ (rigid particles), $\tau_1 = \tau_2 = 0$ and $G' = 0$, whereas

$$\frac{G''}{\omega} = \eta_0 = \eta_c \left[1 + \frac{5}{2}\phi + O(\phi^2) \right] \tag{15.20}$$

Figure 15.2 shows the effect of particle rigidity on the storage and loss moduli of a dispersion of soft elastic particles. The plots are generated from the Frohlich and Sack model (Equation 15.18 and Equation 15.19). With the increase in particle rigidity (G_d), the plateau value of the storage modulus of the dispersion increases. Also, the frequency at which the terminal zone (low-frequency zone where $G' \propto \omega^2$) begins is shifted to higher values with increase in G_d. This is because the relaxation time of particles decreases with an increase in G_d. Particle rigidity appears to have a small effect on the loss modulus of the dispersion.

The effect of continuous-medium viscosity η_c on the storage and loss moduli is shown in Figure 15.3. With the increase in η_c, the plateau value of the storage modulus of the dispersion is unaffected. However, the frequency at which the terminal zone begins is shifted to lower values with an increase in η_c. This is due to an increase in the relaxation time of particles with an increase in η_c. The loss modulus–frequency plots shift toward the left, the increase in η_c indicating an increase in loss modulus.

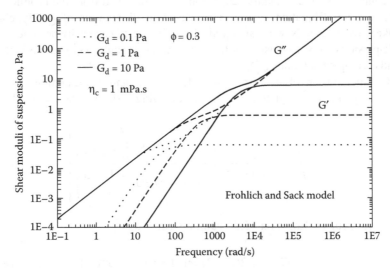

FIGURE 15.2 Effect of particle rigidity (G_d) on the storage and loss moduli of a dispersion of soft elastic particles.

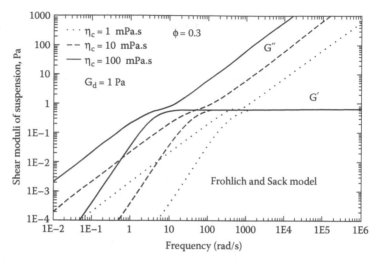

FIGURE 15.3 Effect of continuous medium viscosity (η_c) on the storage and loss moduli of a dispersion of soft elastic particles.

Figure 15.4 shows the effect of dispersed-phase concentration ϕ on the storage and loss moduli of a dispersion. The value of the storage modulus increases with increase in ϕ at all frequencies. The loss modulus increases with increase in ϕ at low frequencies; at high frequencies, a reverse trend is seen in that the loss modulus decreases with increase in ϕ.

15.1.1 GENERALIZATION OF FROHLICH AND SACK MODEL

The complex modulus equation, Equation 15.17, obtained from the Frohlich and Sack equation (Equation 15.11) is restricted to dispersions of purely elastic Hookean particles in Newtonian liquids. Equation 15.17 can easily be generalized to dispersions of viscoelastic particles (complex modulus G_d^*) in viscoelastic liquids (complex modulus G_c^*).

On substitution of the expressions for η_0, τ_1, and τ_2 (Equation 15.12 to Equation 15.14) into Equation 15.17, the following result can be obtained:

$$\frac{G^*}{j\omega\eta_c} = \left(\frac{2+3\phi}{2-2\phi}\right)\left[\frac{1+(3\eta_c j\omega/G_d)\left(\dfrac{1-\phi}{2+3\phi}\right)}{1+(\eta_c j\omega/2G_d)\left(\dfrac{3+2\phi}{1-\phi}\right)}\right] \qquad (15.21)$$

On rearrangement, Equation 15.21 can be written as:

$$\frac{G^*}{j\omega\eta_c} = \left[\frac{2G_d+3j\omega\eta_c+3\phi(G_d-j\omega\eta_c)}{2G_d+3j\omega\eta_c-2\phi(G_d-j\omega\eta_c)}\right] \qquad (15.22)$$

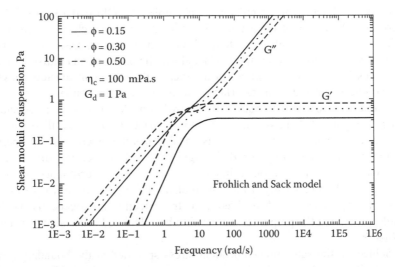

FIGURE 15.4 Effect of dispersed-phase concentration (ϕ) on the storage and loss moduli of a dispersion of soft elastic particles.

Because $G_c^* = j\omega\eta_c$ for Newtonian continuous-phase, and $G_d^* = G_d$ for elastic Hookean particles, Equation 15.22 can be rewritten in a more general form as:

$$\frac{G^*}{G_c^*} = \left[\frac{2G_d^* + 3G_c^* + 3\phi(G_d^* - G_c^*)}{2G_d^* + 3G_c^* - 2\phi(G_d^* - G_c^*)} \right] \tag{15.23}$$

Equation 15.23 is valid for dispersions of viscoelastic solid particles in viscoelastic liquids.

15.2 DISPERSIONS OF DEFORMABLE LIQUID DROPLETS WITH PURE INTERFACE

In the absence of any additives at the interface of droplets, the only mechanism for shape recovery of droplets is the interfacial tension (assuming that the droplets are composed of purely viscous inelastic fluid). When the external liquid (continuous-phase) is subjected to a shear flow, the droplet deforms under the influence of the viscous stresses. The interfacial tension provides a force that tends to restore the droplet shape to its equilibrium (spherical) form. Thus, droplets possess elasticity of shape because of the restoring force of interfacial tension. When the emulsion is subjected to *variable* macroscopic rate of shear, the droplets experience variations in stress and, consequently, the amount of elastic strain energy stored in the emulsion varies. The interchange of energy between internal potential energy and kinetic energy of flow manifests itself as elasticity of shape in the emulsion as a whole [2]. Oldroyd [3] studied the viscoelastic behavior of emulsions. He utilized the approach of Frohlich and Sack [1], that is, the self-consistent effective medium approach, to derive a constitutive equation governing the viscoelastic behavior of an emulsion of one incompressible Newtonian liquid in another with interfacial tension between the two phases providing a restoring force for the droplet shape.

Similar to Frohlich and Sack [1], Oldroyd considers two systems A and B, shown in Figure 15.1. In system A, the entire space $0 < r < b$ is filled with the effective homogeneous medium (emulsion) that has the macroscopic properties we wish to calculate. In system B, a single spherical droplet of radius R made up of incompressible Newtonian liquid of viscosity η_d is surrounded by continuous-phase fluid (viscosity η_c) of the emulsion filling the space $R < r < a$, and the remaining space $r > b$ is filled with the effective homogeneous medium (emulsion) having the properties that need to be calculated. The radius a is equal to $R/(\phi)^{1/3}$.

Subject to the same time-dependent stress specified at the boundary $r = b$, the velocity distributions in the two systems are determined. The governing equations are:

$$\text{System A:} \quad \eta \nabla^2 \vec{u} = \nabla P, \quad \nabla \bullet \vec{u} = 0 \quad \text{for} \quad 0 \leq r \leq b \qquad (15.24)$$

$$\text{System B:} \quad \eta_d \nabla^2 \vec{u}_d = \nabla P_d, \quad \nabla \bullet \vec{u}_d = 0 \quad \text{for} \quad 0 \leq r \leq R \qquad (15.25)$$

$$\eta_c \nabla^2 \vec{u}_c = \nabla P_c, \quad \nabla \bullet \vec{u}_c = 0 \quad \text{for} \quad R \leq r \leq a \qquad (15.26)$$

$$\eta \nabla^2 \vec{u} = \nabla P, \quad \nabla \bullet \vec{u} = 0 \quad \text{for} \quad a \leq r \leq b \qquad (15.27)$$

The boundary conditions to be satisfied by system B are:

1. Continuity of velocity at the surface of the droplets, that is,

$$\vec{u}_c = \vec{u}_d = \frac{\partial \vec{r}}{\partial t} \quad \text{at} \quad r = R \qquad (15.28)$$

2. Normal stress discontinuity across the interface at $r = R$

$$(\overline{\overline{\sigma}}_c - \overline{\overline{\sigma}}_d) : \hat{n}\hat{n} = \gamma \nabla_s \bullet \hat{n} \quad \text{at} \quad r = R \qquad (15.29)$$

where ∇_s is the surface gradient operator given as $\nabla_s = (\overline{\overline{\delta}} - \hat{n}\hat{n}) \bullet \nabla$, γ is the interfacial tension between the two liquids (dispersed and continuous phases of the emulsion), and $\hat{n}$ is the unit outward normal at the surface $r = R$.

3. Continuity of tangential stresses at the interface at $r = R$

$$(\overline{\overline{\sigma}}_c - \overline{\overline{\sigma}}_d) : \hat{n}\hat{t} = 0 \tag{15.30}$$

where $\hat{t}$ is the unit tangential vector at the interface $r = R$.

4. Continuity of stress at $r = a$, that is,

$$(\overline{\overline{\sigma}} - \overline{\overline{\sigma}}_c) \bullet \hat{n} = 0 \quad \text{at} \quad r = a \tag{15.31}$$

5. Continuity of velocity at $r = a$, that is,

$$\vec{u}_c = \vec{u} \quad \text{at} \quad r = a \tag{15.32}$$

The two systems (A and B), subjected to the same boundary stresses at $r = b$, are required to be indistinguishable when observations are made at $r = b$, where b is sufficiently large compared to a. In other words, the following condition must be fulfilled:

$$\vec{u}_A\big|_{r=b} = \vec{u}_B\big|_{r=b} \tag{15.33}$$

Oldroyd's analysis leads to the following constitutive equation, valid at small variable rates of strain, for emulsions of two immiscible Newtonian liquids with constant interfacial tension:

$$\left(1 + \tau_1 \frac{\partial}{\partial t}\right) \overline{\overline{\tau}} = 2\eta_0 \left(1 + \tau_2 \frac{\partial}{\partial t}\right) \overline{\overline{E}} \tag{15.34}$$

where η_0 is the zero-shear viscosity of the emulsion, τ_1 is the relaxation time, and τ_2 is the retardation time. Expressions for η_0, τ_1, and τ_2 are as follows:

$$\eta_0 = \eta_c \left[\frac{10(1+\lambda) + 3\phi(2+5\lambda)}{10(1+\lambda) - 2\phi(2+5\lambda)} \right]$$

$$= \eta_c \left[1 + \frac{5\lambda+2}{2(\lambda+1)}\phi + O(\phi^2) \right] \tag{15.35}$$

$$\tau_1 = \left(\frac{\eta_c R}{4\gamma}\right)\left[\frac{(2\lambda+3)+2\phi(1-\lambda)}{10(\lambda+1)-2\phi(2+5\lambda)}\right](16+19\lambda)$$

$$= \frac{(16+19\lambda)(2\lambda+3)}{40(\lambda+1)}\left(\frac{\eta_c R}{\gamma}\right)\left[1+\phi\frac{(19\lambda+16)}{5(\lambda+1)(2\lambda+3)}+O(\phi^2)\right]$$

(15.36)

$$\tau_2 = \left(\frac{\eta_c R}{4\gamma}\right)\left[\frac{(2\lambda+3)-3\phi(1-\lambda)}{10(\lambda+1)+3\phi(2+5\lambda)}\right](16+19\lambda)$$

$$= \frac{(16+19\lambda)(2\lambda+3)}{40(\lambda+1)}\left(\frac{\eta_c R}{\gamma}\right)\left[1-\phi\frac{3(19\lambda+16)}{10(\lambda+1)(2\lambda+13)}+O(\phi^2)\right]$$

(15.37)

where λ is the viscosity ratio η_d/η_c. Note that the Oldroyd equation (Equation 15.34) has exactly the same form as the Frohlich and Sack equation (Equation 15.11). However, the expressions for η_0, τ_1, and τ_2 are different.

In oscillatory shear flow, the Oldroyd equation (Equation 15.34) in conjunction with $\tau^* = 2G^*e^*$ gives:

$$G^* = j\omega\eta_0\left[\frac{1+j\omega\tau_2}{1+j\omega\tau_1}\right]$$

(15.38)

$$G' = \frac{\eta_0\omega^2(\tau_1-\tau_2)}{1+\omega^2\tau_1^2}$$

(15.39)

$$G'' = \frac{\eta_0\omega(1+\omega^2\tau_1\tau_2)}{1+\omega^2\tau_1^2}$$

(15.40)

Equation 15.38 to Equation 15.40 are the same as Equation 15.17 to Equation 15.19 except that the expressions for η_0, τ_1, and τ_2 are now different.

In the limit $\gamma/R \to \infty$, $\tau_1 = \tau_2 = 0$ and $G' = 0$, whereas

$$\frac{G''}{\omega} = \eta_0 = \eta_c\left[1+\frac{5\lambda+2}{2(\lambda+1)}+O(\phi^2)\right]$$

(15.41)

This gives the well-known Taylor [4] result for the viscosity of dilute emulsions of spherical droplets.

Figure 15.5 shows the effect of viscosity ratio λ on the storage and loss moduli of emulsion. The plots are generated from the Oldroyd model (Equation 15.39 and Equation 15.40). With the increase in λ, the plateau value of the storage modulus decreases whereas the loss modulus shows a slight increase. The effect of interfacial stress γ/R on the storage and loss moduli of emulsion is shown in Figure 15.6. With the increase in γ/R, the plateau value of the storage modulus increases. Also, the frequency at which the terminal zone (low-frequency zone where $G' \propto \omega^2$) begins is shifted to higher values with increase in γ/R. The interfacial stress γ/R appears to have a negligible effect on the loss modulus of the emulsion.

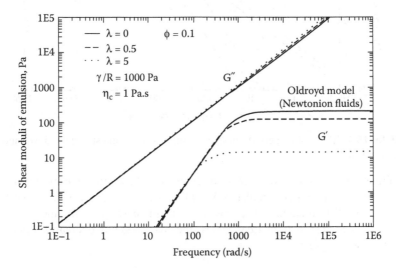

FIGURE 15.5 Effect of viscosity ratio λ on the storage and loss moduli of an emulsion.

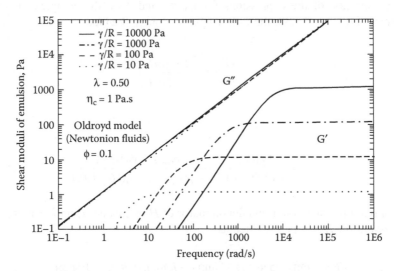

FIGURE 15.6 Effect of interfacial stress γ/R on the storage and loss moduli of an emulsion.

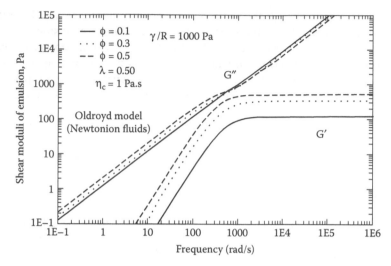

FIGURE 15.7 Effect of dispersed-phase concentration ϕ on the storage and loss moduli of an emulsion.

Figure 15.7 shows the effect of dispersed-phase concentration ϕ on the storage and loss moduli of emulsion. With the increase in ϕ, the storage modulus increases at all frequencies. The loss modulus increases with the increase in ϕ at low frequencies; at high frequencies, the loss modulus decreases somewhat with the increase in ϕ.

15.2.1 GENERALIZATION OF OLDROYD MODEL

On substitution of the expressions for η_0, τ_1, and τ_2 in the complex modulus equation, Equation 15.38, the following equation is obtained:

$$\frac{G^*}{j\omega\eta_c} = \left(\frac{1+3\phi H}{1-2\phi H} \right) \qquad (15.42)$$

where

$$H = \frac{(4\gamma/R)(2+5\lambda) + j\omega\eta_c(\lambda-1)(16+19\lambda)}{(40\gamma/R)(1+\lambda) + j\omega\eta_c(2\lambda+3)(16+19\lambda)} \qquad (15.43)$$

Multiplying the numerator and denominator of H by $j\omega$ and using $\lambda = \eta_d/\eta_c$, the expression for H can be rewritten as:

$$H = \frac{(4\gamma/R)(2j\omega\eta_c + 5j\omega\eta_d) + (j\omega\eta_d - j\omega\eta_c)(16j\omega\eta_c + 19j\omega\eta_d)}{(40\gamma/R)(j\omega\eta_c + j\omega\eta_d) + (2j\omega\eta_d + 3j\omega\eta_c)(16j\omega\eta_c + 19j\omega\eta_d)} \qquad (15.44)$$

The complex modulus equation (Equation 15.42), with H given by Equation 15.44, is restricted to emulsions of two immiscible Newtonian liquids. Equation 15.42 can easily be generalized to emulsions of two viscoelastic liquids with interfacial tension.

As $G_c^* = j\omega\eta_c$ and $G_d^* = j\omega\eta_d$ for Newtonian liquids, Equation 15.42 can be rewritten in a more general form as:

$$\frac{G^*}{G_c^*} = \left(\frac{1+3\phi H}{1-2\phi H}\right) \tag{15.45}$$

where

$$H = \frac{(4\gamma/R)(2G_c^* + 5G_d^*) + (G_d^* - G_c^*)(16G_c^* + 19G_d^*)}{(40\gamma/R)(G_c^* + G_d^*) + (2G_d^* + 3G_c^*)(16G_c^* + 19G_d^*)} \tag{15.46}$$

Equation 15.45, with H given by Equation 15.46, is valid for emulsions of two immiscible viscoelastic liquids with interfacial tension. A more formal derivation of Equation 15.45 is given by Palierne [5]. In the limit $\phi \to 0$, Equation 15.45 reduces to:

$$\frac{G^*}{G_c^*} = 1 + 5\phi H \tag{15.47}$$

Figure 15.8 shows the storage and loss moduli of emulsion of two immiscible viscoelastic liquids with interfacial tension. The data are generated from Equation 15.45. The two liquids (continuous-phase and dispersed-phase) are assumed to follow the Maxwell model, that is:

$$G_c' = \frac{G_c\omega^2\tau_c^2}{1+\omega^2\tau_c^2}, \quad G_d' = \frac{G_d\omega^2\tau_d^2}{1+\omega^2\tau_d^2} \tag{15.48a}$$

$$G_c'' = \frac{G_c\omega\tau_c}{1+\omega^2\tau_c^2}, \quad G_d'' = \frac{G_d\omega\tau_d}{1+\omega^2\tau_d^2} \tag{15.48b}$$

where G_c and G_d are high-frequency limiting values of storage moduli of continuous-phase and dispersed-phase liquids, respectively, and τ_c and τ_d are the relaxation times of the continuous-phase and dispersed-phase, respectively. At high frequencies, the dynamic moduli (G' and G'') of the emulsion are similar to those of the matrix (continuous-phase). But at low frequencies, the storage modulus of the emulsion deviates from that of the matrix. The values of emulsion G' are much

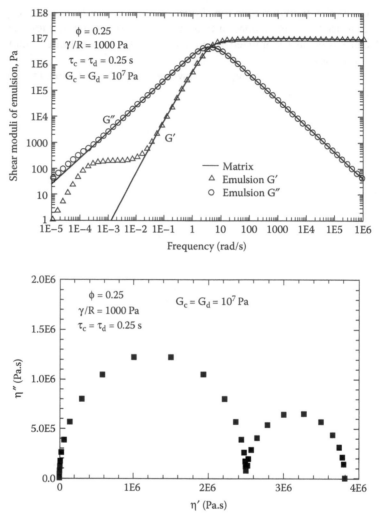

FIGURE 15.8 Dynamic moduli and Cole–Cole diagram for an emulsion of two immiscible viscoelastic liquids with interfacial tension.

higher than those of the matrix in the low-frequency range. It should be further noted that the data for emulsion G' exhibit two plateaus: *upper plateau* at high frequencies (overlapping with the matrix plateau) and *lower plateau* at low frequencies. The Cole–Cole diagram, where $\eta''(= G'/\omega)$ is plotted as a function of $\eta'(= G''/\omega)$, is also shown in Figure 15.8. The Cole–Cole diagram clearly shows two frequency domains; the right-hand side semicircle corresponds to low-frequency behavior and the left-hand side semicircle corresponds to high-frequency behavior. The elastic behavior in the low-frequency domain is due to the interfacial-tension effect (elasticity of the shape of the droplets). In the high-frequency domain, the elastic behavior is due to the elasticity of the matrix.

Figure 15.9 to Figure 15.12 show the effects of various parameters on the storage and loss moduli of emulsions. With the increase in the dispersed-phase concentration ϕ of an emulsion, the lower plateau value of the storage modulus of the emulsion increases (see Figure 15.9). The loss modulus increases with increase in ϕ only in the low-frequency range. With an increase in the interfacial stress γ/R, the lower plateau value of the storage modulus of the emulsion increases (see Figure 15.10). However, the frequency at which the terminal zone (low-frequency zone

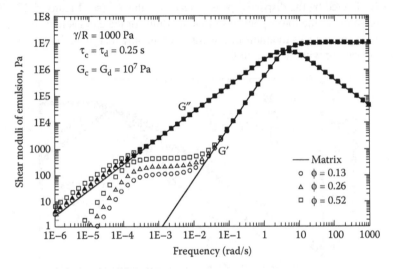

FIGURE 15.9 Effect of dispersed-phase concentration ϕ on the storage and loss moduli of an emulsion of two immiscible viscoelastic liquids.

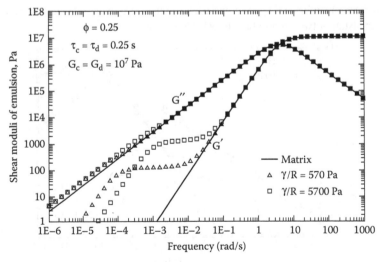

FIGURE 15.10 Effect of interfacial stress γ/R on the storage and loss moduli of an emulsion of two immiscible viscoelastic liquids.

where $G' \propto \omega^2$) begins is shifted to higher values with an increase in γ/R. The effect of γ/R on the loss modulus of an emulsion is small. With an increase in the dispersed-phase relaxation time τ_d, the lower plateau value of the storage modulus of the emulsion decreases (see Figure 15.11). At low frequencies (less than $1/\tau_c$), the loss modulus of the emulsion increases slightly with increase in τ_d. At high frequencies ($>1/\tau_c$), the opposite effect is seen, that is, the loss modulus of the emulsion tends to decrease with the increase in τ_d. The lower plateau value of the storage modulus is strongly affected by the dispersed-phase shear modulus G_d (see Figure 15.12). With an increase in G_d, the lower plateau storage modulus of the emulsion decreases. The effect of G_d on the loss modulus of the emulsion is only marginal; the loss modulus tends to increase with increase in G_d.

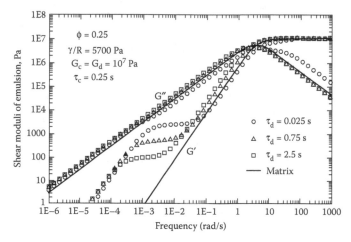

FIGURE 15.11 Effect of dispersed-phase relaxation time τ_d on the storage and loss moduli of an emulsion of two immiscible viscoelastic liquids.

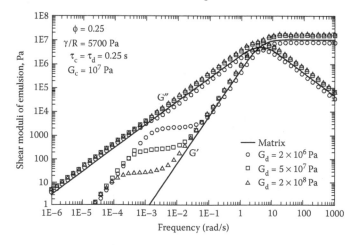

FIGURE 15.12 Effect of dispersed-phase shear modulus G_d on the storage and loss moduli of an emulsion of two immiscible viscoelastic liquids.

15.3 DISPERSIONS OF DEFORMABLE LIQUID DROPLETS COATED WITH ADDITIVE

The rheological behavior of dispersions of deformable liquid droplets coated with some additive (surfactant, polymer, etc.) is dependent on the rheological characteristics of the film formed by the additive. For example, if the interfacial film is purely viscous in nature, with no elasticity of shape or area, its presence has a very different effect on the rheological behavior of an emulsion as compared to when the interfacial film is purely elastic in nature.

15.3.1 DROPLETS COATED WITH PURELY VISCOUS FILMS

For an infinitely thin (two-dimensional) purely viscous film, the surface-stress tensor $(\bar{\bar{P}}_s)$ is related to the surface rate-of-strain tensor $(\bar{\bar{E}}_s)$ as [6]:

$$\bar{\bar{P}}_s = \gamma \bar{\bar{\delta}}_s + 2\eta_s \bar{\bar{E}}_s + \left(\eta_s^k - \eta_s\right)\left(\bar{\bar{\delta}}_s : \bar{\bar{E}}_s\right)\bar{\bar{\delta}}_s \tag{15.49}$$

where γ is the interfacial tension, η_s is the surface shear viscosity, η_s^k is the surface dilational viscosity, and $\bar{\bar{\delta}}_s$ is the surface unit tensor $(= \bar{\bar{\delta}} - \hat{n}\hat{n})$.

Emulsions of liquid droplets coated with purely viscous films were treated by Oldroyd [6]. He developed the following equation to describe the viscoelastic behavior of such emulsions:

$$\left(1 + \tau_1 \frac{\partial}{\partial t}\right)\bar{\bar{\tau}} = 2\eta_0\left(1 + \tau_2 \frac{\partial}{\partial t}\right)\bar{\bar{E}} \tag{15.50}$$

where η_0, τ_1, and τ_2 are given as:

$$\eta_0 = \eta_c\left[\frac{10(1+\lambda)+(8N_{Bo}^{\eta})+(12N_{Bo}^k)+3\phi\{2+5\lambda+(4N_{Bo}^{\eta})+(6N_{Bo}^k)\}}{10(1+\lambda)+(8N_{Bo}^{\eta})+(12N_{Bo}^k)-2\phi\{2+5\lambda+(4N_{Bo}^{\eta})+(6N_{Bo}^k)\}}\right] \tag{15.51}$$

$$\tau_1 = \left(\frac{\alpha_1 - 2\phi\alpha_2}{\beta_1 - 2\phi\beta_2}\right)\left(\frac{\eta_c R}{4\gamma}\right) \tag{15.52}$$

$$\tau_2 = \left(\frac{\alpha_1 + 3\phi\alpha_2}{\beta_1 + 3\phi\beta_2}\right)\left(\frac{\eta_c R}{4\gamma}\right) \tag{15.53}$$

$$\alpha_1 = [(3+2\lambda)(16+19\lambda)+4(12+13\lambda)(N_{Bo}^\eta)$$
$$+ 2(32+23\lambda)(N_{Bo}^k)+32N_{Bo}^k N_{Bo}^\eta] \tag{15.54}$$

$$\alpha_2 = [(\lambda-1)(16+19\lambda)+2(8+13\lambda)(N_{Bo}^\eta)$$
$$+ (23\lambda-16)(N_{Bo}^k)+16N_{Bo}^k N_{Bo}^\eta] \tag{15.55}$$

$$\beta_1 = 10(1+\lambda)+8N_{Bo}^\eta+12N_{Bo}^k \tag{15.56}$$

$$\beta_2 = 2+5\lambda+4N_{Bo}^\eta+6N_{Bo}^k \tag{15.57}$$

N_{Bo}^η is the shear Boussinesq number defined as $\eta_s/\eta_c R$ and N_{Bo}^k is the dilational Boussinesq number defined as $\eta_s^k/\eta_c R$. Equation 15.50 was developed using the same "self-consistent effective medium approach" used earlier by Oldroyd [3] and Frohlich and Sack [1]. Because of the presence of a film on the droplet, the stress boundary condition at the interface ($r = R$) was modified. The remaining boundary conditions were unchanged. From the force balance at the interface, it can be shown that [7]:

$$\hat{n}\bullet\left(\bar{\bar{\sigma}}_c-\bar{\bar{\sigma}}_d\right)=\left(\nabla_s\bullet\hat{n}\right)\left(\hat{n}\bullet\bar{\bar{P}}_s\right)-\nabla_s\bullet\bar{\bar{P}}_s \tag{15.58}$$

where ∇_s is the surface gradient operator ($=(\bar{\bar{\delta}}-\hat{n}\hat{n})\bullet\nabla$). As the surface stress tensor ($\bar{\bar{P}}_s$) does not have any component in the direction of $\hat{n}$, $\hat{n}\bullet\bar{\bar{P}}_s=0$. Thus,

$$\hat{n}\bullet\left(\bar{\bar{\sigma}}_c-\bar{\bar{\sigma}}_d\right)=-\nabla_s\bullet\bar{\bar{P}}_s \tag{15.59}$$

From Equation 15.59, the normal stress discontinuity across the interface at $r = R$ is:

$$\hat{n}\bullet\left(\bar{\bar{\sigma}}_c-\bar{\bar{\sigma}}_d\right)\bullet\hat{n}=-\left(\nabla_s\bullet\bar{\bar{P}}_s\right)\bullet\hat{n} \tag{15.60}$$

The tangential stress discontinuity at the interface at $r = R$ is:

$$\hat{n}\bullet\left(\bar{\bar{\sigma}}_c-\bar{\bar{\sigma}}_d\right)\bullet\bar{\bar{\delta}}_s=-\left(\nabla_s\bullet\bar{\bar{P}}_s\right)\bullet\bar{\bar{\delta}}_s \tag{15.61}$$

where $\bar{\bar{P}}_s$ is given by Equation 15.49.

When the surface dilational and surface shear viscosities are both zero, Equation 15.51 to Equation 15.57 reduce to Equation 15.35 to Equation 15.37. In oscillatory shear flow, Equation 15.50 to Equation 15.53 give:

$$\frac{G^*}{j\omega\eta_c} = \left(\frac{1+3\phi H}{1-2\phi H} \right) \tag{15.62}$$

where H is now given by

$$H = \left[\frac{(4\gamma/R)\beta_2 + (j\omega\eta_c)\alpha_2}{(4\gamma/R)\beta_1 + (j\omega\eta_c)\alpha_1} \right] \tag{15.63}$$

When the surface dilational and shear viscosities are both zero, the expression for H given by Equation 15.63 becomes the same as that given by Equation 15.43.

Figure 15.13 shows the effect of the shear Boussinesq number on the storage and loss moduli of emulsion of liquid droplets covered with purely viscous film. The plots are generated from Equation 15.62 and Equation 15.63. With an increase in the shear Boussinesq number N_{Bo}^η, the storage modulus of the emulsion decreases significantly. The loss modulus of the emulsion increases with increase in N_{Bo}^η, although the increase is only marginal. The effect of the dilational Boussinesq number N_{Bo}^k on the dynamic moduli of the emulsion is shown in Figure 15.14. Both the storage and loss moduli of the emulsion tend to increase with the increase in N_{Bo}^k.

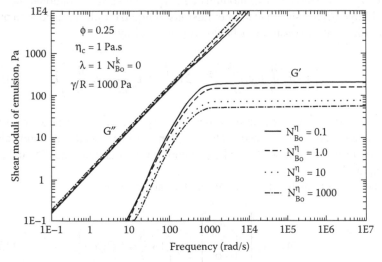

FIGURE 15.13 The effect of the shear Boussinesq number (N_{Bo}^η) on the storage and loss moduli of an emulsion of liquid droplets covered with purely viscous film.

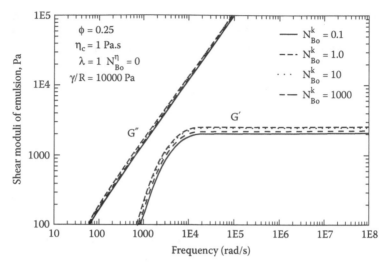

FIGURE 15.14 The effect of the dilational Boussinesq number (N_{Bo}^k) on the storage and loss moduli of an emulsion of liquid droplets covered with purely viscous film.

15.3.2 DROPLETS COATED WITH PURELY ELASTIC FILMS

For purely elastic two-dimensional membranes, the surface-stress tensor $(\bar{\bar{P}}_s)$ is related to the surface-strain tensor $\bar{\bar{e}}_s$ as [6]:

$$\bar{\bar{P}}_s = \gamma \bar{\bar{\delta}}_s + 2G_s \bar{\bar{e}}_s + (K_s - G_s)(\bar{\bar{\delta}}_s : \bar{\bar{e}}_s)\bar{\bar{\delta}}_s \qquad (15.64)$$

where G_s is the surface shear modulus and K_s is the surface dilational modulus.

Oldroyd [6] has shown that for emulsions of liquid droplets coated with purely elastic films, the viscoelastic behavior is governed by:

$$\left(1 + \tau_1 \frac{\partial}{\partial t} + v_1 \frac{\partial^2}{\partial t^2}\right)\bar{\bar{\tau}} = 2\eta_0 \left(1 + \tau_2 \frac{\partial}{\partial t} + v_2 \frac{\partial^2}{\partial t^2}\right)\bar{\bar{E}} \qquad (15.65)$$

where τ_1 and $\sqrt{v_1}$ are relaxation times, and τ_2 and $\sqrt{v_2}$ are retardation times. Thus, to describe the viscoelastic behavior of emulsions coated with purely elastic films, two relaxation times (τ_1 and $\sqrt{v_1}$) and two retardation times (τ_2 and $\sqrt{v_2}$) are required.

The expressions for η_0, τ_1, τ_2, v_1, and v_2 are given as:

$$\eta_0 = \eta_c \left[\frac{1 + \frac{3}{2}\phi}{1 - \phi}\right] = \eta_c \left[1 + \frac{5}{2}\phi + O(\phi^2)\right] \qquad (15.66)$$

$$\tau_1 = \left(\frac{\alpha_1 - 2\phi\alpha_2}{\beta_1 - 2\phi\beta_2}\right)\left(\frac{\eta_c R}{8\gamma}\right) \tag{15.67}$$

$$\tau_2 = \left(\frac{\alpha_1 + 3\phi\alpha_2}{\beta_1 + 3\phi\beta_2}\right)\left(\frac{\eta_c R}{8\gamma}\right) \tag{15.68}$$

$$v_1 = \left(\frac{\alpha_1' - 2\phi\alpha_2'}{\beta_1 - 2\phi\beta_2}\right)\left(\frac{\eta_c^2 R^2}{8\gamma}\right) \tag{15.69}$$

$$v_2 = \left(\frac{\alpha_1' + 3\phi\alpha_2'}{\beta_1 + 3\phi\beta_2}\right)\left(\frac{\eta_c^2 R^2}{8\gamma}\right) \tag{15.70}$$

$$\alpha_1 = \left[40(1+\lambda)\gamma + 4(12+13\lambda)G_s + 2(32+23\lambda)K_s\right] \tag{15.71}$$

$$\alpha_2 = \left[4(2+5\lambda)\gamma + 2(8+13\lambda)G_s + (23\lambda-16)K_s\right] \tag{15.72}$$

$$\beta_1 = 2[2G_s + 3K_s + (2G_sK_s/\gamma)] \tag{15.73}$$

$$\beta_2 = [2G_s + 3K_s + (2G_sK_s/\gamma)] \tag{15.74}$$

$$\alpha_1' = (3+2\lambda)(16+19\lambda) \tag{15.75}$$

$$\alpha_2' = (\lambda-1)(16+19\lambda) \tag{15.76}$$

In oscillatory shear flow, Equation 15.65 to Equation 15.70 give:

$$\frac{G^*}{j\omega\eta_c} = \left(\frac{1+3\phi H}{1-2\phi H}\right) \tag{15.77}$$

where H is now given by:

$$H = \frac{(8\gamma/R)\,\beta_2 + (j\omega\eta_c)\alpha_2 - (R\omega^2\eta_c^2)\,\alpha_2'}{(8\gamma/R)\,\beta_1 + (j\omega\eta_c)\alpha_1 - (R\omega^2\eta_c^2)\alpha_1'} \tag{15.78}$$

Figure 15.15 and Figure 15.16 show the dynamic moduli of the emulsion predicted from Equation 15.77 and Equation 15.78. The corresponding Cole–Cole diagrams are also shown. Clearly, the emulsion exhibits two frequency domains (the right-hand side semicircle of the Cole–Cole plot corresponds to low-frequency behavior and the left-hand side semicircle corresponds to high-frequency behavior). The elastic behavior in the low-frequency domain is due to the interfacial relaxation effect, whereas in the high-frequency domain, the elastic behavior is due to form relaxation (the deformed droplets tend to attain their original spherical shape).

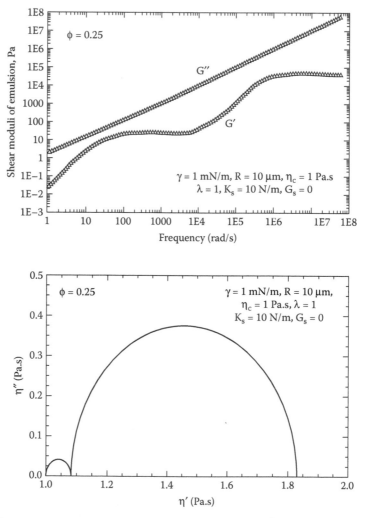

FIGURE 15.15 Dynamic moduli and Cole–Cole diagram for an emulsion of liquid droplets covered with purely elastic film.

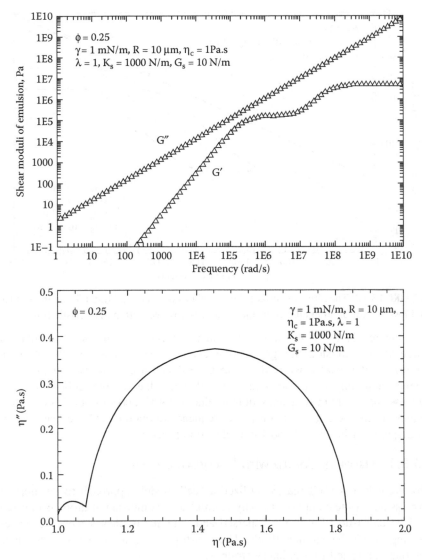

FIGURE 15.16 Dynamic moduli and Cole–Cole diagram for an emulsion of liquid droplets covered with purely elastic film.

Figure 15.17 to Figure 15.19 show the effects of various parameters on the dynamic moduli of an emulsion of liquid droplets coated with purely elastic films, as predicted from Equation 15.77 and Equation 15.78. With an increase in G_s (surface shear modulus of the interfacial film), the upper plateau value of the storage modulus of an emulsion increases; the frequency at which the upper plateau begins also increases with the increase in G_s (see Figure 15.17). With the increase in K_s (surface dilational modulus), a similar behavior is observed

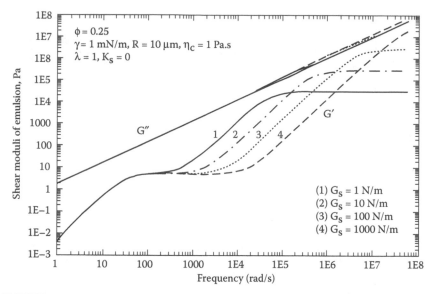

FIGURE 15.17 The effect of surface shear modulus of interfacial film (G_s) on the storage and loss moduli of an emulsion of liquid droplets covered with purely elastic film.

(see Figure 15.18). With the decrease in interfacial tension, the upper plateau value of the emulsion storage modulus decreases (see Figure 15.19). At high values of interfacial tension, the Cole–Cole diagram exhibits two distinct frequency domains — the low-frequency domain due to interfacial-relaxation process and the high-frequency domain due to form-relaxation process. At low values of interfacial tension, only one frequency domain is observed because of overlapping of interfacial and form relaxation processes.

15.3.3 DROPLETS COATED WITH VISCOELASTIC FILMS

Oosterbroek and Mellema [8] utilized a "cell model" approach to develop an expression for the complex viscosity of emulsions composed of droplets covered with viscoelastic films. The dispersed and continuous phases were treated as incompressible Newtonian fluids. The constitutive equation for a viscoelastic interface subjected to oscillatory strain is:

$$\overline{\overline{P}}_s^* = \gamma \overline{\overline{\delta}}_s + 2G_s^* \overline{\overline{e}}_s^* + \left(K_s^* - G_s^*\right)\left(\overline{\overline{\delta}}_s : \overline{\overline{e}}_s^*\right)\overline{\overline{\delta}}_s \tag{15.79}$$

where $\overline{\overline{P}}_s^*$ is the complex surface-stress tensor, $\overline{\overline{e}}_s^*$ is the complex surface-strain tensor, G_s^* is the complex surface-shear modulus, and K_s^* is the complex surface-dilational modulus. G_s^* and K_s^* are defined as:

$$G_s^* = G_s + j\eta_s\omega \tag{15.80}$$

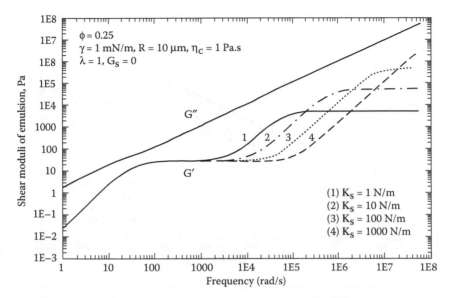

FIGURE 15.18 The effect of surface dilational modulus of interfacial film (K_s) on the storage and loss moduli of an emulsion of liquid droplets covered with purely elastic film.

$$K_s^* = K_s + j\eta_s^k \omega \qquad (15.81)$$

In general, G_s, η_s, K_s, and η_s^k are functions of frequency. The complex viscosity expression developed by Oosterbroek and Mellema is as follows:

$$\eta^* = \eta_c \left[1 + \phi \left(\frac{\alpha_0 + \beta_0 H + \nu_0 H^2}{\alpha_0' + \beta_0' H + \nu_0' H^2} \right) \right] \qquad (15.82)$$

where $H = jR\eta_c\omega/\gamma$, and $\alpha_0, \alpha_0', \beta_0, \beta_0', \nu_0,$ and ν_0' are given as follows:

$$\alpha_0 = A_{10}(\phi^{-10/3}) + A_3(\phi^{-1});$$
$$\alpha_0' = A_{10}'(\phi^{-10/3}) + A_7'(\phi^{-7/3}) + A_5'(\phi^{-5/3}) + A_3'(\phi^{-1}) + A_0' \qquad (15.83)$$

$$\beta_0 = B_{10}(\phi^{-10/3}) + B_3(\phi^{-1});$$
$$\beta_0' = B_{10}'(\phi^{-10/3}) + B_7'(\phi^{-7/3}) + B_5'(\phi^{-5/3}) + B_3'(\phi^{-1}) + B_0' \qquad (15.84)$$

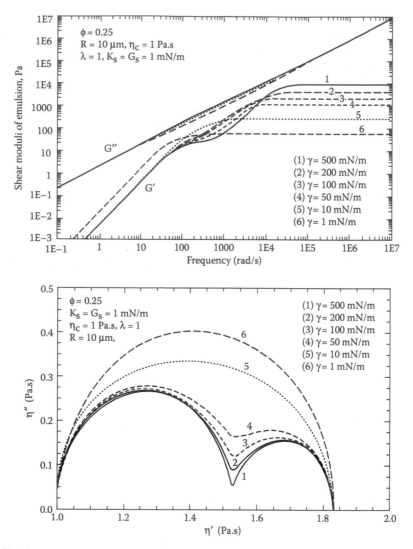

FIGURE 15.19 The effect of interfacial tension (γ) on the dynamic moduli and Cole–Cole plot of an emulsion of liquid droplets covered with purely elastic film.

$$v_0 = C_{10}(\phi^{-10/3}) + C_3(\phi^{-1});$$

$$v_0' = C_{10}'(\phi^{-10/3}) + C_7'(\phi^{-7/3}) + C_5'(\phi^{-5/3}) + C_3'(\phi^{-1}) + C_0' \qquad (15.85)$$

The coefficients (A, B, and C etc.) in Equation 15.83 to Equation 15.85 are given in terms of the variables $\overline{A}$, $\overline{B}$, and $\overline{C}$ defined as:

$$\overline{A} = (3K_s/\gamma) + (2G_s/\gamma) + (2K_sG_s/\gamma^2) \qquad (15.86)$$

$$\bar{B} = (24\eta_s^k/R\eta_c) + (20\lambda) + (16\eta_s/R\eta_c) + (23K_s\lambda/\gamma) \quad (15.87)$$
$$+ (16K_s\eta_s/R\eta_c\gamma) + (16G_s\eta_s^k/R\eta_c\gamma) + (26\lambda G_s/\gamma)$$

$$\bar{C} = (23\eta_s^k\lambda/R\eta_c) + (16\eta_s\eta_s^k/R^2\eta_c^2) + (19\lambda^2) + (26\lambda\eta_s/R\eta_c) \quad (15.88)$$

The coefficients are:

$$A_3 = -80\bar{A}; \ A_{10} = 80\bar{A}; \ A_0' = 32\bar{A}; \ A_3' = -200\bar{A};$$
$$A_5' = 336\bar{A}; \ A_7' = -200\bar{A}; \ A_{10}' = 32\bar{A} \quad (15.89a)$$

$$B_3 = -10\left[\bar{B} - 20 - 23(K_s/\gamma) - 26(G_s/\gamma)\right];$$
$$B_{10} = 10\left[\bar{B} + 8 - 16(K_s/\gamma) + 16(G_s/\gamma)\right] \quad (5.89b)$$

$$B_0' = -4\left[\bar{B} - 20 - 23(K_s/\gamma) - 26(G_s/\gamma)\right];$$
$$B_3' = -25\left[\bar{B} - 8 + 4(K_s/\gamma) + 25(G_s/\gamma)\right] \quad (15.89c)$$

$$B_5' = 42\left[\bar{B} - 8(K_s/\gamma) + 24(G_s/\gamma)\right];$$
$$B_7' = -25\left[\bar{B} + 8 - 16(K_s/\gamma) + 16(G_s/\gamma)\right] \quad (15.89d)$$

$$B_{10}' = 4\left[\bar{B} + 20 + 32(K_s/\gamma) + (G_s/\gamma)\right] \quad (15.89e)$$

$$C_3 = -10\left[\bar{C} - 23(\eta_s^k/R\eta_c) - 38\lambda - 26(\eta_s/R\eta_c) + 19\right] \quad (15.89f)$$

$$C_{10} = 10\left[\bar{C} - 16(\eta_s^k/R\eta_c) - 3\lambda + 16(\eta_s/R\eta_c) - 16\right] \quad (15.89g)$$

$$C_0' = 4\left[\bar{C} - 23(\eta_s^k/R\eta_c) - 38\lambda - 26(\eta_s/R\eta_c) + 19\right] \quad (15.89h)$$

$$C_3' = -25\left[\bar{C} + 4(\eta_s^k/R\eta_c) - \lambda + 24(\eta_s/R\eta_c) - 18\right] \quad (15.89i)$$

$$C_5' = 42\left[\bar{C} - 8(\eta_s^k/R\eta_c) - 3\lambda - 24(\eta_s/R\eta_c) - 16\right] \quad (15.89j)$$

$$C_7' = -25\left[\bar{C} - 16(\eta_s^k/R\eta_c) - 3\lambda + 16(\eta_s/R\eta_c) - 16\right] \qquad (15.89k)$$

$$C_{10}' = 2\left[2\bar{C} + 64(\eta_s^k/R\eta_c) + 89\lambda + 48(\eta_s/R\eta_c) + 48\right] \qquad (15.89l)$$

Figure 15.20 shows the dynamic moduli and Cole–Cole plot of an emulsion predicted from Equation 15.82. As expected, the emulsion exhibits two frequency

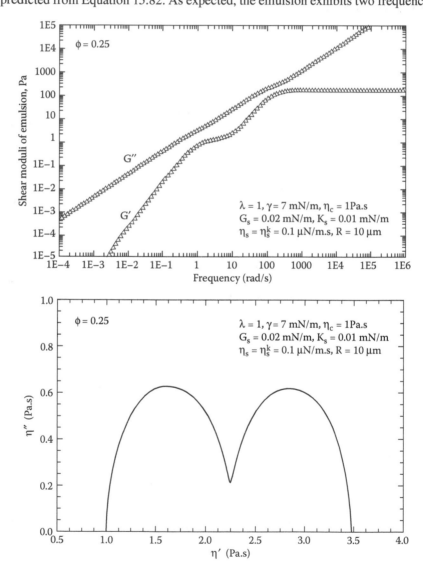

FIGURE 15.20 Dynamic moduli and Cole–Cole diagram for an emulsion of liquid droplets covered with viscoelastic film.

domains corresponding to two relaxation processes. The high-frequency domain is due to the form relaxation of droplets and the low-frequency domain is due to the interfacial relaxation process.

15.3.3.1 Palierne Model

Palierne [5] has developed a fairly general model for the complex modulus of dispersions of liquid droplets coated with viscoelastic films. The dispersed and continuous phases were treated as incompressible viscoelastic liquids. A self-consistent, effective medium approach similar to that of Frohlich and Sack [1] and Oldroyd [3] was utilized.

For dispersions of uniform-size droplets, the complex modulus expression developed by Palierne is given as:

$$G^* = G_c^* \left[\frac{1 + 3\phi H}{1 - 2\phi H} \right] \qquad (15.90)$$

where

$$H = \frac{Q}{2W} \qquad (15.91)$$

and Q and W are given as:

$$Q = \left[2(G_d^* - G_c^*)(19G_d^* + 16G_c^*) \right] + \left[48\gamma K_s^*/R^2 \right] \qquad (15.92)$$

$$+ \left[32G_s^*(\gamma + K_s^*)/R^2 \right] + \left[(8\gamma/R)(5G_d^* + 2G_c^*) \right]$$

$$+ \left[(2K_s^*/R)(23G_d^* - 16G_c^*) \right] + \left[(4G_s^*/R)(13G_d^* + 8G_c^*) \right]$$

$$W = \left[(2G_d^* + 3G_c^*)(19G_d^* + 16G_c^*) \right] + \left[(40\gamma/R)(G_d^* + G_c^*) \right] \qquad (15.93)$$

$$+ \left[\frac{2K_s^*}{R}(23G_d^* + 32G_c^*) \right] + \left[\frac{4G_s^*}{R}(13G_d^* + 12G_c^*) \right]$$

$$+ \left[48K_s^*\gamma/R^2 \right] + \left[32G_s^*(\gamma + K_s^*)/R^2 \right]$$

15.3.4 DROPLETS COATED WITH FILMS POSSESSING BENDING RIGIDITY

Thus far, we have considered films with negligible bending rigidity. The bending or flexural rigidity (M) is defined as [8–12]:

$$M = \frac{Eh^3}{12(1-v^2)} \tag{15.94}$$

where E is the Young's modulus, v is the Poisson ratio, and h is the film thickness. The *bending rigidity* (also referred to as *curvature modulus*) of a membrane can be significant if the thickness of the film is not extremely small.

When a droplet of radius R is subjected to a shear flow, the shear stress that tends to elongate the droplet is $\eta_c \dot{\gamma}$, where $\dot{\gamma}$ is the shear rate. If the droplet is coated with a film of bending rigidity M, an opposing stress of magnitude M/R^3 tends to maintain the droplet in a spherical shape. Assuming that the opposing stress to droplet deformation is due to only bending rigidity, the shape of the droplet is governed by a dimensionless group $\eta_c \dot{\gamma} R^3/M$, which is analogous to capillary number $\eta_c \dot{\gamma} R/\gamma$ in systems in which interfacial tension plays a dominant role.

Seki and Komura [12] have studied the viscoelastic behavior of emulsions composed of droplets coated with elastic membranes of nonnegligible bending rigidity and nonzero dilational modulus. They developed an expression for the bulk stress of a dilute emulsion. Both the continuous phase and the dispersed phase were assumed to be incompressible and Newtonian. The effect of interfacial tension was also considered in the analysis. They first calculated the flow perturbation due to the presence of a single droplet in the matrix (continuous phase) subjected to an oscillatory motion. As nearly spherical droplets were considered, the rotational part of the shear flow was neglected and only the pure straining contribution of the flow was considered. At large distances from the droplet, the velocity field prescribed was as follows:

$$\vec{u}_\infty = \overline{\overline{\Gamma}}_\infty \bullet \vec{r} \tag{15.95}$$

where $\overline{\overline{\Gamma}}_\infty$ is a second-order symmetric tensor given in terms of the elongation rate $\dot{\varepsilon}$ as:

$$\overline{\overline{\Gamma}}_\infty = \frac{\dot{\varepsilon}}{2} \begin{pmatrix} -1 & 0 & 0 \\ 0 & -1 & 0 \\ 0 & 0 & 2 \end{pmatrix} e^{j\omega t} \tag{15.96}$$

The fluid motion outside and inside the droplet was described by the Stokes equations:

$$\nabla^2 \vec{u}_c = \frac{1}{\eta_c} \nabla P_c \quad \text{for} \quad r > R \tag{15.97}$$

$$\nabla \bullet \vec{u}_c = 0 \quad \text{for} \quad r > R \tag{15.98}$$

$$\nabla^2 \vec{u}_d = \frac{1}{\eta_d} \nabla P_d \quad \text{for} \quad r < R \tag{15.99}$$

$$\nabla \bullet \vec{u}_d = 0 \quad \text{for} \quad r < R \tag{15.100}$$

where subscripts c and d refer to the continuous phase and the dispersed phase, respectively. Equation 15.97 to Equation 15.100 were solved for velocity and pressure fields inside and outside the droplet. The boundary conditions were:

(1) Far away from the droplet ($r \to \infty$)

$$\vec{u}_c \to \vec{u}_\infty \tag{15.101}$$

(2) Continuity of velocity at the surface of the droplet, that is,

$$\vec{u}_c = \vec{u}_d = \vec{u}_m = \frac{\partial \vec{r}}{\partial t} \quad \text{at } r = R \tag{15.102}$$

where $\vec{u}_m$ is the membrane velocity.
(3) Force balance at the interface ($r = R$)

$$\left(\bar{\bar{\sigma}}_c - \bar{\bar{\sigma}}_d \right) \bullet \hat{n} + \vec{F}_n + \vec{F}_t = 0 \tag{15.103}$$

where $\hat{n}$ is the unit outer normal to the particle surface, $\vec{F}_n$ is the normal restoring force per unit area due to the membrane, and $\vec{F}_t$ is the tangential restoring force per unit area due to the membrane.

From flow field calculations, Seki and Komura [12] determined the dipole strength $\bar{\bar{S}}^0$ of a single droplet located in an infinite matrix fluid. The dipole strength $\bar{\bar{S}}^0$ is defined as (see Chapter 3):

$$\bar{\bar{S}}^0 = \int_{A_0} \left[\bar{\bar{\sigma}} \bullet \hat{n}\vec{x} - \frac{1}{3} (\vec{x} \bullet \bar{\bar{\sigma}} \bullet \hat{n})\bar{\bar{\delta}} - \eta_c (\vec{u}\hat{n} + \hat{n}\vec{u}) \right] dA \tag{15.104}$$

where $\bar{\bar{\sigma}}$ and $\bar{u}$ are evaluated at the surface of the droplet and A_0 is the surface area of the droplet.

The average or bulk stress in a dispersion of identical particles is given by (see Chapter 3):

$$\langle\bar{\bar{\sigma}}\rangle = -\langle P\rangle\bar{\bar{\delta}}+2\,\eta_c\,\langle\bar{\bar{E}}\rangle + \frac{N}{V}\langle\bar{\bar{S}}\rangle \qquad (15.105)$$

where N/V is the number density of the particles in the dispersion given as $3\phi/4\pi R^3$, $\langle\bar{\bar{E}}\rangle$ is the average rate of deformation tensor, and $\langle\bar{\bar{S}}\rangle$ is the average value of $\bar{\bar{S}}$ for a reference particle. As the interaction between the particles is negligible in the case of a dilute dispersion, the average $\langle\bar{\bar{S}}\rangle$ is the same as the $\bar{\bar{S}}^0$ value of a single particle in an infinite matrix.

The average stress in a dispersion can also be expressed as:

$$\langle\bar{\bar{\sigma}}\rangle = -\langle P\rangle\bar{\bar{\delta}}+2\,\eta^*\,\langle\bar{\bar{E}}\rangle \qquad (15.106)$$

where η^* is the viscosity (complex viscosity in the present situation) of the dispersion. From Equation 15.105 and Equation 15.106, it follows that:

$$\eta^* = \eta_c\left[1+\frac{3\,\phi}{8\,\pi\,R^3\,\eta_c}\left(\frac{\bar{\bar{S}}}{\bar{\bar{E}}}\right)\right] \qquad (15.107)$$

To simplify the notation, the brackets $\langle\,\rangle$ have been dropped. On substituting the expression for the term $\bar{\bar{S}}/\bar{\bar{E}}$ in Equation 15.107, Seki and Komura [12] obtained the following formula for the complex viscosity of a dilute emulsion composed of droplets coated with films characterized by interfacial tension, dilational modulus, and bending rigidity:

$$\eta^* = \eta_c\left[1+5\,\phi\,H\right] \qquad (15.108)$$

where H is given by:

$$H = Y/Z \qquad (15.109a)$$

and Y and Z are given as:

$$Y = [24K_sB_s]+[(23\lambda-16)K_s+4(5\lambda+2)B_s]f$$
$$+[(\lambda-1)(19\lambda+16)f^2] \qquad (15.109b)$$

$$Z = [48K_sB_s]+[2(23\lambda+32)K_s+40(\lambda+1)B_s]f$$
$$+[(2\lambda+3)(19\lambda+16)f^2] \qquad (15.109c)$$

In Equation 15.109b and Equation 15.109c, K_s is the dilational modulus, λ is the ratio of dispersed-phase viscosity to continuous-phase viscosity, f is $j\omega R\eta_c$, and B_s is given as:

$$B_s = \gamma + \frac{M}{R^2}\left(6 - 2 J_0 R + \frac{1}{2} J_0^2 R^2\right)$$ (15.110)

where γ is the interfacial tension, M is the bending rigidity, and J_0 is the spontaneous curvature of the film.

Equation 15.108 could also be expressed in terms of complex modulus because $\eta^* = G^*/j\omega$:

$$\frac{G^*}{j\omega\eta_c} = 1 + 5\phi H$$ (15.111)

Under conditions of negligible bending rigidity ($M = 0$) and negligible surface-dilational modulus ($K_s = 0$), Equation 15.111 reduces to the Oldroyd equation (Equation 15.42) in the limit $\phi \to 0$.

When $K_s = 0$ and $B_s > 0$, Equation 15.108 reduces to the following expressions for low-frequency and high-frequency limiting viscosities:

$$\eta_0 = \eta_c\left[1 + \frac{5\lambda+2}{2(\lambda+1)}\phi\right]$$ (15.112)

$$\eta_\infty = \eta_c\left[1 + \frac{5(\lambda-1)}{(2\lambda+3)}\phi\right]$$ (15.113)

Equation 15.112 is the well-known Taylor law [4] for the viscosity of dilute emulsions of spherical droplets.

It should also be noted that as long as the film is characterized by nonzero K_s and B_s, the low-frequency limiting viscosity is always given by:

$$\eta_0 = \eta_c\left[1 + \frac{5}{2}\phi\right]$$ (15.114)

Equation 15.114 is the well-known Einstein law [13,14] for the viscosity of dilute suspensions of rigid spheres.

Figure 15.21 shows the dynamic moduli and Cole–Cole plot of an emulsion predicted from Equation 15.108 for different values of bending rigidity M of

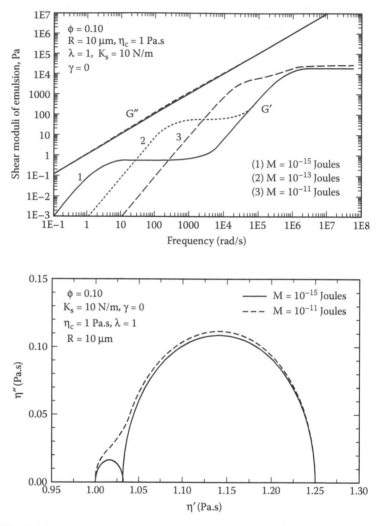

FIGURE 15.21 Dynamic moduli and Cole–Cole diagram for an emulsion of liquid droplets covered with film possessing bending rigidity.

the film (J_0 is taken to be zero). The emulsion exhibits two relaxation processes. The elastic behavior in the high-frequency domain is due to form relaxation; this is governed by interfacial tension and bending rigidity M. In the low-frequency domain, the elastic behavior of the emulsion is due to interfacial relaxation governed by the surface dilational modulus (K_s is 10 N/m for the example under consideration). With an increase in the bending rigidity M of the film, the lower plateau value of the emulsion storage modulus G' increases. The upper plateau value of the storage modulus (G') of the emulsion and the value of the loss modulus G'' of the emulsion are not affected significantly when the bending rigidity of the film is increased.

15.3.5 DROPLETS COATED WITH LIQUID FILMS

Oosterbroek and Mellema [8] studied the linear viscoelastic behavior of emulsions composed of droplets covered with films of another immiscible liquid. Figure 15.22 shows a droplet whose interfacial film consists of a liquid and has a finite thickness. The inner core drop has a radius of a_c and the liquid film thickness is $(b - a_c)$. The external continuous phase (fluid 1) has a viscosity of η_1, the film liquid (fluid 2) has a viscosity of η_2, and the inner droplet liquid (fluid 3) has a viscosity of η_3. The interfacial tensions between different fluids are γ_{12} between fluids 1 and 2, and γ_{23} between fluids 2 and 3.

Using a cell model approach, Oosterbroek and Mellema [8] derived an expression for the complex viscosity of emulsions composed of droplets covered with films of another immiscible liquid. For dilute emulsions, the complex viscosity expression developed by Oosterbroek and Mellema [8] is as follows:

$$\eta^* = \eta_c \left[1 + \phi \left(\frac{\alpha_0 + \beta_0 H + \mu_0 H^2}{\alpha_0' + \beta_0' H + \mu_0' H^2} \right) \right] \tag{15.115}$$

where

$$\alpha_0 = 5\, FD^2 \left[96\, \lambda_{21} + (160\, \lambda_{31} - 17856\, \lambda_{21} + 64)\, D \right] \tag{15.116}$$

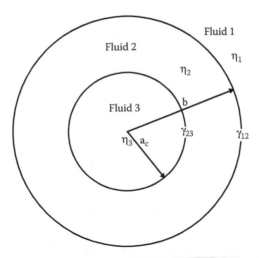

FIGURE 15.22 Droplet of one liquid covered with a film of another immiscible liquid.

$$\alpha_0' = 2\,FD^2\,[96\,\lambda_{21} + (160\,\lambda_{31} - 17856\,\lambda_{21} + 169)\,D] \qquad (15.117)$$

$$\beta_0 = \left[40(F+1)\left(\frac{5}{2}\lambda_{31} + 1\right)\lambda_{21}\right] + 5[-(3748 + 3740F)\lambda_{21}\lambda_{31}$$
$$+ 40(F+1)\lambda_{31} + 88(F+1)\lambda_{21}^2 - (1504 + 1568F)]D \qquad (15.118)$$

$$\beta_0' = [40(F+1)\,(\lambda_{31} + 1)\,\lambda_{21}] + 2\,[-(3748 + 3740\,F)\,\lambda_{21}\,\lambda_{31}$$
$$+ 100\,(F+1)\,\lambda_{31} + 88\,(F+1)\lambda_{21}^2 - (3760 + 3728\,F)]\,D \qquad (15.119)$$

$$\mu_0 = 5[19\lambda_{31}^2\lambda_{21} - 3\lambda_{31}\lambda_{21} - 16\lambda_{21} + D(-3572\lambda_{31}^2\lambda_{21}$$
$$+ 38\lambda_{31}^2 + 95\lambda_{31}\lambda_{21}^2 + 543\lambda_{31}\lambda_{21} - 80\lambda_{31}$$
$$- 32\lambda_{21}^2 - 3008\lambda_{21})] \qquad (15.120)$$

$$\mu_0' = 2\left[19\lambda_{31}^2\lambda_{21} + \frac{89}{2}\lambda_{31}\lambda_{21} + 24\lambda_{21} + D(-3572\lambda_{31}^2\lambda_{21}\right.$$
$$+ 95\lambda_{31}^2 + 95\lambda_{31}\lambda_{21}^2 - \frac{16717}{2}\lambda_{21}\lambda_{31} + 120\lambda_{31} \qquad (15.121)$$
$$\left. + 120\lambda_{21}^2 - 4512\lambda_{21}\right]$$

$$\lambda_{21} = \eta_2/\eta_1 \qquad (15.122)$$

$$\lambda_{31} = \eta_3/\eta_1 \qquad (15.123)$$

$$F = \gamma_{23}/\gamma_{12} \qquad (15.124)$$

$$D = (b - a_c)/b \qquad (15.125)$$

$$H = j\,b\,\eta_1\,\omega/\gamma_{12} \qquad (15.126)$$

The complex viscosity expression (Equation 15.115) is an approximate solution, valid for small values of the relative film thickness D and viscosity ratio λ_{21}. In the limit $D \to 0$, that is, the film is of very small thickness, the low-frequency viscosity of the emulsion becomes:

$$\eta_0 = \eta_c \left[1 + \phi \left(\frac{\alpha_0}{\alpha_0'} \right) \right] = \eta_c \left[1 + \frac{5}{2} \phi \right] \qquad (15.127)$$

Thus, at low frequencies, emulsions of droplets covered with thin films of another immiscible liquid behave as suspensions of rigid particles regardless of the viscosity of the inner droplet liquid.

Figure 15.23 shows the dynamic moduli of the emulsion predicted from Equation 15.115. The corresponding Cole–Cole diagram is also shown. Clearly,

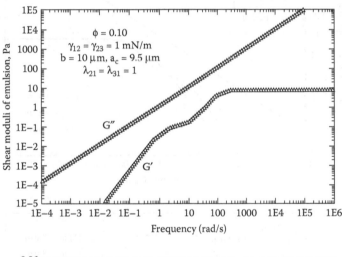

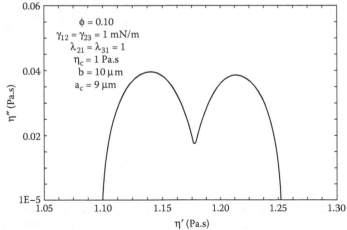

FIGURE 15.23 Dynamic moduli and Cole–Cole diagram for an emulsion of droplets of one liquid covered with a film of another immiscible liquid.

the emulsion exhibits two relaxation processes — one corresponds to relaxation of the thin interfacial film surrounding the droplet and the other corresponds to relaxation of the shape of the droplet. Figure 15.24 shows the effect of film thickness on the dynamic behavior of the emulsion. With an increase in the film thickness, that is, decrease in a_c keeping b constant, the larger of the two relaxation times decreases; the smaller relaxation time appears to be unaffected by the film thickness. The Cole–Cole plot indicates a merging of the two frequency domains with the increase in the film thickness.

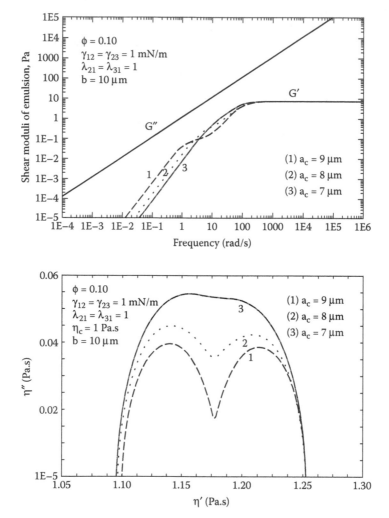

FIGURE 15.24 The effect of film thickness on the dynamic moduli and Cole–Cole plot of an emulsion of droplets of one liquid covered with a film of another immiscible liquid.

NOTATION

A	surface area
A_0	surface area of a particle
$\bar{A}$	parameter given by Equation 15.86
A_i	coefficients given by Equation 15.89a
A_i'	coefficients given by Equation 15.89a
a	radius defined as $R/(\phi)^{1/3}$, where R is the particle radius and ϕ is the volume fraction of particles in a dispersion (see Figure 15.1)
a_c	radius of the core of an emulsion droplet covered with film of another immiscible liquid
$\bar{B}$	parameter given by Equation 15.87
B_i	coefficients given by Equation 15.89b
B_i'	coefficients given by Equation 15.89c to Equation 15.89e
B_s	parameter given by Equation 15.110
b	radius of a large spherical region filled with effective homogeneous medium whose macroscopic properties are to be calculated (see Figure 15.1); also outer radius of interfacial film (Figure 15.22)
$\bar{C}$	parameter given by Equation 15.88
C_i	coefficients given by Equation 15.89f and Equation 15.89g
C_i'	coefficients given by Equation 15.89h to Equation 15.89l
D	relative film thickness defined by Equation 15.125
E	Young's modulus of a film
$\bar{\bar{E}}$	rate of strain tensor
$\bar{\bar{E}}_s$	surface rate of strain tensor
e_0	oscillatory strain amplitude
e^*	complex strain
$\bar{\bar{e}}$	strain tensor
$\bar{\bar{e}}_d$	strain tensor inside the particle
$\bar{\bar{e}}_s$	surface strain tensor
$\bar{\bar{e}}_s^*$	complex surface strain tensor
F	ratio of interfacial tensions; see Equation 15.124
$\vec{F}_n$	normal restoring force per unit area due to the membrane
$\vec{F}_t$	tangential restoring force per unit area due to the membrane
f	$= j\omega R\eta_c$; see Equation 15.109
G_d	shear modulus of a particle
G_s	surface shear modulus
G^*	complex shear modulus
G_c^*	complex shear modulus of continuous phase
G_d^*	complex shear modulus of dispersed phase
G_s^*	complex surface shear modulus
G'	storage (shear) modulus

G''	loss (shear) modulus
h	film thickness
J_0	spontaneous curvature of a film; see Equation 15.110
j	imaginary number
K_s	surface dilational modulus
K_s^*	complex surface dilational modulus
M	bending or flexural rigidity of a film; see Equation 15.94
N	number of particles
$\hat{n}$	outward unit normal vector
N_{Bo}^k	dilational Boussinesq number
N_{Bo}^{η}	shear Boussinesq number
P	pressure
P_c	pressure in continuous phase (matrix)
P_d	pressure in dispersed phase (particle/droplet)
$\bar{\bar{P}}_s$	surface stress tensor
$\bar{\bar{P}}_s^*$	complex surface stress tensor
Q	parameter given by Equation 15.92
R	particle or droplet radius
r	radial distance
$\vec{r}$	radial position vector
$\bar{\bar{S}}$	dipole strength of a particle or droplet
$\bar{\bar{S}}^0$	dipole strength of a single particle or droplet located in an infinite matrix fluid
t	time
$\hat{t}$	unit tangent vector at the interface
$\vec{u}$	velocity vector
$\vec{u}_c$	velocity vector in continuous phase (matrix)
$\vec{u}_d$	velocity vector in dispersed phase (particle or droplet)
$\vec{u}_m$	membrane velocity vector
$\vec{u}_\infty$	velocity vector of unperturbed flow far away from the particle or droplet
V	volume
W	parameter given by Equation 15.93
$\vec{x}$	position vector
Y	parameter given by Equation 15.109b
Z	parameter given by Equation 15.109c

GREEK

γ	interfacial tension
γ_{12}	interfacial tension between fluid 1 and fluid 2
γ_{23}	interfacial tension between fluid 2 and fluid 3
$\dot{\gamma}$	shear rate
$\overline{\overline{\Gamma}}_{\infty}$	velocity gradient tensor of unperturbed flow far away from the particle or droplet
∇	del operator
∇_s	surface gradient operator
∇^2	Laplacian
$\overline{\overline{\delta}}$	unit tensor
$\overline{\overline{\delta}}_s$	surface unit tensor
$\dot{\varepsilon}$	elongation rate
η	viscosity
η_1	viscosity of fluid 1
η_2	viscosity of fluid 2
η_3	viscosity of fluid 3
η_c	continuous-phase viscosity
η_d	dispersed-phase viscosity
η_0	zero shear or low-frequency limiting viscosity
η_s	surface shear viscosity
η_{∞}	high-frequency or high-shear limiting viscosity
η_s^k	surface dilational viscosity
η^*	complex viscosity
λ	viscosity ratio
λ_{21}	viscosity ratio defined as η_2/η_1
λ_{31}	viscosity ratio defined as η_3/η_1
v_1	square of relaxation time; see Equation 15.65
v_2	square of retardation time; see Equation 15.65
$\overline{\overline{\sigma}}$	total stress tensor
$\overline{\overline{\sigma}}_c$	total stress tensor in continuous phase (matrix)
$\overline{\overline{\sigma}}_d$	total stress tensor in dispersed phase (particle or droplet)
τ_1	relaxation time
τ_2	retardation time
τ_c	relaxation time of continuous phase
τ_d	relaxation time of dispersed phase
$\overline{\overline{\tau}}$	deviatoric stress tensor
$\overline{\overline{\tau}}_d$	deviatoric stress tensor inside the particle
ϕ	volume fraction of particles or droplets
ω	frequency of oscillation of strain or stress
$\langle X \rangle$	volume average of quantity X

REFERENCES

1. H. Frohlich, R. Sack, *Proc. R. Soc. Lond. A*, 185: 415–430, 1946.
2. J.G. Oldroyd, in *Rheology of Disperse Systems*, C.C. Mill, Ed., Pergamon Press, London, 1959, pp. 1–15.
3. J.G. Oldroyd, *Proc. R. Soc. Lond. A*, 218: 122–132, 1953.
4. G.I. Taylor, *Proc. R. Soc. Lond. A*, 138: 41–48, 1932.
5. J.F. Palierne, *Rheol. Acta*, 29, 204–214, 1990.
6. J.G. Oldroyd, *Proc. R. Soc. Lond. A*, 232: 567–577, 1955.
7. A. Nadim, *Chem. Eng. Commun.*, 148–150: 391–407, 1996.
8. M. Oosterbroek. J. Mellema, *J. Colloid Interface Sci.*, 84: 14–35, 1981.
9. R.A. de Bruijn, J. Mellema, *Rheol. Acta*, 24: 159–174, 1985.
10. J.B.A.F. Smenlders, C. Blom, J. Mellema, *Phys. Rev. A*, 42: 3483–3498, 1990.
11. J.B.A.F. Smenlders, J. Mellema, C. Blom, *Phys. Rev. A*, 46: 7708–7722, 1992.
12. K. Seki, S. Komura, *Physica A*, 219: 253–289, 1995.
13. A. Einstein, *Ann. Phys.*, 19: 289–306, 1906.
14. A. Einstein, *Ann. Phys.*, 34: 591–592, 1911.

16 Dynamic Viscoelastic Behavior of Composites

16.1 INTRODUCTION

Composite materials often contain polymer as one of the two phases (dispersed phase and matrix phase). The polymeric phase generally imparts viscoelastic behavior to the composite [1]. For example, the EPDM rubber (EPDM refers to ethylene-propylene-diene monomer) is being increasingly used in many applications because of its excellent oxidation resistance and electrical properties [2]. As the EPDM rubber, similar to many other polymeric materials, is viscoelastic in nature, the composites prepared from EPDM rubber as one of the phases are also viscoelastic. Thus, a good understanding of the effective properties of viscoelastic composites is of significant industrial interest [3–14].

In this chapter, the macroscopic dynamic behavior of viscoelastic composites is described in terms of the effective complex moduli.

16.2 ELASTIC-VISCOELASTIC CORRESPONDENCE PRINCIPLE

The elastic-viscoelastic correspondence principle states that the expressions for the effective complex moduli of a viscoelastic heterogeneous material can be obtained from the corresponding expressions for the effective elastic moduli of an associated heterogeneous elastic material (with identical phase geometry) simply by replacing phase elastic moduli by phase complex moduli [6,15]. The correspondence principle is based on the fact that the governing equations and the boundary conditions for elastic and viscoelastic problems are the same; only the stress–strain relations in the two cases are different.

16.3 COMPLEX MODULI OF VISCOELASTIC PARTICULATE COMPOSITES

Using the elastic-viscoelastic correspondence principle, the expressions for the effective elastic moduli of particulate composites (given in Chapter 12) can be converted into the corresponding expressions for the complex moduli of viscoelastic particulate composites.

For dilute composites of spherical particles, the results are as follows:

$$\frac{G^*}{G_c^*} = 1 + \left[\frac{15(1 - v_c^*)(G_d^* - G_c^*)}{2G_d^*(4 - 5v_c^*) + G_c^*(7 - 5v_c^*)} \right] \phi \qquad (16.1)$$

$$\frac{K^*}{K_c^*} = 1 + \left[\left(\frac{3K_c^* + 4G_c^*}{3K_c^*} \right) \left(\frac{3K_d^* - 3K_c^*}{3K_d^* + 4G_c^*} \right) \right] \phi \tag{16.2}$$

where G^*, G_c^*, and G_d^* are the complex shear moduli of composite, continuous phase, and dispersed phase, respectively, v_c^* is the complex Poisson's ratio of the continuous phase, and K^*, K_c^*, and K_d^* are the complex bulk moduli of composite, continuous phase, and dispersed phase, respectively. The definitions of complex shear modulus, complex Poisson's ratio, and complex bulk modulus are:

$$G^* = G' + jG'' \tag{16.3}$$

$$v^* = v' + jv'' \tag{16.4}$$

$$K^* = K' + jK'' \tag{16.5}$$

where single-primed quantities are the storage components, double-primed quantities are the loss components, and j is the imaginary number ($\sqrt{-1}$).

For the case where (1) the particles are purely elastic and (2) the matrix is elastic in dilation and viscoelastic in shear, one can write:

$$G_d^* = G_d \tag{16.6}$$

$$K_d^* = K_d \tag{16.7}$$

$$K_c^* = K_c \tag{16.8}$$

$$G_c^* = G_c' + jG_c'' \tag{16.9}$$

Introducing these expressions for G_d^*, K_d^*, K_c^*, and G_c^* into Equation 16.1 and Equation 16.2, and making the additional assumption that $\tan^2 \delta_{Gc} = (G_c''/G_c')^2 \ll 1$, the following results are obtained for the bulk moduli [6]:

$$K' = K_c + \frac{(K_d - K_c)(4G_c' + 3K_c)}{(4G_c' + 3K_d)} \phi \tag{16.10a}$$

$$K'' = 12 \left[\frac{K_d - K_c}{4G_c' + 3K_d} \right]^2 G_c'' \phi \tag{16.10b}$$

where K' and K'' are storage and loss bulk moduli, respectively, of the composite. Interestingly, the composite is viscoelastic in dilation (K' and K'' are nonzero) even though the dispersed phase and matrix are purely elastic in dilation. The shear viscoelasticity of the matrix introduces dilational viscoelasticity in the composite.

For the case where the particles are elastic and the matrix is incompressible and viscoelastic in shear, the expressions for the complex bulk and shear moduli are as follows:

$$K^* = K' + jK'' = \frac{3K_d + 4(G'_c + jG''_c)}{3\phi} \tag{16.11}$$

$$K' = \frac{3K_d + 4G'_c}{3\phi} \tag{16.12a}$$

$$K'' = \frac{4G''_c}{3\phi} \tag{16.12b}$$

$$\frac{G^*}{G^*_c} = 1 + \frac{5}{2}\left[\frac{G_d - G^*_c}{G_d + \frac{3}{2}G^*_c}\right]\phi \tag{16.13}$$

Figure 16.1 shows the shear moduli of the matrix and dilational (bulk) moduli of the composite. The dilational moduli are calculated from Equation 16.12a and Equation 16.12b. The matrix is assumed to follow the Maxwell model, that is:

$$G'_c = G_c\omega^2\tau^2_{gc}/(1 + \omega^2\tau^2_{gc}) \tag{16.14}$$

$$G''_c = G_c\omega\tau_{gc}/(1 + \omega^2\tau^2_{gc}) \tag{16.15}$$

where G_c is the high-frequency limiting value of shear modulus, ω is the frequency, and τ_{gc} is the shear relaxation time of the matrix. Again, shear viscoelasticity of matrix results in dilational viscoelasticity of the composite.

When the particles are viscoelastic in both shear and dilation, and the matrix is incompressible and viscoelastic in shear, the expressions for the complex bulk and shear moduli of the composite become:

$$K^* = \frac{3K^*_d + 4G^*_c}{3\phi} \tag{16.16}$$

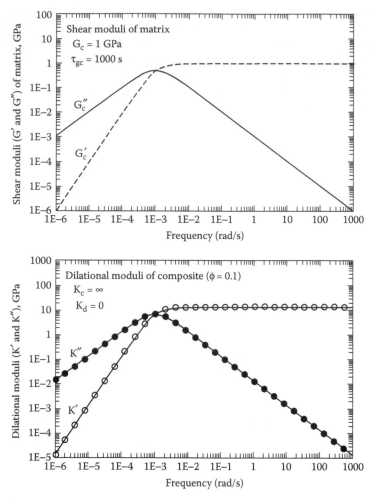

FIGURE 16.1 Shear viscoelasticity of the matrix results in dilational viscoelasticity of the composite.

$$K' = \frac{3K_d' + 4G_c'}{3\phi} \tag{16.17}$$

$$K'' = \frac{3K_d'' + 4G_c''}{3\phi} \tag{16.18}$$

$$\frac{G^*}{G_c^*} = 1 + \frac{5}{2}\left[\frac{G_d^* - G_c^*}{G_d^* + \frac{3}{2}G_c^*}\right]\phi \tag{16.19}$$

From Equation 16.19, the expressions for the storage and loss shear moduli (G' and G'') of the composite can be obtained:

$$G' = (1+\phi A_1)G'_c - \phi A_2 G''_c \tag{16.20}$$

$$G'' = \phi A_2 G'_c + (1+\phi A_1)G''_c \tag{16.21}$$

where A_1 and A_2 are given as:

$$A_1 = 5\left[\frac{ac+bd}{c^2+d^2}\right] \tag{16.22}$$

$$A_2 = 5\left[\frac{bc-ad}{c^2+d^2}\right] \tag{16.23}$$

$$a = G'_d - G'_c, \quad b = G''_d - G''_c \tag{16.24}$$

$$c = 2G'_d + 3G'_c, \quad d = 2G''_d + 3G''_c \tag{16.25}$$

Figure 16.2a and Figure 16.2b show the plots of bulk moduli of the composite calculated from Equation 16.17 and Equation 16.18. The shear moduli of the

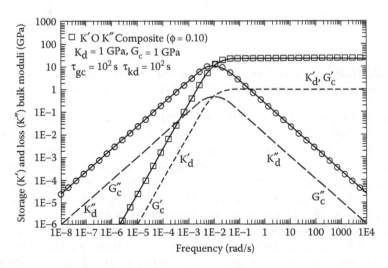

FIGURE 16.2a Bulk moduli of the composite when the shear relaxation time of the matrix (τ_{gc}) is equal to the dilational relaxation time of the particles (τ_{kd}).

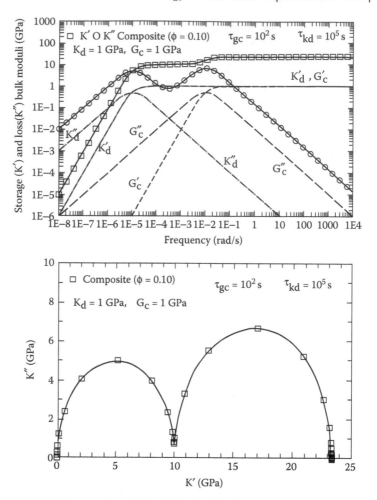

FIGURE 16.2b Bulk moduli and Cole–Cole diagram for composite when the shear relaxation time of the matrix and dilational relaxation time of the particles are different.

matrix phase and the bulk moduli of the matrix phase are assumed to follow the Maxwell model. When the shear relaxation time of the matrix (τ_{gc}) and the dilational relaxation time of the particles (τ_{kd}) are equal, the composite exhibits a single relaxation time (see Figure 16.2a). When the shear relaxation time of the matrix and dilational relaxation time of the particles are different, the composite exhibits two relaxation times as can be seen in Figure 6.2b. The Cole–Cole plot (K'' vs. K' plot) clearly shows two frequency domains.

Figure 16.3 shows the effects of K_d (high-frequency limiting value of the dispersed phase bulk modulus) and G_c (high-frequency limiting value of the matrix shear modulus) on the bulk moduli of the composite. The shear relaxation time of the matrix is taken to be equal to the dilational relaxation time of the particulate

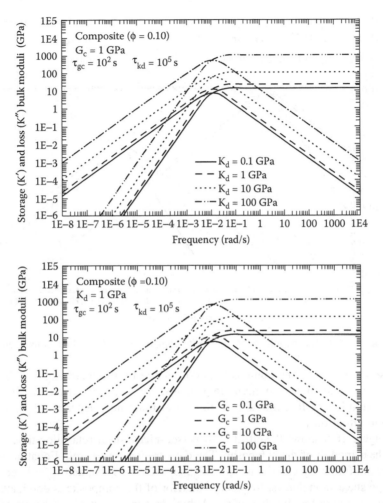

FIGURE 16.3 The effects of K_d (dispersed-phase bulk modulus at high frequency) and G_c (matrix shear modulus at high frequency) on the bulk moduli of the composite.

phase. With an increase in K_d, the bulk moduli K' and K'' of the composite increase as shown in the top portion of Figure 16.3. The bulk moduli of the composite (K' and K'') also increase with the increase in shear modulus of the matrix G_c, as can be seen in the bottom portion of Figure 16.3.

Figure 16.4 shows the plots of storage and loss shear moduli of viscoelastic composite of Maxwellian components obtained from Equation 16.19 under the following conditions: $\phi = 0.10$, $G_c = 10^{10}$ Pa, $G_d = 10^7$ Pa, $\tau_{gc} = 100$ sec, and $\tau_{gd} = 10^{-5}$ sec. Clearly, the composite response is dominated by the matrix material properties; the storage modulus values are closer to that of the matrix. However,

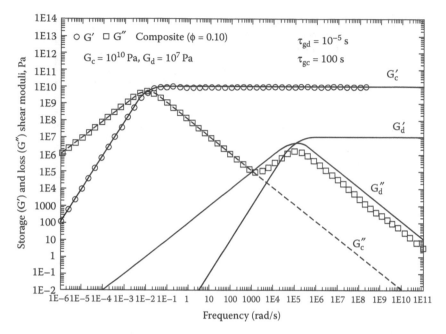

FIGURE 16.4 Shear moduli of the composite composed of Maxwellian components.

the loss modulus of the composite is dominated by the matrix only at low to moderate frequencies; at high frequencies, the composite loss modulus is dominated by the loss modulus of the softer material (dispersed phase in the present situation).

Figure 16.5 shows the effect of dispersed-phase shear relaxation time τ_{gd} on the shear moduli of the composite. The data are generated from Equation 16.19 under the following conditions: $\phi = 0.10$, $G_c = G_d = 10^8$ Pa, and $\tau_{gc} = 1$ sec. The storage shear modulus vs. frequency behavior of the composite is dominated by the matrix phase for all values of τ_{gd}. At high frequencies, the loss shear modulus of the composite deviates from the loss shear modulus of the matrix phase. With a decrease in τ_{gd}, the deviation of the composite loss modulus from the matrix loss modulus increases; the frequency at which the deviation occurs also increases with decrease in τ_{gd}.

The effect of G_d (dispersed-phase shear modulus) on the shear moduli of the composite is shown in Figure 16.6 under the following conditions: $\phi = 0.10$, $G_c = 10^5$ Pa, and $\tau_{gc} = \tau_{gd} = 1$ sec. With an increase in G_d, the storage shear modulus vs. frequency plot shifts toward higher modulus values. The loss modulus vs. frequency plot also shifts toward higher modulus values. Thus, both storage and loss shear moduli increase with increase in G_d.

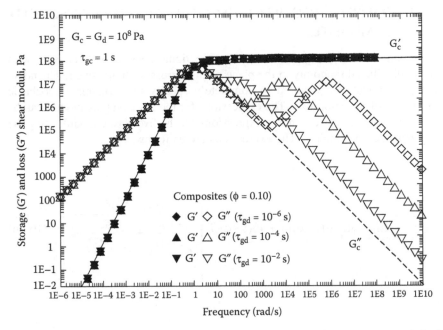

FIGURE 16.5 The effect of dispersed-phase relaxation time (τ_{gd}) on the shear moduli of the composite.

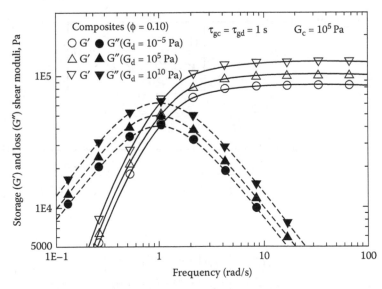

FIGURE 16.6 Effect of dispersed-phase shear modulus (G_d) on the shear moduli of the composite.

16.4 COMPLEX MODULI OF VISCOELASTIC FIBER-REINFORCED COMPOSITES

Assuming that the composite is of a unidirectional continuous-fiber type, with fibers distributed randomly in the plane of their cross-sections, five independent complex moduli are required to completely specify the dynamic viscoelastic properties of the composite. The expressions for the five complex moduli can be obtained from the corresponding expressions for the five elastic moduli using the correspondence principle. The expressions are as follows:

$$E_{11}^* = \phi E_f^* + (1-\phi)E_c^* + 4\phi(1-\phi)G_c^* \left[\dfrac{(v_f^* - v_c^*)^2}{\dfrac{(1-\phi)G_c^*}{K_f^* + \dfrac{G_f^*}{3}} + \dfrac{\phi G_c^*}{K_c^* + \dfrac{G_c^*}{3}} + 1} \right] \quad (16.26)$$

$$v_{12}^* = \phi v_f^* + (1-\phi)v_c^* + \dfrac{\phi(1-\phi)(v_f^* - v_c^*)\left[\dfrac{G_c^*}{K_c^* + \dfrac{G_c^*}{3}} - \dfrac{G_c^*}{K_f^* + \dfrac{G_f^*}{3}} \right]}{\left[\dfrac{(1-\phi)G_c^*}{K_f^* + \dfrac{G_f^*}{3}} + \dfrac{\phi G_c^*}{K_c^* + \dfrac{G_c^*}{3}} + 1 \right]} \quad (16.27)$$

$$K_{23}^* = K_c^* + \dfrac{G_c^*}{3} + \dfrac{\phi}{\left[\dfrac{1}{K_f^* - K_c^* + \dfrac{1}{3}(G_f^* - G_c^*)} + \dfrac{1-\phi}{K_c^* + \dfrac{4}{3}G_c^*} \right]} \quad (16.28)$$

$$\dfrac{G_{12}^*}{G_c^*} = \left[\dfrac{G_f^*(1+\phi) + G_c^*(1-\phi)}{G_f^*(1-\phi) + G_c^*(1+\phi)} \right] \quad (16.29)$$

$$\frac{G_{23}^*}{G_c^*} = 1 + \left[\left(\frac{G_c^*}{G_f^* - G_c^*} \right) + \frac{\phi}{\dfrac{\left(K_c^* + \dfrac{7}{3} G_c^* \right)}{\left(2K_c^* + \dfrac{8}{3} G_c^* \right)} (1 - \phi)} \right] \qquad (16.30)$$

where E_{11}^* is the complex longitudinal Young's modulus, v_{12}^* is the complex longitudinal Poisson's ratio, K_{23}^* is the complex transverse bulk modulus, G_{12}^* is the complex longitudinal shear modulus, and G_{23}^* is the complex transverse shear modulus. The quantities with subscript c are continuous-phase properties, and those with subscript f are fiber properties.

For the case where the fibers are purely elastic and the matrix (continuous phase) is elastic in dilation and viscoelastic in shear:

$$G_f^* = G_f \qquad (16.31a)$$

$$K_f^* = K_f \qquad (16.31b)$$

$$E_f^* = E_f \qquad (16.31c)$$

$$v_f^* = v_f \qquad (16.31d)$$

$$K_c^* = K_c \qquad (16.32a)$$

$$G_c^* = G_c' + jG_c'' \qquad (16.32b)$$

$$E_c^* = E_c' + jE_c'' \qquad (16.32c)$$

$$v_c^* = v_c' + jv_c'' \qquad (16.32d)$$

Introducing the preceding expressions for G_f^*, K_f^*, E_f^*, v_f^*, K_c^*, G_c^*, E_c^*, and v_c^* into Equation 16.26 to Equation 16.30 and making some additional approximations, the following results are obtained [15]:

$$E_{11}^* = E_{11}' + jE_{11}'' = \phi E_f + (1-\phi)E_c^* \qquad (16.33a)$$

$$E_{11}' = \phi E_f + (1-\phi)E_c' \qquad (16.33b)$$

$$E_{11}'' = (1-\phi)E_c'' \qquad (16.33c)$$

$$\tan \delta_{E11} = \frac{E_{11}''}{E_{11}'} = \frac{E_c''(1-\phi)}{\phi E_f + (1-\phi)E_c'}$$

$$= \frac{\tan \delta_{Ec}}{1 + \left(\dfrac{\phi}{1-\phi}\right)\left(\dfrac{E_f}{E_c'}\right)} \qquad (16.34)$$

$$v_{12}^* = v_{12}' + jv_{12}'' = \phi v_f + (1-\phi)v_c^* \qquad (16.35a)$$

$$v_{12}' = \phi v_f + (1-\phi)v_c' \qquad (16.35b)$$

$$v_{12}'' = \phi v_f + (1-\phi)v_c'' \qquad (16.35c)$$

$$v_c^* = \frac{3K_c - 2G_c^*}{2[3K_c + G_c^*]} \qquad (16.36a)$$

$$v_c' = \frac{(3K_c - 2G_c')(3K_c + G_c') - 2(G_c'')^2}{2[(3K_c + G_c')^2 + (G_c'')^2]} \qquad (16.36b)$$

$$v_c'' = \frac{-G_c''[(3K_c - 2G_c') + 2(3K_c + G_c')]}{2[(3K_c + 2G_c')^2 + (G_c'')^2]}$$

(16.36c)

$$K_{23}' = \left(K_c + \frac{G_c'}{3} \right) \left[1 + \frac{\phi}{\dfrac{\left(K_c + \dfrac{G_c'}{3} \right)}{\left(K_f + \dfrac{G_f}{3} \right) - \left(K_c + \dfrac{G_c'}{3} \right)} + \dfrac{\left(K_c + \dfrac{G_c'}{3} \right)(1-\phi)}{\left(K_c + \dfrac{4}{3}G_c' \right)}} \right]$$

(16.37a)

$$K_{23}'' = \frac{1}{3}G_c'' \left[1 - \frac{\left(K_c + \dfrac{4}{3}G_c' \right)^2 - 4(1-\phi)\left(K_f - K_c - \dfrac{G_c'}{3} \right)^2}{\left\{ K_c + \dfrac{4}{3}G_c' + (1-\phi)\left(K_f - K_c - \dfrac{G_c'}{3} \right) \right\}^2} \phi \right]$$

(16.37b)

$$\frac{G_{12}'}{G_c'} = \left[\frac{\dfrac{G_f}{G_c'}(1-\phi) + (1-\phi)}{\dfrac{G_f}{G_c'}(1-\phi) + (1+\phi)} \right]$$

(16.38a)

$$\frac{G_{12}''}{G_c''} = \left[\frac{\left(1 + \dfrac{G_f}{G_c'} \right)^2 + \phi\left(\dfrac{G_f}{G_c'} - 1 \right)^2}{\left\{ \dfrac{G_f}{G_c'}(1-\phi) + (1+\phi) \right\}^2} \right](1-\phi)$$

(16.38b)

$$G_{23}' = G_c' \left[1 + \frac{a_1\phi}{1-\phi} \right] - \left(G_c'' \right)^2 \left[\frac{a_2\phi}{1-\phi} \right]$$

(16.39a)

$$G_{23}'' = G_c'' \left[1 + \left(\frac{a_1 \phi}{1 - \phi} \right) + \left(\frac{a_2 \phi}{1 - \phi} \right) G_c' \right] \tag{16.39b}$$

where

$$a_1 = \left[\frac{(6K_c + 8G_c')(3K_c + 7G_c') + 56(G_c'')^2}{(3K_c + 7G_c')^2 + (7G_c'')^2} \right] \tag{16.39c}$$

$$a_2 = \left[\frac{8(3K_c + 7G_c') - 7(6K_c + 8G_c')}{(3K_c + 7G_c')^2 + (7G_c'')^2} \right] \tag{16.39d}$$

In arriving at Equation 16.33a, the third term on the right-hand side of Equation 16.26 is neglected. The third term is generally insignificant in comparison with the first two terms of Equation 16.26. If fibers are much stiffer than the matrix, that is, if E_f/E_c' is large, Equation 16.34 indicates that $\tan \delta_{E11} \ll \tan \delta_{Ec}$. This implies that the viscoelastic effect in axial stressing is insignificant if E_f/E_c' is large. In arriving at Equation 16.35a, the third term on the right-hand side of Equation 16.27 is neglected. Equation 16.37a and Equation 16.37b involve the assumption that $\tan^2 \delta_{Gc} \ll 1$. The approximation $\tan^2 \delta_{Gc} \ll 1$ is also made in arriving at Equation 16.38a and Equation 16.38b. Equation 16.39a to Equation 16.39d are obtained from Equation 16.30 assuming that the fibers are rigid ($K_f \rightarrow \infty$, $G_f \rightarrow \infty$). Note that when the fibers and matrix are both purely elastic in dilation, the composite is viscoelastic in dilation in that the storage and loss components of the transverse bulk modulus K_{23}^* are not zero. Clearly, the shear viscoelasticity of the matrix introduces dilational viscoelasticity in the composite.

As an example, consider a matrix which is elastic in dilation with $K_c = 1$ GPa and Maxwellian viscoelastic in shear with G_c (the high-frequency limiting value of shear modulus) of 10 GPa and relaxation time of $\tau_{gc} = 1000$ sec. The frequency response of this matrix material is shown in Figure 16.7. The Poisson ratio as well as the Young modulus of the matrix exhibit viscoelastic behavior as can be seen in the figure. The frequency response of a fiber-reinforced composite prepared from this matrix at a fiber volume fraction of 0.1 is shown in Figure 16.8. The fibers are purely elastic, with K_f and G_f values as indicated in the figure. The composite values are calculated from the preceding equations. All the rheological properties of the composite (longitudinal Young moduli, longitudinal shear moduli, longitudinal Poisson ratio, transverse bulk moduli, and transverse shear moduli) exhibit viscoelastic behavior.

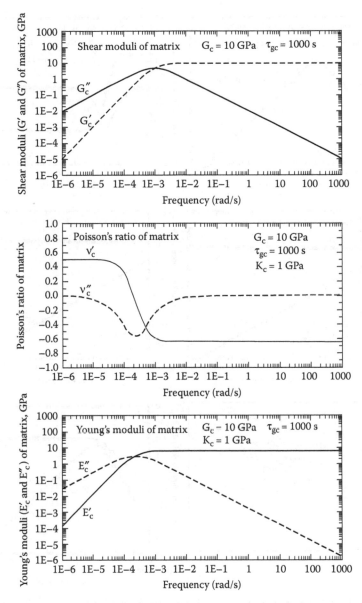

FIGURE 16.7 Frequency response of a matrix that is elastic in dilation and viscoelastic in shear.

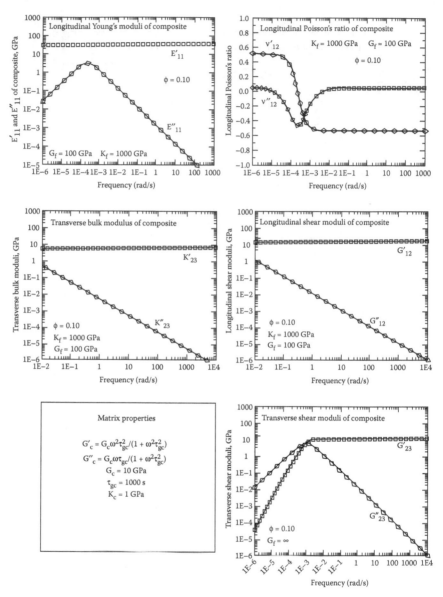

FIGURE 16.8 Frequency response of a fiber-reinforced composite prepared from a matrix that is elastic in dilation and viscoelastic in shear (fibers are assumed to be elastic in both dilation and shear).

NOTATION

E	Young's modulus
E^*	complex Young's modulus
E'	storage Young's modulus
E''	loss Young's modulus
E_{11}^*	complex longitudinal Young's modulus
E_{11}'	storage longitudinal Young's modulus
E_{11}''	loss longitudinal Young's modulus
G	shear modulus
G^*	complex shear modulus
G'	storage shear modulus
G''	loss shear modulus
G_{12}^*	complex longitudinal shear modulus
G_{12}'	storage longitudinal shear modulus
G_{12}''	loss longitudinal shear modulus
G_{23}^*	complex transverse shear modulus
G_{23}'	storage transverse shear modulus
G_{23}''	loss transverse shear modulus
K	bulk (dilational) modulus
K^*	complex bulk (dilational) modulus
K'	storage bulk (dilational) modulus
K''	loss bulk (dilational) modulus
K_{23}^*	complex transverse bulk (dilational) modulus
K_{23}'	storage transverse bulk (dilational) modulus
K_{23}''	loss transverse bulk (dilational) modulus

GREEK

δ_{E11}	$\tan^{-1}(E_{11}''/E_{11}')$
δ_{Ec}	$\tan^{-1}(E_c''/E_c')$
δ_G	$\tan^{-1}(G''/G')$
δ_{Gc}	$\tan^{-1}(G_c''/G_c')$
ϕ	volume fraction of the dispersed phase
ν^*	complex Poisson's ratio
ν'	storage Poisson's ratio
ν''	loss Poisson's ratio
ν_{12}^*	complex longitudinal Poisson's ratio
ν_{12}'	storage longitudinal Poisson's ratio
ν_{12}''	loss longitudinal Poisson's ratio
ω	frequency

τ	relaxation time
τ_g	shear relaxation time
τ_k	bulk (dilational) relaxation time

SUBSCRIPTS

c	continuous phase
d	dispersed phase
f	fiber

REFERENCES

1. R.M. Christensen, *Mechanics of Composite Materials*, John Wiley & Sons, New York, 1979.
2. M.S. Sohn, K.S. Kim, S.H. Hong, J.K. Kim, *J. Appl. Polym. Sci.*, 87: 1595–1601, 2003.
3. L.C. Brinson, W.S. Lin, *Comp. Struct.*, 41: 353–367, 1998.
4. R.F. Landel, T.L. Smith, *ARS J.*, 31: 599–608, 1961.
5. Z. Hashin, *Appl. Mech., Trans. ASME*, 32E, 630–636, 1965.
6. Z. Hashin, *Int. J. Solids Struct.*, 6: 539–552, 1970.
7. T.B. Lewis, L.E. Nielsen, *J. Appl. Polym. Sci.*, 14: 1449–1471, 1970.
8. P. Mete, N.D. Alberola, *Comp. Sci. Technol.*, 56: 849–853, 1996.
9. L.V. Gibiansky, R. Lakes, *Mech. Mater.*, 25: 79–95, 1997.
10. S. Nikkeski, M. Kudo, T. Masuki, *J. Appl. Polym. Sci.*, 69: 2593–2598, 1998.
11. J.Z. Liang, R.K.Y. Li, S.C. Tjong, *Polym. Int.*, 48: 1068–1072, 1999.
12. A.D. Drozdov, *Math. Comp. Modeling*, 29: 11–25, 1999.
13. J.Z. Liang, R.K.Y. Li, S.C. Tjong, *Polym. Test.*, 19: 213–220, 2000.
14. V.P. Privalko, V.F. Shumsky, E.G. Privalko, V.M. Karaman, R. Walter, K. Friedrich, M.Q. Zhang, M.Z. Rong, *Sci. Technol. Adv. Mater.*, 3: 111–116, 2002.
15. Z. Hashin, *Int. J. Solids Struct.*, 6: 797–807, 1970.

Part VI Appendices

Part VI consists of four appendices. Appendix A gives the equations of fluid mechanics and solid mechanics in different coordinate systems. Appendix B reviews vector and tensor operations. Appendix C describes the Gauss divergence theorem, and Appendix D gives the definition of symmetric deviator of fourth order tensor.

Appendix A

Equations Related to Mechanics of Fluids and Solids in Different Coordinate Systems

A.1 EQUATIONS RELATED TO FLUID MECHANICS

A.1.1 CONTINUITY EQUATION IN DIFFERENT COORDINATE SYSTEMS

($\dfrac{\partial \rho}{\partial t} + \nabla \bullet \rho \vec{u} = 0$, where ρ is density and $\vec{u}$ is velocity vector)

Rectangular coordinates (x, y, z):

$$\frac{\partial \rho}{\partial t} + \frac{\partial}{\partial x}(\rho u_x) + \frac{\partial}{\partial y}(\rho u_y) + \frac{\partial}{\partial z}(\rho u_z) = 0$$

Cylindrical coordinates (r, θ, z):

$$\frac{\partial \rho}{\partial t} + \frac{1}{r}\frac{\partial}{\partial x}(\rho r u_r) + \frac{1}{r}\frac{\partial}{\partial \theta}(\rho u_\theta) + \frac{\partial}{\partial z}(\rho u_z) = 0$$

Spherical coordinates (r, θ, ϕ):

$$\frac{\partial \rho}{\partial t} + \frac{1}{r^2}\frac{\partial}{\partial r}(\rho r^2 u_r) + \frac{1}{r\sin\theta}\frac{\partial}{\partial \theta}(\rho u_\theta \sin\theta) + \frac{1}{r\sin\theta}\frac{\partial}{\partial \phi}(\rho u_\phi) = 0$$

A.1.2   EQUATIONS OF MOTION IN DIFFERENT COORDINATE SYSTEMS

$\left(\rho\left(\dfrac{\partial \vec{u}}{\partial t} + \vec{u} \bullet \nabla \vec{u} \right) = -\nabla P + \rho \vec{g} + \nabla \bullet \overline{\overline{\tau}} \right.$, where ρ is density, $\vec{u}$ is the velocity vector, P is pressure, $\vec{g}$ is acceleration due to gravity, and $\overline{\overline{\tau}}$ is deviatoric stress tensor).

Rectangular coordinates (x, y, z):

x-Direction

$$\rho\left(\frac{\partial u_x}{\partial t} + u_x \frac{\partial u_x}{\partial x} + u_y \frac{\partial u_x}{\partial y} + u_z \frac{\partial u_x}{\partial z} \right) = -\frac{\partial P}{\partial x} + \rho g_x + \left(\frac{\partial \tau_{xx}}{\partial x} + \frac{\partial \tau_{yx}}{\partial y} + \frac{\partial \tau_{zx}}{\partial z} \right)$$

y-Direction

$$\rho\left(\frac{\partial u_y}{\partial t} + u_x \frac{\partial u_y}{\partial x} + u_y \frac{\partial u_y}{\partial y} + u_z \frac{\partial u_y}{\partial z} \right) = -\frac{\partial P}{\partial y} + \rho g_y + \left(\frac{\partial \tau_{xy}}{\partial x} + \frac{\partial \tau_{yy}}{\partial y} + \frac{\partial \tau_{zy}}{\partial z} \right)$$

z-Direction

$$\rho\left(\frac{\partial u_z}{\partial t} + u_x \frac{\partial u_z}{\partial x} + u_y \frac{\partial u_z}{\partial y} + u_z \frac{\partial u_z}{\partial z} \right) = -\frac{\partial P}{\partial z} + \rho g_z + \left(\frac{\partial \tau_{xz}}{\partial x} + \frac{\partial \tau_{yz}}{\partial y} + \frac{\partial \tau_{zz}}{\partial z} \right)$$

Cylindrical coordinates (r, θ, z):

r-Direction

$$\rho\left(\frac{\partial u_r}{\partial t} + u_r \frac{\partial u_r}{\partial r} + \frac{u_\theta}{r} \frac{\partial u_r}{\partial \theta} - \frac{u_\theta^2}{r} + u_z \frac{\partial u_r}{\partial z} \right) = -\frac{\partial P}{\partial r}$$

$$+ \rho g_r + \left[\frac{1}{r} \frac{\partial}{\partial r}(r\tau_{rr}) + \frac{1}{r} \frac{\partial \tau_{\theta r}}{\partial \theta} - \frac{\tau_{\theta\theta}}{r} + \frac{\partial \tau_{zr}}{\partial z} \right]$$

θ-Direction

$$\rho\left(\frac{\partial u_\theta}{\partial t} + u_r \frac{\partial u_\theta}{\partial r} + \frac{u_\theta}{r} \frac{\partial u_\theta}{\partial \theta} + \frac{u_r u_\theta}{r} + u_z \frac{\partial u_\theta}{\partial z} \right) = -\frac{1}{r} \frac{\partial P}{\partial \theta}$$

$$+ \rho g_\theta + \left[\frac{1}{r^2} \frac{\partial}{\partial r}(r^2 \tau_{r\theta}) + \frac{1}{r} \frac{\partial \tau_{\theta\theta}}{\partial \theta} + \frac{\partial \tau_{z\theta}}{\partial z} \right]$$

z-Direction

$$\rho\left(\frac{\partial u_z}{\partial t}+u_r\frac{\partial u_z}{\partial r}+\frac{u_\theta}{r}\frac{\partial u_z}{\partial \theta}+u_z\frac{\partial u_z}{\partial z}\right)=-\frac{\partial P}{\partial z}+\rho g_z$$

$$+\left[\frac{1}{r}\frac{\partial}{\partial r}(r\tau_{rz})+\frac{1}{r}\frac{\partial \tau_{\theta z}}{\partial \theta}+\frac{\partial \tau_{zz}}{\partial z}\right]$$

Spherical coordinates (r, θ, ϕ):

r-Direction

$$\rho\left(\frac{\partial u_r}{\partial t}+u_r\frac{\partial u_r}{\partial r}+\frac{u_\theta}{r}\frac{\partial u_r}{\partial \theta}+\frac{u_\phi}{r\sin\theta}\frac{\partial u_r}{\partial \phi}-\frac{u_\theta^2+u_\phi^2}{r}\right)=-\frac{\partial P}{\partial r}+\rho g_r$$

$$+\left[\frac{1}{r^2}\frac{\partial}{\partial r}(r^2\tau_{rr})+\frac{1}{r\sin\theta}\frac{\partial}{\partial \theta}(\tau_{\theta r}\sin\theta)+\frac{1}{r\sin\theta}\frac{\partial \tau_{\phi r}}{\partial \phi}-\frac{\tau_{\theta\theta}+\tau_{\phi\phi}}{r}\right]$$

θ-Direction

$$\rho\left(\frac{\partial u_\theta}{\partial t}+u_r\frac{\partial u_\theta}{\partial r}+\frac{u_\theta}{r}\frac{\partial u_\theta}{\partial \theta}+\frac{u_\phi}{r\sin\theta}\frac{\partial u_\theta}{\partial \phi}+\frac{u_r u_\theta}{r}-\frac{u_\phi^2\cot\theta}{r}\right)=-\frac{1}{r}\frac{\partial P}{\partial r}+\rho g_\theta$$

$$+\left[\frac{1}{r^2}\frac{\partial}{\partial r}(r^2\tau_{r\theta})+\frac{1}{r\sin\theta}\frac{\partial}{\partial \theta}(\tau_{\theta\theta}\sin\theta)+\frac{1}{r\sin\theta}\frac{\partial \tau_{\phi\theta}}{\partial \phi}+\frac{\tau_{r\theta}}{r}-\frac{\tau_{\phi\phi}\cot\theta}{r}\right]$$

φ-Direction

$$\rho\left(\frac{\partial u_\phi}{\partial t}+u_r\frac{\partial u_\phi}{\partial r}+\frac{u_\theta}{r}\frac{\partial u_\phi}{\partial \theta}+\frac{u_\phi}{r\sin\theta}\frac{\partial u_\phi}{\partial \phi}+\frac{u_\phi u_r}{r}+\frac{u_\theta u_\phi\cot\theta}{r}\right)=-\frac{1}{r\sin\theta}\frac{\partial P}{\partial \phi}$$

$$+\rho g_\phi+\left[\frac{1}{r^2}\frac{\partial}{\partial r}(r^2\tau_{r\phi})+\frac{1}{r}\frac{\partial \tau_{\theta\phi}}{\partial \theta}+\frac{1}{r\sin\theta}\frac{\partial \tau_{\phi\phi}}{\partial \phi}+\frac{\tau_{r\phi}}{r}+\frac{2\tau_{\theta\phi}\cot\theta}{r}\right]$$

A.1.3 RATE OF STRAIN TENSOR $(\overline{\overline{E}})$ IN DIFFERENT COORDINATE SYSTEMS

($\overline{\overline{E}} = \dfrac{1}{2}\left[\nabla\vec{u}+(\nabla\vec{u})^{T}\right]$, where $\vec{u}$ is velocity vector)

Rectangular coordinates (x, y, z):

$$E_{xx} = \frac{\partial u_{x}}{\partial x}, \qquad E_{xy} = E_{yx} = \frac{1}{2}\left(\frac{\partial u_{x}}{\partial y}+\frac{\partial u_{y}}{\partial x}\right)$$

$$E_{yy} = \frac{\partial u_{y}}{\partial y}, \qquad E_{yz} = E_{zy} = \frac{1}{2}\left(\frac{\partial u_{y}}{\partial z}+\frac{\partial u_{z}}{\partial y}\right)$$

$$E_{zz} = \frac{\partial u_{z}}{\partial z}, \qquad E_{zx} = E_{xz} = \frac{1}{2}\left(\frac{\partial u_{z}}{\partial x}+\frac{\partial u_{x}}{\partial z}\right)$$

Cylindrical coordinates (r, θ, z):

$$E_{rr} = \frac{\partial u_{r}}{\partial r}, \qquad E_{r\theta} = E_{\theta r} = \frac{1}{2}\left[r\frac{\partial}{\partial r}\left(\frac{u_{\theta}}{r}\right)+\frac{1}{r}\frac{\partial u_{r}}{\partial\theta}\right]$$

$$E_{\theta\theta} = \frac{1}{r}\frac{\partial u_{\theta}}{\partial\theta}+\frac{u_{r}}{r}, \qquad E_{\theta z} = E_{z\theta} = \frac{1}{2}\left(\frac{\partial u_{\theta}}{\partial z}+\frac{1}{r}\frac{\partial u_{z}}{\partial\theta}\right)$$

$$E_{zz} = \frac{\partial u_{z}}{\partial z}, \qquad E_{zr} = E_{rz} = \frac{1}{2}\left(\frac{\partial u_{z}}{\partial r}+\frac{\partial u_{r}}{\partial z}\right)$$

Spherical coordinates (r, θ, ϕ):

$$E_{rr} = \frac{\partial u_{r}}{\partial r}, \qquad E_{r\theta} = E_{\theta r} = \frac{1}{2}\left[r\frac{\partial}{\partial r}\left(\frac{u_{\theta}}{r}\right)+\frac{1}{r}\frac{\partial u_{r}}{\partial\theta}\right]$$

$$E_{\theta\theta} = \frac{1}{r}\frac{\partial u_{\theta}}{\partial\theta}+\frac{u_{r}}{r}, \quad E_{\theta\phi} = E_{\phi\theta} = \frac{1}{2}\left[\frac{\sin\theta}{r}\frac{\partial}{\partial\theta}\left(\frac{u_{\phi}}{\sin\theta}\right)+\frac{1}{r\sin\theta}\frac{\partial u_{\theta}}{\partial\phi}\right]$$

$$E_{\phi\phi} = \left(\frac{1}{r\sin\theta}\frac{\partial u_{\phi}}{\partial\phi}+\frac{u_{r}}{r}+\frac{u_{\theta}\cot\theta}{r}\right), \quad E_{\phi r} = E_{r\phi} = \frac{1}{2}\left[\frac{1}{r\sin\theta}\frac{\partial u_{r}}{\partial\phi}+r\frac{\partial}{\partial r}\left(\frac{u_{\phi}}{r}\right)\right]$$

A.1.4 SHEAR RATE TENSOR $(\bar{\bar{\dot{\gamma}}})$ IN DIFFERENT COORDINATE SYSTEMS

$(\bar{\bar{\dot{\gamma}}} = \nabla \vec{u} + (\nabla \vec{u})^T$, where $\vec{u}$ is velocity vector)

Rectangular coordinates (x, y, z):

$$\dot{\gamma}_{xx} = 2\frac{\partial u_x}{\partial x}, \qquad \dot{\gamma}_{xy} = \dot{\gamma}_{yx} = \left(\frac{\partial u_x}{\partial y} + \frac{\partial u_y}{\partial x}\right)$$

$$\dot{\gamma}_{yy} = 2\frac{\partial u_y}{\partial y}, \qquad \dot{\gamma}_{yz} = \dot{\gamma}_{zy} = \left(\frac{\partial u_y}{\partial z} + \frac{\partial u_z}{\partial y}\right)$$

$$\dot{\gamma}_{zz} = 2\frac{\partial u_z}{\partial z}, \qquad \dot{\gamma}_{zx} = \dot{\gamma}_{xz} = \left(\frac{\partial u_z}{\partial x} + \frac{\partial u_x}{\partial z}\right)$$

Cylindrical coordinates (r, θ, z):

$$\dot{\gamma}_{rr} = 2\frac{\partial u_r}{\partial r}, \qquad \dot{\gamma}_{r\theta} = \dot{\gamma}_{\theta r} = \left[r\frac{\partial}{\partial r}\left(\frac{u_\theta}{r}\right) + \frac{1}{r}\frac{\partial u_r}{\partial \theta}\right]$$

$$\dot{\gamma}_{\theta\theta} = 2\left(\frac{1}{r}\frac{\partial u_\theta}{\partial \theta} + \frac{u_r}{r}\right), \qquad \dot{\gamma}_{\theta z} = \dot{\gamma}_{z\theta} = \left(\frac{\partial u_\theta}{\partial z} + \frac{1}{r}\frac{\partial u_z}{\partial \theta}\right)$$

$$\dot{\gamma}_{zz} = 2\frac{\partial u_z}{\partial z}, \qquad \dot{\gamma}_{zr} = \dot{\gamma}_{rz} = \left(\frac{\partial u_z}{\partial r} + \frac{\partial u_r}{\partial z}\right)$$

Spherical coordinates (r, θ, ϕ):

$$\dot{\gamma}_{rr} = 2\frac{\partial u_r}{\partial r}, \qquad \dot{\gamma}_{r\theta} = \dot{\gamma}_{\theta r} = \left[r\frac{\partial}{\partial r}\left(\frac{u_\theta}{r}\right) + \frac{1}{r}\frac{\partial u_r}{\partial \theta}\right]$$

$$\dot{\gamma}_{\theta\theta} = 2\left(\frac{1}{r}\frac{\partial u_\theta}{\partial \theta} + \frac{u_r}{r}\right), \qquad \dot{\gamma}_{\theta\phi} = \dot{\gamma}_{\phi\theta} = \left[\frac{\sin\theta}{r}\frac{\partial}{\partial \theta}\left(\frac{u_\phi}{\sin\theta}\right) + \frac{1}{r\sin\theta}\frac{\partial u_\theta}{\partial \phi}\right]$$

$$\dot{\gamma}_{\phi\phi} = 2\left(\frac{1}{r\sin\theta}\frac{\partial u_\phi}{\partial \phi} + \frac{u_r}{r} + \frac{u_\theta\cot\theta}{r}\right), \qquad \dot{\gamma}_{\phi r} = \dot{\gamma}_{r\phi} = \left[\frac{1}{r\sin\theta}\frac{\partial u_r}{\partial \phi} + r\frac{\partial}{\partial r}\left(\frac{u_\phi}{r}\right)\right]$$

A.1.5   Vorticity Tensor ($\bar{\bar{\Omega}}$) in Different Coordinate Systems

($\bar{\bar{\Omega}} = \dfrac{1}{2}\left[(\nabla\vec{u})^T - \nabla\vec{u}\right]$, where $\vec{u}$ is velocity vector)

Rectangular coordinates (x, y, z):

$$\Omega_{xy} = -\Omega_{yx} = -\frac{1}{2}\left(\frac{\partial u_y}{\partial x} - \frac{\partial u_x}{\partial y}\right)$$

$$\Omega_{yz} = -\Omega_{zy} = -\frac{1}{2}\left(\frac{\partial u_z}{\partial y} - \frac{\partial u_y}{\partial z}\right)$$

$$\Omega_{zx} = -\Omega_{xz} = -\frac{1}{2}\left(\frac{\partial u_x}{\partial z} - \frac{\partial u_z}{\partial x}\right)$$

Cylindrical coordinates (r, θ, z):

$$\Omega_{r\theta} = -\Omega_{\theta r} = -\frac{1}{2}\left[\frac{1}{r}\frac{\partial}{\partial r}(ru_\theta) - \frac{1}{r}\frac{\partial u_r}{\partial \theta}\right]$$

$$\Omega_{\theta z} = -\Omega_{z\theta} = -\frac{1}{2}\left[\frac{1}{r}\frac{\partial u_z}{\partial \theta} - \frac{\partial u_\theta}{\partial z}\right]$$

$$\Omega_{zr} = -\Omega_{rz} = -\frac{1}{2}\left[\frac{\partial u_r}{\partial z} - \frac{\partial u_z}{\partial r}\right]$$

Spherical coordinates (r, θ, ϕ):

$$\Omega_{r\theta} = -\Omega_{\theta r} = -\frac{1}{2}\left[\frac{1}{r}\frac{\partial}{\partial r}(ru_\theta) - \frac{1}{r}\frac{\partial u_r}{\partial \theta}\right]$$

$$\Omega_{\theta\phi} = -\Omega_{\phi\theta} = -\frac{1}{2}\left[\frac{1}{r\sin\theta}\frac{\partial}{\partial \theta}(u_\phi \sin\theta) - \frac{1}{r\sin\theta}\frac{\partial u_\theta}{\partial \phi}\right]$$

$$\Omega_{\theta r} = -\Omega_{r\phi} = -\frac{1}{2}\left[\frac{1}{r\sin\theta}\frac{\partial u_r}{\partial \phi} - \frac{1}{r}\frac{\partial}{\partial r}(ru_\phi)\right]$$

A.1.6   Viscous Stress Tensor ($\bar{\bar{\tau}}$) for Newtonian Fluids in Different Coordinate Systems

($\bar{\bar{\tau}} = 2\eta\bar{\bar{E}} + (\eta^k - \frac{2}{3}\eta)(tr\bar{\bar{E}})\bar{\bar{\delta}}$, where η is the shear viscosity, η^k is the dilational viscosity, tr refers to trace, and $\bar{\bar{\delta}}$ is unit tensor)

Rectangular coordinates (x, y, z):

$$\tau_{xx} = 2\eta\left(\frac{\partial u_x}{\partial x}\right) + \left[(\eta^k - \tfrac{2}{3}\eta)\nabla \bullet \vec{u}\right]$$

$$\tau_{yy} = 2\eta\left(\frac{\partial u_y}{\partial y}\right) + \left[(\eta^k - \tfrac{2}{3}\eta)\nabla \bullet \vec{u}\right]$$

$$\tau_{zz} = 2\eta\left(\frac{\partial u_z}{\partial z}\right) + \left[(\eta^k - \tfrac{2}{3}\eta)\nabla \bullet \vec{u}\right]$$

$$\tau_{xy} = \tau_{yx} = \eta\left(\frac{\partial u_x}{\partial y} + \frac{\partial u_y}{\partial x}\right)$$

$$\tau_{yz} = \tau_{zy} = \eta\left(\frac{\partial u_y}{\partial z} + \frac{\partial u_z}{\partial y}\right)$$

$$\tau_{zx} = \tau_{xz} = \eta\left(\frac{\partial u_z}{\partial x} + \frac{\partial u_x}{\partial z}\right)$$

Note that: $\nabla \bullet \vec{u} = \dfrac{\partial u_x}{\partial x} + \dfrac{\partial u_y}{\partial y} + \dfrac{\partial u_z}{\partial z}$

Cylindrical coordinates (r, θ, z):

$$\tau_{rr} = 2\eta\left(\frac{\partial u_r}{\partial r}\right) + \left[(\eta^k - \tfrac{2}{3}\eta)\nabla \bullet \vec{u}\right]$$

$$\tau_{\theta\theta} = 2\eta\left(\frac{1}{r}\frac{\partial u_\theta}{\partial \theta} + \frac{u_r}{r}\right) + \left[(\eta^k - \tfrac{2}{3}\eta)\nabla \bullet \vec{u}\right]$$

$$\tau_{zz} = 2\eta\left(\frac{\partial u_z}{\partial z}\right) + \left[(\eta^k - \tfrac{2}{3}\eta)\nabla \bullet \vec{u}\right]$$

$$\tau_{r\theta} = \tau_{\theta r} = \eta\left[r\frac{\partial}{\partial r}\left(\frac{u_\theta}{r}\right) + \frac{1}{r}\frac{\partial u_r}{\partial \theta}\right]$$

$$\tau_{\theta z} = \tau_{z\theta} = \eta\left(\frac{\partial u_\theta}{\partial z} + \frac{1}{r}\frac{\partial u_z}{\partial \theta}\right)$$

$$\tau_{zr} = \tau_{rz} = \eta\left(\frac{\partial u_z}{\partial r} + \frac{\partial u_r}{\partial z}\right)$$

$$\nabla \bullet \vec{u} = \frac{1}{r}\frac{\partial}{\partial r}(\rho u_r) + \frac{1}{r}\frac{\partial u_\theta}{\partial \theta} + \frac{\partial u_z}{\partial z}$$

Spherical coordinates (r, θ, ϕ):

$$\tau_{rr} = 2\eta\left(\frac{\partial u_r}{\partial r}\right) + \left[(\eta^k - \tfrac{2}{3}\eta)\nabla \bullet \vec{u}\right]$$

$$\tau_{\theta\theta} = 2\eta\left(\frac{1}{r}\frac{\partial u_\theta}{\partial \theta} + \frac{u_r}{r}\right) + \left[(\eta^k - \tfrac{2}{3}\eta)\nabla \bullet \vec{u}\right]$$

$$\tau_{\phi\phi} = 2\eta\left(\frac{1}{r\sin\theta}\frac{\partial u_\phi}{\partial \phi} + \frac{u_r}{r} + \frac{u_\theta \cot\theta}{r}\right) + \left[(\eta^k - \tfrac{2}{3}\eta)\nabla \bullet \vec{u}\right]$$

$$\tau_{r\theta} = \tau_{\theta r} = \eta\left[r\frac{\partial}{\partial r}\left(\frac{u_\theta}{r}\right) + \frac{1}{r}\frac{\partial u_r}{\partial \theta}\right]$$

$$\tau_{\theta\phi} = \tau_{\phi\theta} = \eta\left[\frac{\sin\theta}{r}\frac{\partial}{\partial \theta}\left(\frac{u_\phi}{\sin\theta}\right) + \frac{1}{r\sin\theta}\frac{\partial u_\theta}{\partial \phi}\right]$$

$$\tau_{\phi r} = \tau_{r\phi} = \eta\left[\frac{1}{r\sin\theta}\frac{\partial u_r}{\partial \phi} + r\frac{\partial}{\partial r}\left(\frac{u_\phi}{r}\right)\right]$$

$$\nabla \bullet \vec{u} = \frac{1}{r^2}\frac{\partial}{\partial r}(r^2 u_r) + \frac{1}{r\sin\theta}\frac{\partial}{\partial \theta}(u_\theta \sin\theta) + \frac{1}{r\sin\theta}\frac{\partial u_\phi}{\partial \phi}$$

A.1.7 NAVIER–STOKES EQUATIONS IN DIFFERENT COORDINATE SYSTEMS

$(\rho\left[\dfrac{\partial \vec{u}}{\partial t} + \vec{u} \bullet \nabla\vec{u}\right] = -\nabla P + \rho\vec{g} + \eta\nabla^2\vec{u}$, where ρ is density, $\vec{u}$ is the velocity vector, P is pressure, $\vec{g}$ is acceleration due to gravity, and η is viscosity)

Rectangular coordinates (x, y, z):
x-Direction

$$\rho\left(\frac{\partial u_x}{\partial t} + u_x\frac{\partial u_x}{\partial x} + u_y\frac{\partial u_x}{\partial y} + u_z\frac{\partial u_x}{\partial z}\right) = -\frac{\partial P}{\partial x} + \rho g_x + \eta\left(\frac{\partial^2 u_x}{\partial x^2} + \frac{\partial^2 u_x}{\partial y^2} + \frac{\partial^2 u_x}{\partial z^2}\right)$$

y-Direction

$$\rho\left(\frac{\partial u_y}{\partial t} + u_x\frac{\partial u_y}{\partial x} + u_y\frac{\partial u_y}{\partial y} + u_z\frac{\partial u_y}{\partial z}\right) = -\frac{\partial P}{\partial y} + \rho g_y + \eta\left(\frac{\partial^2 u_y}{\partial x^2} + \frac{\partial^2 u_y}{\partial y^2} + \frac{\partial^2 u_y}{\partial z^2}\right)$$

z-Direction

$$\rho\left(\frac{\partial u_z}{\partial t}+u_x\frac{\partial u_z}{\partial x}+u_y\frac{\partial u_z}{\partial y}+u_z\frac{\partial u_z}{\partial z}\right)=-\frac{\partial P}{\partial z}+\rho g_z+\eta\left(\frac{\partial^2 u_z}{\partial x^2}+\frac{\partial^2 u_z}{\partial y^2}+\frac{\partial^2 u_z}{\partial z^2}\right)$$

Cylindrical coordinates (r, θ, z):

r-Direction

$$\rho\left(\frac{\partial u_r}{\partial t}+u_r\frac{\partial u_r}{\partial r}+\frac{u_\theta}{r}\frac{\partial u_r}{\partial\theta}-\frac{u_\theta^2}{r}+u_z\frac{\partial u_r}{\partial z}\right)=-\frac{\partial P}{\partial r}+\rho g_r$$

$$+\eta\left[\frac{\partial}{\partial r}\left(\frac{1}{r}\frac{\partial}{\partial r}(ru_r)\right)+\frac{1}{r^2}\frac{\partial^2 u_r}{\partial\theta^2}-\frac{2}{r^2}\frac{\partial u_\theta}{\partial\theta}+\frac{\partial^2 u_r}{\partial z^2}\right]$$

θ-Direction

$$\rho\left(\frac{\partial u_\theta}{\partial t}+u_r\frac{\partial u_\theta}{\partial r}+\frac{u_\theta}{r}\frac{\partial u_\theta}{\partial\theta}+\frac{u_r u_\theta}{r}+u_z\frac{\partial u_\theta}{\partial z}\right)=-\frac{1}{r}\frac{\partial P}{\partial\theta}+\rho g_\theta$$

$$+\eta\left[\frac{\partial}{\partial r}\left(\frac{1}{r}\frac{\partial}{\partial r}(ru_\theta)\right)+\frac{1}{r^2}\frac{\partial^2 u_\theta}{\partial\theta^2}+\frac{2}{r^2}\frac{\partial u_r}{\partial\theta}+\frac{\partial^2 u_\theta}{\partial z^2}\right]$$

z-Direction

$$\rho\left(\frac{\partial u_z}{\partial t}+u_r\frac{\partial u_z}{\partial r}+\frac{u_\theta}{r}\frac{\partial u_z}{\partial\theta}+u_z\frac{\partial u_z}{\partial z}\right)=-\frac{\partial P}{\partial z}+\rho g_z$$

$$+\eta\left[\frac{1}{r}\frac{\partial}{\partial r}\left(r\frac{\partial u_z}{\partial r}\right)+\frac{1}{r^2}\frac{\partial^2 u_z}{\partial\theta^2}+\frac{\partial^2 u_z}{\partial z^2}\right]$$

Spherical coordinates (r, θ, ϕ):

r-Direction

$$\rho\left(\frac{\partial u_r}{\partial t}+u_r\frac{\partial u_r}{\partial r}+\frac{u_\theta}{r}\frac{\partial u_r}{\partial\theta}+\frac{u_\phi}{r\sin\theta}\frac{\partial u_r}{\partial\phi}-\frac{u_\theta^2+u_\phi^2}{r}\right)=-\frac{\partial P}{\partial r}+\rho g_r$$

$$+\eta\left(\nabla^2 u_r-\frac{2u_r}{r^2}-\frac{2}{r^2}\frac{\partial u_\theta}{\partial\theta}-\frac{2u_\theta\cot\theta}{r^2}-\frac{2}{r^2\sin\theta}\frac{\partial u_\phi}{\partial\phi}\right)$$

θ-Direction

$$\rho\left(\frac{\partial u_\theta}{\partial t} + u_r \frac{\partial u_\theta}{\partial r} + \frac{u_\theta}{r}\frac{\partial u_\theta}{\partial \theta} + \frac{u_\phi}{r\sin\theta}\frac{\partial u_\theta}{\partial \phi} + \frac{u_r u_\theta}{r} - \frac{u_\phi^2 \cot\theta}{r}\right) = -\frac{1}{r}\frac{\partial P}{\partial \theta}$$

$$+\rho g_\theta + \eta\left(\nabla^2 u_\theta + \frac{2}{r^2}\frac{\partial u_r}{\partial \theta} - \frac{u_\theta}{r^2 \sin^2\theta} - \frac{2\cos\theta}{r^2 \sin^2\theta}\frac{\partial u_\phi}{\partial \phi}\right)$$

φ-Direction

$$\rho\left(\frac{\partial u_\phi}{\partial t} + u_r \frac{\partial u_\phi}{\partial r} + \frac{u_\theta}{r}\frac{\partial u_\phi}{\partial \theta} + \frac{u_\phi}{r\sin\theta}\frac{\partial u_\phi}{\partial \phi} + \frac{u_\phi u_r}{r} + \frac{u_\theta u_\phi}{r}\cot\theta\right) = -\frac{1}{r\sin\theta}\frac{\partial P}{\partial \phi}$$

$$+\rho g_\phi + \eta\left(\nabla^2 u_\phi - \frac{u_\phi}{r^2 \sin^2\theta} + \frac{2}{r^2 \sin\theta}\frac{\partial u_r}{\partial \phi} + \frac{2\cos\theta}{r^2 \sin^2\theta}\frac{\partial u_\theta}{\partial \phi}\right)$$

Note that the Laplacian ∇^2 in spherical coordinates is:

$$\nabla^2 = \frac{1}{r^2}\frac{\partial}{\partial r}\left(r^2\frac{\partial}{\partial r}\right) + \frac{1}{r^2 \sin\theta}\frac{\partial}{\partial \theta}\left(\sin\theta\frac{\partial}{\partial \theta}\right) + \frac{1}{r^2 \sin^2\theta}\left(\frac{\partial^2}{\partial \phi^2}\right)$$

A.2   EQUATIONS RELATED TO MECHANICS OF SOLIDS

A.2.1   INFINITESIMAL STRAIN TENSOR $(\bar{\bar{e}})$ IN DIFFERENT COORDINATE SYSTEMS

$(\bar{\bar{e}} = \frac{1}{2}\left[\nabla\vec{\xi} + (\nabla\vec{\xi})^T\right]$ where $\vec{\xi}$ is displacement vector)

Rectangular coordinates (x, y, z):

$$e_{xx} = \frac{\partial \xi_x}{\partial x}, \qquad e_{xy} = e_{yx} = \frac{1}{2}\left(\frac{\partial \xi_x}{\partial y} + \frac{\partial \xi_y}{\partial x}\right)$$

$$e_{yy} = \frac{\partial \xi_y}{\partial y}, \qquad e_{yz} = e_{zy} = \frac{1}{2}\left(\frac{\partial \xi_y}{\partial z} + \frac{\partial \xi_z}{\partial y}\right)$$

$$e_{zz} = \frac{\partial \xi_z}{\partial z}, \qquad e_{zx} = e_{xz} = \frac{1}{2}\left(\frac{\partial \xi_z}{\partial x} + \frac{\partial \xi_x}{\partial z}\right)$$

Cylindrical coordinates (r, θ, z):

$$e_{rr} = \frac{\partial \xi_r}{\partial r}, \qquad\qquad e_{r\theta} = e_{\theta r} = \frac{1}{2}\left(\frac{1}{r}\frac{\partial \xi_r}{\partial \theta} - \frac{\xi_\theta}{r} + \frac{\partial \xi_\theta}{\partial r}\right)$$

$$e_{\theta\theta} = \frac{1}{r}\left(\frac{\partial \xi_\theta}{\partial \theta} + \xi_r\right), \qquad e_{\theta z} = e_{z\theta} = \frac{1}{2}\left(\frac{\partial \xi_\theta}{\partial z} + \frac{1}{r}\frac{\partial \xi_z}{\partial \theta}\right)$$

$$e_{zz} = \frac{\partial \xi_z}{\partial z}, \qquad\qquad e_{zr} = e_{rz} = \frac{1}{2}\left(\frac{\partial \xi_r}{\partial z} + \frac{\partial \xi_z}{\partial r}\right)$$

Spherical coordinates (r, θ, ϕ):

$$e_{rr} = \frac{\partial \xi_r}{\partial r}, \qquad\qquad e_{r\theta} = e_{\theta r} = \frac{1}{2}\left(\frac{1}{r}\frac{\partial \xi_r}{\partial \theta} + \frac{\partial \xi_\theta}{\partial r} - \frac{\xi_\theta}{r}\right)$$

$$e_{\theta\theta} = \frac{1}{r}\left(\frac{\partial \xi_\theta}{\partial \theta} + \xi_r\right), \qquad e_{\theta\phi} = e_{\phi\theta} = \frac{1}{2r}\left(\frac{1}{\sin\theta}\frac{\partial \xi_\theta}{\partial \phi} + \frac{\partial \xi_\phi}{\partial \theta} - \xi_\phi \cot\theta\right)$$

$$e_{\phi\phi} = \frac{1}{r}\left(\frac{1}{\sin\theta}\frac{\partial \xi_\phi}{\partial \phi} + \xi_\theta \cot\theta + \xi_r\right), \quad e_{\phi r} = e_{r\phi} = \frac{1}{2}\left(\frac{1}{r\sin\theta}\frac{\partial \xi_r}{\partial \phi} + \frac{\partial \xi_\phi}{\partial r} - \frac{\xi_\phi}{r}\right)$$

A.2.2 EQUILIBRIUM EQUATION IN DIFFERENT COORDINATE SYSTEMS

($\nabla \bullet \overline{\overline{\sigma}} + \rho \vec{g} = 0$, where $\overline{\overline{\sigma}}$ is stress tensor, ρ is density, and $\vec{g}$ is acceleration due to gravity)

Rectangular coordinates (x, y, z):

x-Direction

$$\frac{\partial \sigma_{xx}}{\partial x} + \frac{\partial \sigma_{yx}}{\partial y} + \frac{\partial \sigma_{zx}}{\partial z} + \rho g_x = 0$$

y-Direction

$$\frac{\partial \sigma_{xy}}{\partial x} + \frac{\partial \sigma_{yy}}{\partial y} + \frac{\partial \sigma_{zy}}{\partial z} + \rho g_y = 0$$

z-Direction

$$\frac{\partial \sigma_{xz}}{\partial x} + \frac{\partial \sigma_{yz}}{\partial y} + \frac{\partial \sigma_{zz}}{\partial z} + \rho g_z = 0$$

Cylindrical coordinates (r, θ, z):

r-Direction

$$\frac{\partial \sigma_{rr}}{\partial r} + \frac{1}{r}\frac{\partial \sigma_{r\theta}}{\partial \theta} + \frac{\partial \sigma_{rz}}{\partial z} + \frac{\sigma_{rr} - \sigma_{\theta\theta}}{r} + \rho g_r = 0$$

θ-Direction

$$\frac{\partial \sigma_{\theta r}}{\partial r} + \frac{1}{r}\frac{\partial \sigma_{\theta\theta}}{\partial \theta} + \frac{\partial \sigma_{\theta z}}{\partial z} + \frac{2\,\sigma_{r\theta}}{r} + \rho g_\theta = 0$$

z-Direction

$$\frac{\partial \sigma_{zr}}{\partial r} + \frac{1}{r}\frac{\partial \sigma_{z\theta}}{\partial \theta} + \frac{\partial \sigma_{zz}}{\partial z} + \frac{\sigma_{zr}}{r} + \rho g_z = 0$$

Spherical coordinates (r, θ, ϕ):

r-Direction

$$\frac{\partial \sigma_{rr}}{\partial r} + \frac{1}{r}\frac{\partial \sigma_{r\theta}}{\partial \theta} + \frac{1}{r\sin\theta}\frac{\partial \sigma_{r\phi}}{\partial \phi} + \frac{1}{r}\left(2\sigma_{rr} - \sigma_{\theta\theta} - \sigma_{\phi\phi} + \sigma_{r\theta}\cot\theta\right) + \rho g_r = 0$$

θ-Direction

$$\frac{\partial \sigma_{\theta r}}{\partial r} + \frac{1}{r}\frac{\partial \sigma_{\theta\theta}}{\partial \theta} + \frac{1}{r\sin\theta}\frac{\partial \sigma_{\theta\phi}}{\partial \phi} + \frac{1}{r}\left[\left(\sigma_{\theta\theta} - \sigma_{\phi\phi}\right)\cot\theta + 3\sigma_{r\theta}\right] + \rho g_\theta = 0$$

φ-Direction

$$\frac{\partial \sigma_{\theta r}}{\partial r} + \frac{1}{r}\frac{\partial \sigma_{\phi\theta}}{\partial \theta} + \frac{1}{r\sin\theta}\frac{\partial \sigma_{\phi\phi}}{\partial \phi} + \frac{1}{r}\left(3\sigma_{\phi r} + 2\sigma_{\phi\theta}\cot\theta\right) + \rho g_\theta = 0$$

A.2.3 Hooke's Law in Different Coordinate Systems

$(\bar{\bar{\sigma}} = \lambda(\operatorname{tr}\bar{\bar{e}})\bar{\bar{\delta}} + 2G\bar{\bar{e}}$, where $\bar{\bar{\sigma}}$ is the stress tensor, λ and G (shear modulus) are Lamé constants, $\bar{\bar{e}}$ is the infinitesimal strain tensor, and $\bar{\bar{\delta}}$ is the unit tensor)

Rectangular coordinates (x, y, z):

$$\sigma_{xx} = \lambda(\operatorname{tr}\bar{\bar{e}}) + 2G\left(\frac{\partial \xi_x}{\partial x}\right)$$

$$\sigma_{yy} = \lambda(\operatorname{tr}\bar{\bar{e}}) + 2G\left(\frac{\partial \xi_y}{\partial y}\right)$$

$$\sigma_{zz} = \lambda(\operatorname{tr}\bar{\bar{e}}) + 2G\left(\frac{\partial \xi_z}{\partial z}\right)$$

$$\sigma_{xy} = \sigma_{yx} = G\left(\frac{\partial \xi_x}{\partial y} + \frac{\partial \xi_y}{\partial x}\right)$$

$$\sigma_{yz} = \sigma_{zy} = G\left(\frac{\partial \xi_y}{\partial z} + \frac{\partial \xi_z}{\partial y}\right)$$

$$\sigma_{zx} = \sigma_{xz} = G\left(\frac{\partial \xi_z}{\partial x} + \frac{\partial \xi_x}{\partial z}\right)$$

Note that $\operatorname{tr}\bar{\bar{e}} = \dfrac{\partial \xi_x}{\partial x} + \dfrac{\partial \xi_y}{\partial y} + \dfrac{\partial \xi_z}{\partial z}$

Cylindrical coordinates (r, θ, z):

$$\sigma_{rr} = \lambda(\operatorname{tr}\bar{\bar{e}}) + 2G\left(\frac{\partial \xi_r}{\partial r}\right)$$

$$\sigma_{\theta\theta} = \lambda(\operatorname{tr}\bar{\bar{e}}) + \frac{2G}{r}\left(\frac{\partial \xi_\theta}{\partial \theta} + \xi_r\right)$$

$$\sigma_{zz} = \lambda(\operatorname{tr}\bar{\bar{e}}) + 2G\left(\frac{\partial \xi_z}{\partial z}\right)$$

$$\sigma_{r\theta} = \sigma_{\theta r} = G\left(\frac{1}{r}\frac{\partial \xi_r}{\partial \theta} - \frac{\xi_\theta}{r} + \frac{\partial \xi_\theta}{\partial r}\right)$$

$$\sigma_{\theta z} = \sigma_{z\theta} = G\left(\frac{\partial \xi_\theta}{\partial z} + \frac{1}{r}\frac{\partial \xi_z}{\partial \theta}\right)$$

$$\sigma_{zr} = \sigma_{rz} = G\left(\frac{\partial \xi_r}{\partial z} + \frac{\partial \xi_z}{\partial r}\right)$$

Note that $\operatorname{tr}\bar{\bar{e}} = \dfrac{\partial \xi_r}{\partial r} + \dfrac{1}{r}\left(\dfrac{\partial \xi_\theta}{\partial \theta} + \xi_r\right) + \dfrac{\partial \xi_z}{\partial z}$

Spherical coordinates (r, θ, ϕ):

$$\sigma_{rr} = \lambda(\operatorname{tr}\bar{\bar{e}}) + 2G\left(\dfrac{\partial \xi_r}{\partial r}\right)$$

$$\sigma_{\theta\theta} = \lambda(\operatorname{tr}\bar{\bar{e}}) + \dfrac{2G}{r}\left(\dfrac{\partial \xi_\theta}{\partial \theta} + \xi_r\right)$$

$$\sigma_{\phi\phi} = \lambda(\operatorname{tr}\bar{\bar{e}}) + \dfrac{2G}{r}\left(\dfrac{1}{\sin\theta}\dfrac{\partial \xi_\phi}{\partial \phi} + \xi_\theta \cot\theta + \xi_r\right)$$

$$\sigma_{r\theta} = \sigma_{\theta r} = G\left(\dfrac{1}{r}\dfrac{\partial \xi_r}{\partial \theta} + \dfrac{\partial \xi_\theta}{\partial r} - \dfrac{\xi_\theta}{r}\right)$$

$$\sigma_{\theta\phi} = \sigma_{\phi\theta} = \dfrac{G}{r}\left(\dfrac{1}{\sin\theta}\dfrac{\partial \xi_\theta}{\partial \phi} + \dfrac{\partial \xi_\phi}{\partial \theta} - \xi_\phi \cot\theta\right)$$

$$\sigma_{\phi r} = \sigma_{r\phi} = G\left(\dfrac{1}{r\sin\theta}\dfrac{\partial \xi_r}{\partial \phi} + \dfrac{\partial \xi_\phi}{\partial r} - \dfrac{\xi_\phi}{r}\right)$$

Note that $\operatorname{tr}\bar{\bar{e}} = \dfrac{\partial \xi_r}{\partial r} + \dfrac{1}{r}\dfrac{\partial \xi_\theta}{\partial \theta} + \dfrac{2\xi_r}{r} + \dfrac{1}{\sin\theta}\dfrac{\partial \xi_\phi}{\partial \phi} + \dfrac{\xi_\theta}{r}\cot\theta$

A.2.4 Navier–Cauchy Equations in Different Coordinate Systems

$(\rho\vec{g} + (\lambda + G)\nabla(\nabla \bullet \vec{\xi}) + G\nabla^2\vec{\xi} = 0$, where ρ is density, $\vec{g}$ is acceleration due to gravity, λ and G (shear modulus) are Lamé constants, and $\vec{\xi}$ is the displacement vector)

Rectangular coordinates (x, y, z):

x-Direction

$$\rho g_x + (\lambda + G)\dfrac{\partial}{\partial x}\left(\dfrac{\partial \xi_x}{\partial x} + \dfrac{\partial \xi_y}{\partial y} + \dfrac{\partial \xi_z}{\partial z}\right) + G\left(\dfrac{\partial^2 \xi_x}{\partial x^2} + \dfrac{\partial^2 \xi_x}{\partial y^2} + \dfrac{\partial^2 \xi_x}{\partial z^2}\right) = 0$$

y-Direction

$$\rho g_y + (\lambda + G)\dfrac{\partial}{\partial y}\left(\dfrac{\partial \xi_x}{\partial x} + \dfrac{\partial \xi_y}{\partial y} + \dfrac{\partial \xi_z}{\partial z}\right) + G\left(\dfrac{\partial^2 \xi_y}{\partial x^2} + \dfrac{\partial^2 \xi_y}{\partial y^2} + \dfrac{\partial^2 \xi_y}{\partial z^2}\right) = 0$$

z-Direction

$$\rho g_z + (\lambda + G)\frac{\partial}{\partial z}\left(\frac{\partial \xi_x}{\partial x} + \frac{\partial \xi_y}{\partial y} + \frac{\partial \xi_z}{\partial z}\right) + G\left(\frac{\partial^2 \xi_z}{\partial x^2} + \frac{\partial^2 \xi_z}{\partial y^2} + \frac{\partial^2 \xi_z}{\partial z^2}\right) = 0$$

Cylindrical coordinates (r, θ, z):

r-Direction

$$\rho g_r + (\lambda + G)\frac{\partial}{\partial r}\left(\frac{\partial \xi_r}{\partial r} + \frac{\xi_r}{r} + \frac{1}{r}\frac{\partial \xi_\theta}{\partial \theta} + \frac{\partial \xi_z}{\partial z}\right)$$

$$+ G\left(\frac{\partial^2 \xi_r}{\partial r^2} + \frac{1}{r}\frac{\partial \xi_r}{\partial r} + \frac{1}{r^2}\frac{\partial^2 \xi_r}{\partial \theta^2} + \frac{\partial^2 \xi_r}{\partial z^2} - \frac{2}{r^2}\frac{\partial \xi_\theta}{\partial \theta} - \frac{\xi_r}{r^2}\right) = 0$$

θ-Direction

$$\rho g_\theta + (\lambda + G)\frac{1}{r}\frac{\partial}{\partial \theta}\left(\frac{\partial \xi_r}{\partial r} + \frac{\xi_r}{r} + \frac{1}{r}\frac{\partial \xi_\theta}{\partial \theta} + \frac{\partial \xi_z}{\partial z}\right)$$

$$+ G\left(\frac{\partial^2 \xi_\theta}{\partial r^2} + \frac{1}{r}\frac{\partial \xi_\theta}{\partial r} + \frac{1}{r^2}\frac{\partial^2 \xi_\theta}{\partial \theta^2} + \frac{\partial^2 \xi_\theta}{\partial z^2} + \frac{2}{r^2}\frac{\partial \xi_r}{\partial \theta} - \frac{\xi_\theta}{r^2}\right) = 0$$

z-Direction

$$\rho g_z + (\lambda + G)\frac{\partial}{\partial z}\left(\frac{\partial \xi_r}{\partial r} + \frac{\xi_r}{r} + \frac{1}{r}\frac{\partial \xi_\theta}{\partial \theta} + \frac{\partial \xi_z}{\partial z}\right)$$

$$+ G\left(\frac{\partial^2 \xi_z}{\partial r^2} + \frac{1}{r}\frac{\partial \xi_z}{\partial r} + \frac{1}{r^2}\frac{\partial^2 \xi_z}{\partial \theta^2} + \frac{\partial^2 \xi_z}{\partial z^2}\right) = 0$$

Spherical coordinates (r, θ, ϕ):

r-Direction

$$\rho g_r + (\lambda + G)\frac{\partial}{\partial r}\left(\frac{\partial \xi_r}{\partial r} + \frac{1}{r}\frac{\partial \xi_\theta}{\partial \theta} + \frac{1}{r\sin\theta}\frac{\partial \xi_\phi}{\partial \phi} + \frac{2\xi_r}{r} + \frac{\xi_\theta}{r}\cot\theta\right)$$

$$+ G\left[\frac{1}{r^2}\frac{\partial}{\partial r}\left(r^2\frac{\partial \xi_r}{\partial r}\right) + \frac{1}{r^2 \sin\theta}\frac{\partial}{\partial \theta}\left(\sin\theta\frac{\partial \xi_r}{\partial \theta}\right) + \frac{1}{r^2 \sin\theta}\frac{\partial^2 \xi_r}{\partial \phi^2}\right.$$

$$\left. - \frac{2}{r^2}\left(\xi_r + \frac{1}{\sin\theta}\frac{\partial}{\partial \theta}(\xi_\theta \sin\theta) + \frac{1}{\sin\theta}\frac{\partial \xi_\phi}{\partial \phi}\right)\right] = 0$$

θ-Direction

$$\rho g_\theta + (\lambda + G) \frac{1}{r} \frac{\partial}{\partial \theta} \left(\frac{\partial \xi_r}{\partial r} + \frac{1}{r} \frac{\partial \xi_\theta}{\partial \theta} + \frac{1}{r \sin \theta} \frac{\partial \xi_\phi}{\partial \phi} + \frac{2}{r} \xi_r + \frac{\xi_\theta}{r} \cot \theta \right)$$

$$+ G \left[\frac{1}{r^2} \frac{\partial}{\partial r} \left(r^2 \frac{\partial \xi_\theta}{\partial r} \right) + \frac{1}{r^2 \sin \theta} \frac{\partial}{\partial \theta} \left(\sin \theta \frac{\partial \xi_\theta}{\partial \theta} \right) + \frac{1}{r^2 \sin \theta} \frac{\partial^2 \xi_\theta}{\partial \phi^2} \right.$$

$$\left. + \frac{2}{r^2} \left(\frac{\partial \xi_r}{\partial \theta} - \frac{\xi_\theta}{2 \sin^2 \theta} - \frac{\cos \theta}{\sin^2 \theta} \frac{\partial \xi_\phi}{\partial \phi} \right) \right] = 0$$

φ-Direction

$$\rho g_\phi + (\lambda + G) \frac{1}{r \sin \theta} \frac{\partial}{\partial \phi} \left(\frac{\partial \xi_r}{\partial r} + \frac{1}{r} \frac{\partial \xi_\theta}{\partial \theta} + \frac{1}{r \sin \theta} \frac{\partial \xi_\phi}{\partial \phi} + \frac{2}{r} \xi_r + \frac{\xi_\theta}{r} \cot \theta \right)$$

$$+ G \left[\frac{1}{r^2} \frac{\partial}{\partial r} \left(r^2 \frac{\partial \xi_\phi}{\partial r} \right) + \frac{1}{r^2 \sin \theta} \frac{\partial}{\partial \theta} \left(\sin \theta \frac{\partial \xi_\phi}{\partial \theta} \right) + \frac{1}{r^2 \sin^2 \theta} \frac{\partial^2 \xi_\phi}{\partial \phi^2} \right.$$

$$\left. + \frac{2}{r^2 \sin \theta} \left(\frac{\partial \xi_r}{\partial \phi} + \cot \theta \frac{\partial \xi_\theta}{\partial \phi} - \frac{\xi_\phi}{2 \sin \theta} \right) \right] = 0$$

Appendix B

Vector and Tensor Operations in Index Notation

B.1 VECTOR OPERATIONS

Dot product of two unit vectors

The dot product (or scalar product) of two unit vectors $\hat{\delta}_i$ and $\hat{\delta}_j$ is given as:

$$\hat{\delta}_i \bullet \hat{\delta}_j = \delta_{ij}$$

where δ_{ij} is the Kronecker delta $(\delta_{ij} = 1$ when $i = j$; $\delta_{ij} = 0$ when $i \neq j)$

Dot product of two vectors

Let vector $\vec{A}$ be $\sum\limits_{i=1}^{3} A_i \hat{\delta}_i$ and vector $\vec{B} = \sum\limits_{j=1}^{3} B_j \hat{\delta}_j$. The dot product of $\vec{A}$ and $\vec{B}$ is:

$$\vec{A} \bullet \vec{B} = \sum_{i=1}^{3} \sum_{j=1}^{3} A_i B_j \left(\hat{\delta}_i \bullet \hat{\delta}_j \right) = \sum_{i=1}^{3} \sum_{j=1}^{3} A_i B_j \delta_{ij} = \sum_{i=1}^{3} A_i B_i$$

Cross product of two unit vectors

The cross product (or vector product) of two unit vectors $\hat{\delta}_i$ and $\hat{\delta}_j$ is given as:

$$\hat{\delta}_i \times \hat{\delta}_j = \sum_{k=1}^{3} \varepsilon_{ijk} \hat{\delta}_k$$

in which ε_{ijk} is the permutation symbol. $\varepsilon_{ijk} = +1$ if $ijk = 123, 231,$ or 312; $\varepsilon_{ijk} = -1$ if $ijk = 321, 132,$ or 213; $\varepsilon_{ijk} = 0$ if any two indices are alike.

Cross product of two vectors

The cross product of two vectors $\vec{A}$ and $\vec{B}$ is given as:

$$\vec{A} \times \vec{B} = \left\{ \sum_{i=1}^{3} \hat{\delta}_i A_i \right\} \times \left\{ \sum_{j=1}^{3} \hat{\delta}_j B_j \right\} = \sum_{i=1}^{3} \sum_{j=1}^{3} A_i B_i \left(\hat{\delta}_i \times \hat{\delta}_j \right)$$

$$= \sum_{i=1}^{3} \sum_{j=1}^{3} \sum_{k=1}^{3} \varepsilon_{ijk} A_i B_j \hat{\delta}_k = \sum_{i=1}^{3} \sum_{j=1}^{3} \sum_{k=1}^{3} \varepsilon_{ijk} \hat{\delta}_i A_j B_k$$

Dyadic product of two vectors

The dyadic product of two vectors $\vec{A}$ and $\vec{B}$ is given as:

$$\vec{A} \vec{B} = \sum_{i=1}^{3} \sum_{j=1}^{3} A_i B_j \ \hat{\delta}_i \hat{\delta}_j$$

Note that the two vectors are placed next to one another with no sign in-between them.

Divergence of a vector

The divergence of a vector $\vec{A}$ is given as:

$$div \ \vec{A} = \nabla \bullet \vec{A} = \sum_{i=1}^{3} \sum_{j=1}^{3} \left(\frac{\partial A_j}{\partial x_i} \right) \hat{\delta}_i \bullet \hat{\delta}_j = \sum_{i=1}^{3} \sum_{i=j}^{3} \left(\frac{\partial A_j}{\partial x_i} \right) \delta_{ij}$$

$$= \sum_{i=1}^{3} \left(\frac{\partial A_i}{\partial x_i} \right)$$

Curl of a vector

The curl of a vector $\vec{A}$ is given as:

$$curl \ \vec{A} = \nabla \times \vec{A} = \sum_{i=1}^{3} \sum_{j=1}^{3} \left(\frac{\partial A_j}{\partial x_i} \right) \hat{\delta}_i \times \hat{\delta}_j = \sum_{i=1}^{3} \sum_{j=1}^{3} \sum_{k=1}^{3} \varepsilon_{ijk} \left(\frac{\partial A_j}{\partial x_i} \right) \hat{\delta}_k$$

$$= \sum_{i=1}^{3} \sum_{j=1}^{3} \sum_{k=1}^{3} \varepsilon_{ijk} \hat{\delta}_i \left(\frac{\partial A_k}{\partial x_j} \right)$$

B.2   TENSOR OPERATIONS

Transpose of a tensor

The transpose of a tensor $\overline{\overline{T}} = \left(\sum\limits_{i=1}^{3} \sum\limits_{j=1}^{3} T_{ij}\, \hat{\delta}_i\, \hat{\delta}_j \right)$, denoted by $\overline{\overline{T}}^T$, is given as:

$$\overline{\overline{T}}^T = \sum_{i=1}^{3} \sum_{j=1}^{3} T_{ji}\, \hat{\delta}_i\, \hat{\delta}_j$$

Dot product of unit tensor $(\overline{\overline{\delta}})$ and vector $(\vec{A})$

$$\overline{\overline{\delta}} \bullet \vec{A} = \left(\sum_{i=1}^{3} \sum_{i=j}^{3} \delta_{ij}\, \hat{\delta}_i\hat{\delta}_j \right) \bullet \left(\sum_{k=1}^{3} A_k\, \hat{\delta}_k \right) = \sum_{i=1}^{3} \sum_{j=1}^{3} \sum_{k=1}^{3} \delta_{ij} A_k \hat{\delta}_i \hat{\delta}_{jk}$$

$$= \sum_{i=1}^{3} A_i\, \hat{\delta}_i = \vec{A}$$

Dot product of tensor $(\overline{\overline{T}})$ and vector $(\vec{A})$

$$\overline{\overline{T}} \bullet \vec{A} = \left(\sum_{i=1}^{3} \sum_{j=1}^{3} T_{ij}\, \hat{\delta}_i\hat{\delta}_j \right) \bullet \left(\sum_{k=1}^{3} A_k\hat{\delta}_k \right) = \sum_{i=1}^{3} \sum_{j=1}^{3} \sum_{k=1}^{3} T_{ij} A_k\, \hat{\delta}_i\, \hat{\delta}_{jk}$$

$$= \sum_{i=1}^{3} \sum_{j=1}^{3} T_{ij} A_j\, \hat{\delta}_i$$

Dot product of unit tensor $(\overline{\overline{\delta}})$ and tensor $(\overline{\overline{T}})$

$$\overline{\overline{\delta}} \bullet \overline{\overline{T}} = \sum_{i=1}^{3} \sum_{j=1}^{3} \delta_{ij}\, \hat{\delta}_i\hat{\delta}_j \bullet \sum_{k=1}^{3} \sum_{\ell=1}^{3} T_{k\ell}\hat{\delta}_k\hat{\delta}_\ell = \sum_{i=1}^{3} \sum_{j=1}^{3} \sum_{k=1}^{3} \sum_{\ell=1}^{3} \delta_{ij} T_{k\ell} \delta_{jk} \hat{\delta}_i \hat{\delta}_\ell$$

$$= \sum_{i=1}^{3} \sum_{\ell=1}^{3} T_{i\ell}\hat{\delta}_i\hat{\delta}_\ell = \overline{\overline{T}}$$

Dot product of two tensors

$$
\bar{\bar{S}} \bullet \bar{\bar{T}} = \left\{ \sum_{i=1}^{3} \sum_{j=1}^{3} S_{ij}\hat{\delta}_i\hat{\delta}_j \right\} \bullet \left\{ \sum_{k=1}^{3} \sum_{\ell=1}^{3} T_{k\ell}\hat{\delta}_k\hat{\delta}_\ell \right\} = \sum_{i=1}^{3} \sum_{j=1}^{3} \sum_{k=1}^{3} \sum_{\ell=1}^{3} S_{ij}T_{k\ell}\hat{\delta}_i \left(\hat{\delta}_j \bullet \hat{\delta}_k \right) \hat{\delta}_\ell
$$

$$
= \sum_{i=1}^{3} \sum_{j=1}^{3} \sum_{k=1}^{3} \sum_{\ell=1}^{3} S_{ij}T_{k\ell}\delta_{jk}\hat{\delta}_i\hat{\delta}_\ell
$$

$$
= \sum_{i=1}^{3} \sum_{j=1}^{3} \sum_{\ell=1}^{3} S_{ij}T_{j\ell}\hat{\delta}_i\hat{\delta}_\ell
$$

Double dot product of two tensors

$$
\bar{\bar{S}} : \bar{\bar{T}} = \left[\sum_{i=1}^{3} \sum_{j=1}^{3} S_{ij}\hat{\delta}_i\hat{\delta}_j \right] : \left[\sum_{k=1}^{3} \sum_{\ell=1}^{3} T_{k\ell}\hat{\delta}_k\hat{\delta}_\ell \right]
$$

$$
= \sum_{i=1}^{3} \sum_{j=1}^{3} \sum_{k=1}^{3} \sum_{\ell=1}^{3} S_{ij}T_{k\ell} \left(\hat{\delta}_j \bullet \hat{\delta}_k \right) \left(\hat{\delta}_i \bullet \hat{\delta}_\ell \right) = \sum_{i=1}^{3} \sum_{j=1}^{3} \sum_{k=1}^{3} \sum_{\ell=1}^{3} S_{ij}T_{k\ell}\delta_{jk}\delta_{il}
$$

$$
= \sum_{i=1}^{3} \sum_{j=1}^{3} S_{ij}T_{ji}
$$

Trace of a tensor

$$
\mathrm{tr}\,(\bar{\bar{T}}) = \bar{\bar{\delta}} : \bar{\bar{T}} = \sum_{i=1}^{3} \sum_{j=1}^{3} \sum_{k=1}^{3} \sum_{\ell=1}^{3} \delta_{ij} T_{k\ell} \left(\hat{\delta}_j \bullet \hat{\delta}_k \right) \left(\hat{\delta}_i \bullet \hat{\delta}_\ell \right)
$$

$$
= \sum_{i=1}^{3} \sum_{j=1}^{3} \sum_{k=1}^{3} \sum_{\ell=1}^{3} \delta_{ij}\,\delta_{jk}\,\delta_{i\ell}\,T_{k\ell} = \sum_{i=1}^{3} T_{ii}
$$

$$div\ \bar{\bar{T}} = \nabla \bullet \bar{\bar{T}} = \left[\sum_{i=1}^{3} \hat{\delta}_i \frac{\partial}{\partial x_i}\right] \bullet \left[\sum_{j=1}^{3}\sum_{k=1}^{3} T_{jk}\, \hat{\delta}_j\, \hat{\delta}_k\right]$$

$$= \sum_{i=1}^{3}\sum_{j=1}^{3}\sum_{k=1}^{3} \left(\frac{\partial T_{jk}}{\partial x_i}\right)\left(\hat{\delta}_i \bullet \hat{\delta}_j\right)\hat{\delta}_k$$

$$= \sum_{i=1}^{3}\sum_{k=1}^{3} \left(\frac{\partial T_{ik}}{\partial x_i}\right)\hat{\delta}_k$$

B.3 SOME USEFUL RELATIONS

$\left(\bar{\bar{A}} + \bar{\bar{B}}\right)^T = \bar{\bar{A}}^T + \bar{\bar{B}}^T$ (Superscript T refers to transpose)

$(\bar{\bar{A}}\,\bar{\bar{B}}) = \bar{\bar{B}}^T\,\bar{\bar{A}}^T$

$\left(\bar{\bar{A}}^T\right)^T = \bar{\bar{A}}$

$\vec{A} \bullet \bar{\bar{T}} = \bar{\bar{T}}^T \bullet \vec{A}$

$\vec{A} \bullet \bar{\bar{T}}^T = \bar{\bar{T}} \bullet \vec{A}$

$\left(\vec{A}\vec{B}\right)^T = \vec{B}\vec{A}$

$\bar{\bar{T}} \bullet \vec{A} = \vec{A} \bullet \bar{\bar{T}}$ if $\bar{\bar{T}}$ is symmetric, that is, $\bar{\bar{T}} = \bar{\bar{T}}^T$

$\nabla\vec{x} = \left(\nabla\vec{x}\right)^T = \bar{\bar{\delta}}$

$\bar{\bar{\delta}} \bullet \bar{\bar{T}} = \bar{\bar{T}} \bullet \bar{\bar{\delta}} = \bar{\bar{T}}$

$\nabla \bullet \left(\bar{\bar{T}}\vec{x}\right) = \bar{\bar{T}}^T + \left(\nabla \bullet \bar{\bar{T}}\right)\vec{x}$

$\nabla \bullet \left(\bar{\bar{T}}^T\vec{x}\right) = \bar{\bar{T}} + \left(\nabla \bullet \bar{\bar{T}}^T\right)\vec{x}$

$\left(\nabla \bullet (\bar{\bar{T}}\vec{x})\right)^T = \bar{\bar{T}} + \vec{x}\left(\nabla \bullet \bar{\bar{T}}\right)$

$\bar{\bar{T}} = \nabla \bullet (\bar{\bar{T}}\vec{x}) = (\nabla \bullet (\bar{\bar{T}}\vec{x}))^T$ if $\bar{\bar{T}}$ is symmetric, $(\bar{\bar{T}} = \bar{\bar{T}}^T)$ and divergence free $\left(\nabla \bullet \bar{\bar{T}} = 0\right)$

$\nabla_s \bullet \bar{\bar{\delta}}_s = -\left(\nabla_s \bullet \hat{n}\right)\hat{n}$ where $\bar{\bar{\delta}}_s = \bar{\bar{\delta}} - \hat{n}\hat{n}$ and $\nabla_s = \bar{\bar{\delta}}_s \bullet \nabla$

$-\nabla_s \bullet \bar{\bar{P}}_s = \gamma\hat{n}\left(\nabla_s \bullet \hat{n}\right)$ where $\bar{\bar{P}}_s = \gamma\bar{\bar{\delta}}_s$ and γ is a constant

Appendix C

Gauss Divergence Theorem

The Gauss divergence theorem relates a volume integral to a surface integral. Let V be a closed region in space bounded by a surface S. Let $\vec{G}$ be a vector function that is continuous and has continuous first-order partial derivatives in some domain containing V. Then,

$$\int_V (\nabla \bullet \vec{G}) dV = \int_S d\vec{S} \bullet \vec{G} = \int_S (\hat{n} \bullet \vec{G}) dS$$

where $\hat{n}$ is the outwardly directed unit normal vector of surface S.

The Gauss divergence theorem can be extended to scalar and tensor fields as:

$$\int_V (\nabla Q) dV = \int_S (\hat{n} Q) dS$$

$$\int_V (\nabla \bullet \bar{\bar{T}}) dV = \int_S (\hat{n} \bullet \bar{\bar{T}}) dS$$

where Q is some scalar field and $\bar{\bar{T}}$ is some tensor field.

Some other useful extensions of the Gauss divergence theorem are:

$$\int_V (\nabla \bullet \vec{A}\vec{B}) dV = \int_S (\hat{n} \bullet \vec{A}\vec{B}) dS$$

$$\int_V (\nabla \bullet \bar{\bar{T}}\vec{A})^T dV = \int_S (\hat{n} \bullet \bar{\bar{T}}\vec{A})^T dS$$

Appendix D

Symmetric Deviator of Fourth-Order Tensor

For any fourth-order tensor H_4 with components H_{ijkl} (there are 81 components), the symmetric deviator of fourth-order $Sd_4(H_4)$ is defined as:

$$Sd_4(H_4)_{ijkl} = \frac{1}{8} \text{ [the sum of 24 permutations of } H_{ijkl}] - \frac{1}{28} \{\text{the sum of 6}$$

$$\text{permutations of } [\delta_{ij}(\text{the sum of 12 permutations of } H_{klmm})]\}$$

$$+ \frac{1}{35} (\delta_{ij}\delta_{kl} + \delta_{ik}\delta_{jl} + \delta_{il}\delta_{jk})(H_{mmpp} + H_{mpmp} + H_{mppm})$$

where summation signs for all repeated indices are understood.

Index

A

Acrivos, Barthes-Biesel and, studies, 183
Acrivos, Chen and, studies, 259
Acrivos, Frankel and, studies, 177
Acrivos, Hinch and, studies, 166
Acrivos, Russel and, studies, 292
additive coatings
    bending rigidity, film coatings, 342–346, *346*
    deformable liquid droplets, 329–350
    dynamic viscoelastic behavior, 329–350
    elastic film coatings, 332–336, *334–338*
    fundamentals, 329
    liquid film coatings, *347,* 347–350, *349–350*
    Palierne model, 341
    viscoelastic film coatings, 336–341, *340*
    viscous film coatings, 329–331, *331–332*
Almog and Frankel studies, 140
applications
    biomedical applications, 5–7, *6*
    biotechnological applications, 7, *7*
    ceramics, 3
    coarse-particle composites, 14–15
    composites, 13–17
    cosmetics, 12
    fiber-reinforced composites, 16–17
    fine-particle composites, 15–16
    food applications, 9, *10*
    geological applications, 12–13, *13*
    marine oil spills, 8–9
    materials, pipeline transportation, 9–10
    nanotechnological applications, 5
    paints, 8
    particulate recording media, *4,* 4–5
    particulate composites, 14–16
    particulate dispersions, 3–13
    petroleum production, 10–11
    pipeline transportation, materials, 9–10
    toiletries, 12
automobile tires, 15

axisymmetric particles
    Brownian motion, 111–121
    non-Brownian motion, 102–111

B

Barthes-Biesel and Acrivos studies, 183
Barthes-Biesel and Chhim studies, 220
Barthes-Biesel and Rallison studies, 223
Barthes-Biesel studies, 220
Batchelor studies, 109
bending rigidity, film coatings, 342–346, *346*
Bergenholtz, Zackrisson and, studies, 80, 234, 236, 238
biomedical applications, 5–7, *6*
biotechnological applications, 7, *7*
Blawzdziewicz studies, 189
body force, 32
Boltzmann constant
    axisymmetric particles, 112
    electrically charged solid particles, 84
Booth studies, 90
bounds, elastic properties
    fiber-reinforced composites, 290–291
    particulate composites, 269–271, *272–275*
Boussinesq number
    bubbly suspensions, 206
    surfactants effect, 186, 190
    viscous film coating, 330–331
breakup, liquid droplets, 165–168, *166–167*
Brenner, Davis and, studies, 234, 241
Brenner, Happel and, studies, 77
Brenner and Weissman studies, 133, 137–138
Brenner studies, 130
Brinkman equations
    porous rigid particles, 78
    solid core-hairy shell particles, 235
broths, 7
Brownian characteristics
    electrically charged solid particles, 86
    nonhydrodynamic effects, 63
    nonspherical dipolar particles, 139–140